National Center for Construction Education and Research

Safety Technology

Annotated Instructor's Guide

PEARSON
Prentice Hall

Upper Saddle River, New Jersey
Columbus, Ohio

contren™
Learning Series

National Center for Construction Education and Research

President: Don Whyte
Director of Curriculum Revision and Development: Daniele Dixon
Director of Safety and Management Education: Gary Wilson
Safety Project Manager and Editor: Tara Cohen
Production Manager: Debie Ness
Quality Assurance Coordinator: Jessica Martin
Desktop Publishers: Rachel Ivines and Laura Parker

The NCCER would like to acknowledge the contract service provider for this curriculum:
Topaz Publications, Liverpool, New York.

10 9 8
ISBN 0-13-106259-X

Preface

This volume was developed by the National Center for Construction Education and Research (NCCER) in response to the training needs of the construction, maintenance, and pipeline industries. It is one of many in NCCER's *Contren™ Learning Series*. The program, covering training for close to 40 construction and maintenance areas, and including skills assessments, safety training, and management education, was developed over a period of years by industry and education specialists.

NCCER also maintains a National Registry that provides transcripts, certificates, and wallet cards to individuals who have successfully completed modules of NCCER's *Contren™ Learning Series*, when the training program is delivered by an NCCER Accredited training Sponsor.

The NCCER is a not-for-profit 501(c)(3) education foundation established in 1995 by the world's largest and most progressive construction companies and national construction associations. It was founded to address the severe workforce shortage facing the industry and to develop a standardized training process and curricula. Today, NCCER is supported by hundreds of leading construction and maintenance companies, manufacturers, and national associations, including the following partnering organizations:

PARTNERING ASSOCIATIONS

- American Fire Sprinkler Association
- American Petroleum Institute
- American Society for Training & Development
- American Welding Society
- Associated Builders & Contractors, Inc.
- Association for Career and Technical Education
- Associated General Contractors of America
- Carolinas AGC, Inc.
- Carolinas Electrical Contractors Association
- Citizens Democracy Corps
- Construction Industry Institute
- Construction Users Roundtable
- Design-Build Institute of America
- Merit Contractors Association of Canada
- Metal Building Manufacturers Association
- National Association of Minority Contractors
- National Association of State Supervisors for Trade and Industrial Education
- National Association of Women in Construction
- National Insulation Association
- National Ready Mixed Concrete Association
- National Systems Contractors Association
- National Utility Contractors Association
- National Vocational Technical Honor Society
- North American Crane Bureau
- Painting & Decorating Contractors of America
- Plumbing-Heating-Cooling Contractors National Association
- Portland Cement Association
- SkillsUSA
- Steel Erectors Association of America
- Texas Gulf Coast Chapter ABC
- U.S. Army Corps of Engineers
- University of Florida
- Women Construction Owners & Executives, USA

Some features of NCCER's *Contren™ Learning Series* are:

- An industry-proven record of success
- Curricula developed by the industry for the industry
- National standardization providing portability of learned job skills and educational credits
- Credentials for individuals through NCCER's National Registry
- Compliance with Apprenticeship, Training, Employer, and Labor Services (ATELS) requirements for related classroom training (CFR 29:29)
- Well-illustrated, up-to-date, and practical information

The Contren™ Safety Learning Series

Welcome to the *Contren™ Safety Learning Series*. This systematic approach to safety education and training provides a standardized curriculum in a modularized form that allows organizations to design custom programs to fit their specific needs. The series is composed of four independent titles: *Safety Orientation*, *Field Safety*, *Safety Technology*, and *Safety Management*. While none of these books serve as mandatory prerequisites for the others, they are listed in a logical education path.

The National Center for Construction Education and Research (NCCER) offers individual recognition for successful completion of all programs, ranging from certificates of completion to Safety Instructor Certification. NCCER's *Contren™ Safety Learning Series* provides unique recognition, the education necessary to gain opportunities for career advancement, and the skills needed to build a safe workplace.

In addition to NCCER's certifications, the *Contren™ Safety Learning Series* has been recognized by the Council on Certification of Health, Environmental, and Safety Technologists (CCHEST) as exam preparation for their Safety Trained Supervisor (STS) and Construction Health and Safety Technician (CHST) certifications. CCHEST is a joint venture of the Board of Certified Safety Professionals and the American Board of Industrial Hygiene.

For more information about courses, certifications, or any other related issue, please contact NCCER Customer Service at (352) 334-0920; visit us online at www.nccer.org; or email your questions to info@nccer.org.

Features of This Book

Capitalizing on a well-received campaign to redesign our textbooks, NCCER is publishing select textbooks in a two-column format. *Safety Technology*, a part of the *Contren™ Safety Learning Series*, incorporates the design and layout of our full-color books along with special pedagogical features. The features augment the technical material to maintain the participants' interest and foster a deeper appreciation of the trade.

The **Think About It** feature uses "what if?" questions to help participants apply theory to real-world experiences and put ideas into action.

The **Case History** feature demonstrates the significance of adhering to safety guidelines and best practices by focusing on the importance of job-site safety. Participants read about causes and repercussions of real-life job-site hazards, incidents, accidents, and near-misses.

The **Profile in Success** feature presents one-page biographies of successful industry professionals and shares their related experiences and advice.

We're excited to be able to offer you these improvements and hope they lead to a more rewarding learning experience. As always, your feedback is welcome. Please let us know how we are doing by visiting NCCER at www.nccer.org or e-mailing us at info@nccer.org.

Acknowledgments

This curriculum was developed as a result of the vision and leadership of NCCER's National Safety Committee and those who served as subject matter experts:

Jim Humphry, Quanta Services
Art Deleon, Underground Construction Co.
Frank McDaniel, Casey Industries

For the long-term professional services provided to NCCER on all safety-related subjects, including the development of the *Contren™ Safety Learning Series*, a sincere thanks is extended to **Steven Pereira, Professional Safety Associates**, Denham Springs, LA.

Contents

Module 75201-03

Introduction to Safety Technology

NCCER STANDARDIZED CRAFT TRAINING PROGRAM

The National Center for Construction Education and Research (NCCER) provides a standardized national program of accredited craft training. Key features of the program include instructor certification, competency-based training, and performance testing. The program provides trainees, instructors, and companies with a standard form of recognition through a National Craft Training Registry. The program is described in full in the *Guidelines for Accreditation*, published by the NCCER. For more information on standardized craft training, contact the NCCER by writing us at P.O. Box 141104, Gainesville, FL 32614-1104; calling 352-334-0911; or e-mailing info@nccer.org. More information may be found at our Web site, www.nccer.org.

HOW TO USE THIS ANNOTATED INSTRUCTOR'S GUIDE

Each page presents two sections of information. The larger section displays each page exactly as it appears in the Participant Module. The narrow column ties suggested participant and instructor actions to each page and provides icons (detailed below) to call your attention to material, safety, audiovisual, or testing requirements. The bottom of each page includes space for your notes.

The **Audiovisual** icon indicates an appropriate time to show a transparency or other audiovisual aid.

The **Classroom** icon prompts you to define a term, stress a point, ask participants to explain a concept, or give examples.

The **Demonstration** icon directs you to show participants how to perform tasks.

The **Examination** icon tells you to administer the written module examination.

The **Homework** icon is placed where you may wish to assign reading for the next class, to assign a project, or to advise participants to prepare for an examination.

The **Laboratory** icon is used when participants are to practice performing tasks.

The **Materials** icon is a reminder for you to gather materials needed for classes, labs, and testing.

The **Performance Testing** icon tells you to administer a performance test or a portion thereof.

The **Safety** icon is used to emphasize safety issues. It is often keyed to *Caution* and *Warning* statements in the Participant Module.

The **Teaching Tip** icon indicates additional guidance is available, such as how to conduct an exercise, get the most educational value from a field trip, or encourage class participation. Teaching Tips may expand on a feature (*Think About It, Did You Know?*) or provide Quick Quizzes or similar exercises. You will be referred to the Teaching Tips section at the back of the module if there is additional material.

The **Combination** icon indicates that the laboratory listed corresponds with a performance task. If desired, you can note the proficiency of the participants during the laboratory and use it to satisfy performance testing requirements.

PREPARATION

Before teaching this module, you should review the Objectives, Performance Tasks, Materials and Equipment List, and the Module Outline. Be sure to allow ample time to prepare your own training or lesson plan and gather all required materials and equipment.

MODULE OVERVIEW

In this module, participants will learn the roles and responsibilities of a safety technician, as well as the basic aspects of safety technology, such as the costs of accidents, basic components of a safety program, and regulatory bodies.

PREREQUISITES

Please refer to the Course Map in the Participant Module. Prior to training with this module, it is recommended that the participant shall have successfully completed the following:

Field Safety

OBJECTIVES

Upon completion of this module, the participant will be able to:

1. Explain the roles and responsibilities of a safety technician.
2. Explain important safety-related terms.
3. Explain the three levels of accident causation.
4. Explain the cost impact of accidents.
5. Describe the basic components of a safety program.
6. Explain the government regulatory requirements that affect the construction industry.

PERFORMANCE TASKS

There are no performance tasks for this module.

MATERIALS AND EQUIPMENT LIST

Overhead projector and screen

Transparencies

Whiteboard/chalkboard

Markers/chalk

Blank acetate sheets

Transparency pens

Pencils and scratch paper

Appropriate personal protective equipment

Copy of *29 CFR 1926, OSHA Construction Industry Regulations*

Construction User's Round Table A-3 report

Module Examinations*

* Located in the Test Booklet.

SAFETY CONSIDERATIONS

Ensure that participants are equipped with appropriate personal protective equipment.

ADDITIONAL RESOURCES

This module is intended to present thorough resources for task training. The following reference works are suggested for both instructors and motivated participants interested in further study. These are optional materials for continued education rather than for task training.

Basic Safety Administration: A Handbook for the New Safety Specialist, 2003. Fred Fanning, CSP. Des Plaines, IL: The American Society of Safety Engineers (ASSE).

Construction Safety Planning, 1995. Avid V. MacCollum, P.E., CSP. Hoboken, NJ: John Wiley & Sons.

Web site: www.osha.gov

Web site: www.msha.gov

Web site: www.dot.gov

Web site: www.epa.gov

Web site: www.asse.org

TEACHING TIME FOR THIS MODULE

An outline for use in developing your lesson plan is presented below. Approximately 2½ hours are suggested to cover *Introduction to Safety Technology*. This includes 10 minutes at the beginning of the session for administrative tasks and one 10-minute break. You may need to adjust the time required for hands-on activities and testing based on your class size and resources. Because laboratories often correspond to Performance Tasks, the proficiency of the participants may be noted during these exercises for Performance Testing purposes.

Topic	Planned Time
A. Introduction	____________
B. The Role and Responsibility of the Safety Technician	____________
C. Important Safety Related Terms	____________
D. Accidents	____________
1. Three Levels of Accident Causation	____________
2. Costs of Accidents	____________
E. Basic Components of a Safety Program	____________
1. Management Support and Policy Statement	____________
2. Policy on Alcohol and Drug Abuse	____________
3. Safety Rules	____________
4. Orientation and Training	____________
5. Safety Meetings and Employee Involvement	____________
6. Emergency Action Plans and Methods for Dealing with the Media	____________
7. Inspections, Employee Observations, and Audits	____________
8. Accident Investigation and Analysis	____________
9. Records	____________
10. Program Evaluation and Follow-Up	____________
11. Nine Industry Best Practices	____________
F. Regulatory Requirements	____________
1. Occupational Safety and Health Administration (OSHA)	____________
2. Mine Safety and Health Administration (MSHA)	____________
3. Environmental Protection Agency (EPA)	____________
4. Driver's License Programs	____________
G. Review	____________
H. Module Examination	____________
1. Participants must score 70% or higher to receive recognition from the NCCER.	
2. Record the testing results on Craft Training Report Form 200 and submit the results to the Training Program Sponsor.	

Module 75201-03

Introduction to Safety Technology

Instructor's Notes:

COURSE MAP

This course map shows all of the modules in Safety Technology. The suggested training order begins at the bottom and proceeds up. The local Training Program Sponsor may adjust the training order.

Assign reading of Module 75201-03.

SAFETY TECHNOLOGY

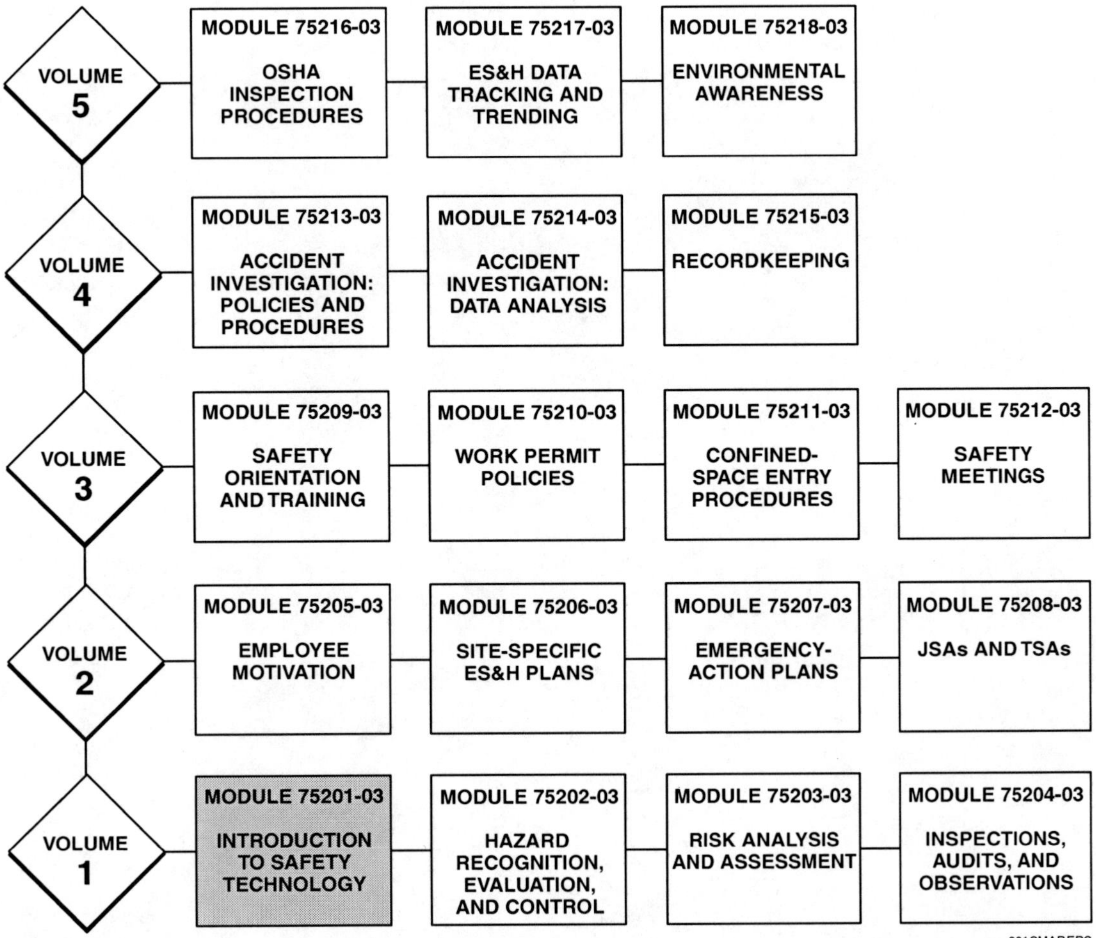

201CMAP.EPS

INTRODUCTION TO SAFETY TECHNOLOGY 1.iii

Instructor's Notes:

MODULE 75201-03 CONTENTS

INTRODUCTION TO SAFETY TECHNOLOGY 1.v

MODULE 75201-03 CONTENTS (Continued)

Figures

Tables

Instructor's Notes:

MODULE 75201-03

Introduction to Safety Technology

Ensure that you have everything you need to teach the course. Check the Materials and Equipment list in the front of this module.

Objectives

When you have completed this module, you will be able to do the following:

1. Explain the roles and responsibilities of a safety technician.
2. Explain important safety-related terms.
3. Explain the three levels of accident causation.
4. Explain the cost impact of accidents.
5. Describe the basic components of a safety program.
6. Explain the government regulatory requirements that affect the construction industry.

Explain that terms shown in bold are defined in the Glossary at the back of this module.

Review the goals of the course and explain what will be expected of the participant.

Show Transparency 1, Objectives.

Prerequisites

Before you begin this module, it is recommended that you successfully complete the following: Field Safety.

Required Materials

1. Pencil and paper
2. Appropriate personal protective equipment
3. Copy of *29 CFR 1926, OSHA Construction Industry Regulations*

1.0.0 ◆ INTRODUCTION

The job of a safety technician is both serious and satisfying. In this role, you have the opportunity to save lives, prevent injuries and disabilities, and help your employer be profitable and competitive. This is important considering that on a typical working day, four to five construction workers will die from on-the-job injuries and nine hundred will be seriously injured. That means more than one thousand construction workers are killed on the job every year.

One way to reduce illness and injury on a site is to be fully aware of the causes and costs of incidents and accidents. Another way is to have a company-wide standardized safety program. This program should provide the basic framework for working safely and efficiently. As a safety technician, you will be an important part of your company's safety program. In fact, some of your main responsibilities include coordinating safety policies and safety procedures, performing audits and inspections, recordkeeping, and risk analysis.

You will also be responsible for knowing the local, state, and national regulations with which your company must comply. These safety regulations are established to achieve injury-free and accident-free worksites. It is important to remember that all of these regulations are minimum standards. Many progressive companies have policies and procedures that are more stringent.

2.0.0 ◆ THE ROLE AND RESPONSIBILITY OF THE SAFETY TECHNICIAN

Explain the roles and responsibilities of a safety technician.

Show Transparency 2, Roles and Responsibilities.

One of the safety technician's most important responsibilities is to serve as a resource to site management on safety, health, and in some cases, environmental regulations. This includes workers, subcontractors, and the public. Being a safety technician is a specific craft. In this role, you are a trainer, motivator, auditor, planner, and advisor. You are a key player in the organization.

Safety technicians also have the following responsibilities:

- In the absence of a site safety manager or supervisor, represent the company during visits by regulatory agencies.
- Provide safety training for both new and experienced workers.

INTRODUCTION TO SAFETY TECHNOLOGY 1.1

Discuss the basic definitions and concepts related to a safety technician's job. Refer the participants to the lists in the Participant Module.

Explain the impact of accidents on a company and its workers.

Describe the accident causation model: Three Levels of Accident Causation.

Show Transparency 3 (Figure 1).

Explain that Level I represents the direct causes of accidents.

Define the term release of energy.

Show Transparency 4, Results of Uncontrolled Releases of Energy.

- Participate in the development, review, or revision of Job Safety Analyses (JSAs), Task Safety Analyses (TSAs), work plans, incident/accident reporting forms, and emergency action plans.
- Audit and inspect the job site or work activities.
- Anticipate, identify, and have management correct safety hazards.
- Audit compliance with regulatory requirements.
- Know how to complete required forms.
- Observe work in progress to make sure safe work methods are used.
- Use proper coaching techniques to correct unsafe behavior and reward safe behavior.
- Conduct safety meetings.
- Audit compliance with work permits.
- Audit permit-required work areas.
- Assist site management in conducting incident investigations.
- Analyze data gathered during accident investigations.
- Manage the site safety and health recordkeeping system.
- Serve as a liaison between the job site and insurance company representatives.
- Provide or coordinate first aid and access to follow-up medical care.

3.0.0 ◆ IMPORTANT SAFETY RELATED TERMS

Here are some of the basic definitions and concepts needed to effectively do your job as a safety technician.

- *Safety* – A general term denoting an acceptable level of risk, or the relative freedom from or low probability of harm. It is the control of recognized hazards to attain an acceptable level of risk.
- *Risk* – A measure of both the probability and the consequences of all hazards of an activity or condition. It is a subjective evaluation of relative failure potential. It is the chance of injury, damage, or loss.
- *Hazards* – Conditions, changing sets of circumstances, or behaviors that present the potential for injury, illness, or property damage.
- *Accidents* – Unplanned and sometimes harmful or damaging events that interrupt the normal progress of a task and invariably preceded by an unsafe act, condition, or a combination of both.
- *Incident/Near Miss* – An undesired event that, under slightly different circumstances, could have resulted in personal harm or property damage or any undesired loss of resources. An incident is sometimes referred to as a near miss or near hit. An example of a near miss would be a falling object that narrowly missed a worker below.
- *Unsafe Acts* – Sometimes called unsafe behavior or placing yourself at risk, are behavioral departures from an accepted, normal, or correct procedure or practice. An unsafe act is an unnecessary exposure to a hazard or conduct that reduces the degree of safety normally present.
- *Unsafe Conditions* – Any physical states that deviate from that which is acceptable, normal, or correct. An unsafe condition is any physical state that results in a reduction in the degree of safety normally present.

It's important to learn these terms and conditions because as a safety technician, you will hear, see, and say them often.

4.0.0 ◆ ACCIDENTS

Accidents can result in property damage, personal injury, or death. Safety rules and guidelines must be followed at all times to minimize the occurrence of accidents. As a safety technician, you must be aware of the causes and costs of accidents as well as how to prevent them.

4.1.0 Three Levels of Accident Causation

The causes of accidents can be classified by three different factors: direct causes, indirect causes, and root causes. These are also called the three levels of accident causation *(Figure 1)*. Each level of causation represents one of these factors.

4.1.1 Level I

Level I represents direct causes of accidents. These are accidents resulting from the uncontrolled **release of energy**. These accidents may or may not cause injury or property damage, but they are still dangerous. When investigating an accident, be sure to look closely at energy sources that may have been released unintentionally. Common examples of the results of uncontrolled releases of energy include:

- Being struck by an object or equipment
- Being caught in between two objects
- Being struck by debris during the detonation of explosives
- Being cut or scraped by a jagged edge

These types of accidents are preventable if equipment is properly maintained and workers follow established safety guidelines.

Instructor's Notes:

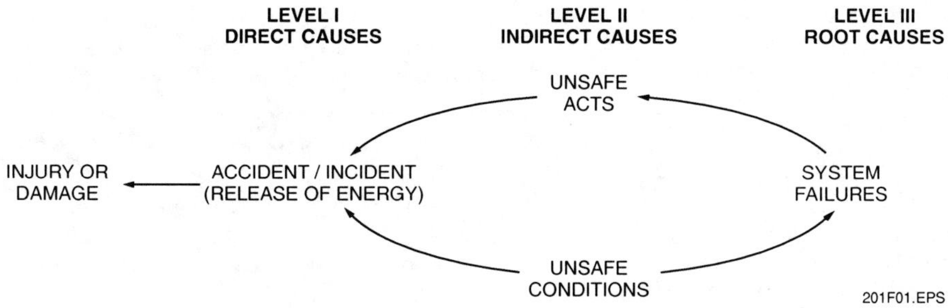

Figure 1 ◈ Three levels of causation.

Explain Level II of the three levels of accident causation.

Discuss the most common types of unsafe behavior found on a site. Refer the participants to the list in the Participant Module.

Describe the most common unsafe conditions found on a site.

4.1.2 Level II

Level II represents indirect causes of accidents. Indirect causes are factors that contribute to an accident but aren't the main cause. Indirect causes are also known as unsafe acts and conditions. In the past, investigators looked only for indirect causes of accidents, not the reasons behind the unsafe act or condition. This proved to be ineffective because indirect causes are often just the symptom of a greater problem.

It's important to be able to recognize when a worker's behavior is unsafe. The following is a list of the most common unsafe acts found on a job site:

- Failing to use personal protective equipment
- Failing to warn co-workers of potentially hazardous conditions or unsafe behaviors
- Failing to follow instructions and procedures
- Using defective equipment
- Lifting improperly
- Taking an improper working position
- Making safety devices inoperable
- Operating equipment at improper speeds
- Operating equipment without authority
- Servicing equipment while it is in motion or energized
- Loading or placing equipment or supplies improperly or dangerously
- Using equipment improperly
- Working while impaired by alcohol or drugs
- Engaging in horseplay

Unsafe conditions are physical conditions that are different from acceptable, normal, or correct conditions. The following is a list of the most common unsafe conditions:

- Congested workplaces
- Defective tools, equipment, or supplies
- Excessive noise
- Fire and explosive hazards
- Hazardous atmospheric conditions
  - Gases
  - Dusts
  - Fumes
  - Vapors
- Inadequate supports or guards
- Inadequate warning systems
- Poor housekeeping
- Poor illumination
- Poor ventilation

Show Transparency 5, Common Unsafe Work Conditions.

Most accidents are caused by unsafe acts and conditions. It's important to be aware of both types of causes and use as many preventive measures as necessary, including proper training, appropriate equipment maintenance, and good work-site housekeeping.

Explain Level III of the three levels of accident causation.

4.1.3 Level III

Level III represents the root causes of accidents. The root or basic causes of accidents are hazardous conditions or unsafe work practices that aren't fixed, even though workers and supervisors know about them. Root causes not only affect single accidents being investigated, they also affect other future accidents and work problems. Once root causes are fixed, the types of accidents that occur because of them will not happen again. Roots causes should be corrected as soon as they are found. This will help prevent accidents and save lives.

4.2.0 Costs of Accidents

Accidents are very costly. When they occur, everyone involved loses, including the injured worker, the employer, and the insurance company. Accident costs are often classified by the term *total incurred.* There are direct (insured) and indirect (uninsured) costs associated with accidents.

INTRODUCTION TO SAFETY TECHNOLOGY 1.3

Explain the direct and indirect costs associated with accidents.

Direct, or insured, costs include medical costs and workers' compensation (WC) insurance benefits, as well as liability and property damage insurance payments. Of the three direct costs, workers' compensation insurance benefits are the most costly. Direct costs of accidents are not generally fixed. Rather, they vary depending upon the severity of the injuries and damages.

Indirect, or uninsured, costs are the hidden costs involved with an accident. Examples include the costs associated with uninsured property damage, equipment damage, production delays, supervisory time, retraining, company image, and worker morale.

Show Transparency 6 (Figure 2).

The costs associated with accidents can be compared to an iceberg, as shown in *Figure 2*. The tip of the iceberg represents the direct costs, which are the costs that can be seen. The larger, indirect costs are unseen. Studies indicate that the indirect costs of accidents usually are two to seven times greater than the direct costs. As a safety technician, you need to be aware of the overall effects of accidents.

Explain how accident costs are calculated.

Describe the relationship between accident experience and future insurance costs.

Explain how to calculate an experience modification rate (EMR).

4.2.1 Calculating Accident Costs

Real dollars are lost when workplace accidents occur. These dollars have a tremendous effect on the company's **profit** margin. For example, consider a company whose **gross income** is 2 million dollars. They operate on a profit margin of 3% of its gross income ($60,000.00). If the company suffers a loss of $50,000 due to accidents, it must increase its gross income by $1,716,667 to make up for the accident. That's almost 2 million dollars more the company will have to earn, just to make up for the $50,000 lost from accidents. This is another reason why preventing accidents is so important.

> **NOTE**
> For more information on calculating accidents costs, please refer to OSHA's web site, www.osha.gov.

4.2.2 Accident Experience vs. Future Insurance Costs

A contractor's accident experience has a significant effect on future insurance costs. Contractors with poor accident experience generally pay more for workers' compensation insurance than those with good records. The difference is in their workers' compensation **experience modification rate (EMR)**.

Experience rating is a method of modifying future workers' compensation insurance premiums by comparing a particular company's actual losses to the losses normally expected for that company's type of work. The average rate for a particular class of work is called the book rate or manual rate. Contractors with better than average loss experience have a modifier/multiplier less than 100%, which is a credit factor. Those with worse than average experience have a modifier of over 100%, which is a penalty factor. *Figure 3* shows an example of this calculation. The EMR is multiplied by the total WC cost to determine what the company will actually pay for WC insurance.

Show Transparency 7 (Figure 3).

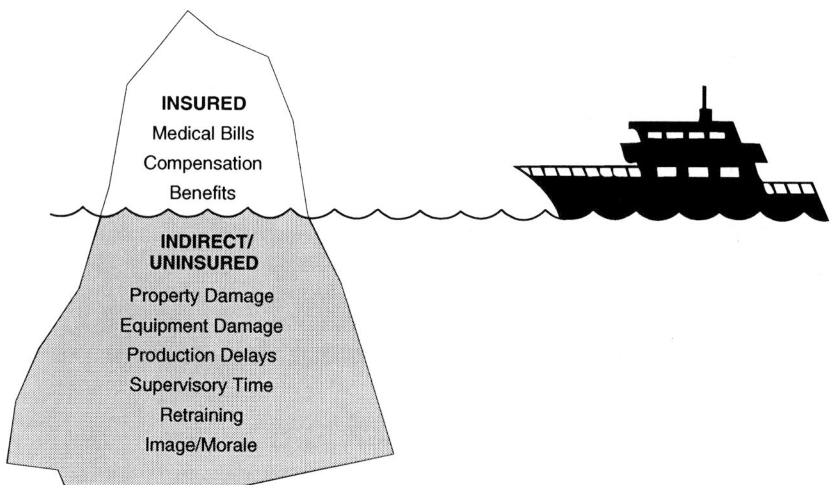

Figure 2 ◆ Costs associated with accidents.

Instructor's Notes:

Insert No. 3

Typical Workers' Compensation Premium Rates for a Louisiana Project (Stock Insurance Companies)

Classification	Payroll	WC Rate [1]	Insurance Cost
Carpenters	$120,000.00	$37.16/100	$44,592.00
Electricians	62,000.00	11.80/100	7,316.00
Pipefitters	61,000.00	9.24/100	5,636.40
Ironworkers	72,000.00	44.39/100	31,960.80
Totals	$315,000.00		$89,505.20

[1] Louisiana Manual Rates as of 9/15/00

Total Workers' Compensation Costs	$89,505.20
EMR	x 1.10
	$98,455.72

Participant Activity

Using the above figures assume that you and two other contractors are bidding the same job. Your labor and materials costs are essentially the same but you have different Workers' Compensation Experience Modifiers.

Company A has an Experience Modifier of 1.0. Company B has an Experience Modifier of 0.8 and your company has an EMR of 1.6. Calculate the insurance costs for:

Company A ____________________
Company B ____________________
Your Company ____________________

As you can see, your EMR can have a major impact on your bid. You should keep in mind that the frequency of job-related injuries generally has a greater impact on your EMR than does severity.

A company's EMR is based on the first three (3) of the last four (4) years' accident experience. In other words, your EMR for the 2003 policy year would be based on your losses in 1999, 2000, and 2001. What you do today will have a significant impact on your operations for years to come.

201F03.EPS

Figure 3 ◆ Experience modifier rate (EMR).

Make the participant activity in Figure 3 a class activity or create your own EMR calculation scenarios based on your own experiences.

Describe the costs and benefits of having an effective safety program. Include statistics from the Construction User's Round Table A-3 Report.

Describe the basic components of a safety program.

Show Transparency 8, Basic Components of a Safety Program.

Describe the components of a management policy statement. Emphasize the importance of a safety technician knowing his or her company's management policy.

Explain the impact alcohol and drug abuse have on a work site.

Describe how many companies deal with substance abuse, including information about employee assistance programs (EAPs).

4.2.3 Cost of Administering an Effective Safety Program

Studies referenced in the **Construction User's Round Table's (CURT)** A-3 report estimate that the cost of maintaining an effective safety program is approximately 2.5% of the **direct labor costs**. This cost includes salaries for safety, medical, and clerical personnel, cost of safety meetings, inspections of tools and equipment, orientation meetings, personal protective equipment, and miscellaneous supplies and equipment. However, most of these items are required by state and federal laws, and contractors are required to pay for them anyway. The likely added cost of a safety program beyond what is legally required is probably closer to 1% of the direct labor costs.

Implementing a comprehensive safety program can actually save your company money. The same studies referenced in the CURT's A-3 report conclude that for every dollar invested in a safety program, the contractor could save as much as $3.20, which means the contractor actually profits by having an effective safety program.

5.0.0 ◈ BASIC COMPONENTS OF A SAFETY PROGRAM

There are certain basic elements common to most effective safety programs. The degree and extent to which these elements are included in your program will vary depending on the size and nature of your operations. The most basic components of a safety program include:

- A management support and policy statement
- A policy on alcohol and drug abuse
- Safety rules
- Orientation and training
- Access to first aid and follow-up medical care
- Pre-project and pre-task safety planning
- Safety meetings and employee involvement
- Sub-contractor management
- Emergency action plans and methods for dealing with the media
- Inspections, employee observations, and audits
- Accident investigation and analysis
- Recordkeeping
- Program evaluation and follow-up

In order to comply with specific OSHA (Occupational Safety and Health Administration) and MSHA (Mine Safety and Health Administration) regulations, the additional safety program elements may be required:

- Hazard communication
- Bloodborne pathogens
- Welding, burning, and cutting
- Respiratory protection
- Personal protective equipment
- Elevated work and fall protection
- Hearing conservation
- Confined-space entry
- Lockout/tagout
- Cranes, hoists, and other lifting devices
- Mobile equipment operations
- Trenching and excavation

5.1.0 Management Support and Policy Statement

A policy statement outlines management's philosophies and goals toward safety and loss prevention. It is signed by the president or chief executive officer (CEO) and communicated to all affected employees.

A good policy statement recognizes management's responsibility for safety and loss prevention without diminishing each employee's role. It expresses a strong, positive, and realistic commitment to safety. Management support and commitment are the most important factors in the success of any safety program. Make sure you know your company's management policy statement and that it is available to all workers.

5.2.0 Policy on Alcohol and Drug Abuse

Most experts agree that at least one out of five, or 20%, of construction workers have an alcohol- or drug-abuse problem. The impact of this problem results in:

- Increased worker's compensation and health-care costs
- Decreased productivity
- Increased tools and materials costs
- Increased absenteeism

Many owners and contractors have recognized this problem and have implemented programs to combat the situation. Make sure you are of aware your company's drug- and alcohol-abuse programs. Be prepared to advise supervisors and other employees of the key elements of your company's program. Some larger companies sponsor employee assistance programs (EAPs). If your company does, make sure the information about the program is available to everyone.

Instructor's Notes:

5.3.0 Safety Rules

Safety rules are a necessary part of any program. They must be logical and they should be prepared and presented in terms that are easily understood. Rules are of no value unless they are communicated and understood, and workers buy into the safety process. Everyone on site must also understand the consequences of breaking safety rules.

Safety rules are generally divided into the following categories:

- General/company safety rules that apply to all employees
- Client-specific, job-specific, or site-specific safety rules
- Craft or special safety rules that apply to a specific type of work operation

Safety rules can be presented as posters on a site, or during orientation, safety meetings, or safety training. Make sure safety rules are available to all workers at all times, so that workers know the rules as well as you do.

Progressive contractors plan safety into the job at the bid stage. This requires a systematic management process that involves the estimator, project manager, and the safety department at a minimum. Likewise, daily pre-task or pre-job safety planning can improve safety, quality, and productivity on any job. As a safety technician, you will be asked to help implement the job-specific safety plan and participate in pre-job or pre-task safety planning.

5.4.0 Orientation and Training

Safety training is an important tool for preventing accidents. Training can be broken down into two major categories: new employee orientation and job-specific safety training. Statistics have shown that employees who have been on the job 30 days or less account for 25% of all construction injuries. This clearly shows the need for an effective orientation program. Safety training should be conducted when:

- A new employee is hired.
- An existing employee is assigned to a new job.
- New jobs are started.

In order for safety training to be effective, it should cover the following:

- Correct work procedures
- Care, use, maintenance, and limitations of any required personal protective equipment
- The hazards and safeguards associated with any harmful materials used
- Where to go for help
- Site emergency reporting and response procedures

Informal safety training will likely be done by the first-line supervisor. Your role in these types of training sessions involves guidance and support. In more formal safety sessions, however, you may be responsible for coordinating or conducting all safety sessions.

> **NOTE**
> Safety orientation and training are covered in more detail in Volume 3, Module 75209-03, *Safety Orientation and Training.*

5.5.0 Safety Meetings and Employee Involvement

Safety meetings are often used to maintain employee interest and involvement. They are a key element in any safety program. Safety meetings can vary from a formal presentation to short five-minute toolbox talks.

Safety meetings, when properly conducted, can be used to do the following:

- Exchange information regarding specific safety matters
- Provide an outlet for workers to discuss critical issues
- Provide a written record of the actions taken
- Establish an effective communications link between management and employees
- Focus on the hazards and safeguards associated with a planned task

Both federal and state governmental regulations require employers to train their employees on the hazards and safeguards associated with the work they are doing. Safety meetings can be used as training sessions.

As a safety technician, you should provide safety meeting topics to crew leaders and supervisors. In some cases, you will plan and conduct safety meetings and critique the quality of the crew leader or supervisor's meeting. Giving feedback allows him or her to adjust safety meetings as needed. This helps to ensure that the message of the meeting, which is always safety, is presented effectively and efficiently. You may also be required to maintain the records of all site safety meetings.

Discuss the importance of including safety rules in a safety program.

Explain the three different categories of safety rules.

Describe the importance of safety orientation and training in an organization.

Describe the characteristics that make training effective.

Show Transparency 9, Components of Effective Safety Training.

Describe how safety meetings, sometimes called toolbox talks, are used to maintain employee interest and involvement in safety.

Show Transparency 10, Goals of a Safety Meeting.

Explain the role of the safety technician in safety meetings.

Explain the requirements of an emergency action plan.

Describe the role of the safety technician in relation to the development of emergency action plans.

Describe the front-line supervisor's and safety technician's roles and responsibilities during inspections, employee observations, and audits.

Explain the tasks the safety technician must complete during the coordination of inspections, observations, and audits. Describe the purpose of each of these functions.

Describe the accident investigation process. Explain why these investigations are important.

Show Transparency 11, Benefits of Accident Investigation.

Explain the roles of both the front-line supervisor and the safety technician during an accident investigation.

> **NOTE**
> Safety and employee involvement are covered in more detail in Volume 2, Module 75205-03, *Employee Motivation* and in Volume 3, Module 75212-03, *Safety Meetings*.

5.6.0 Emergency Action Plans and Methods for Dealing with the Media

Emergency action plans can cut down response time to job-site emergencies. In many cases, advance planning can even reduce the severity of the incident. For example, if the names, addresses, and phone numbers of the nearest medical, fire, police, and emergency response agencies are posted at each job-site phone, it is easier to get the help you need. Also, as part of the emergency action plan, everyone should be familiar with the site emergency reporting and response procedures. Another part of the emergency action plan is the availability of prompt access to first aid and follow-up medical care. Emergency action plans should also require that at least two people on each job site be trained in basic first aid and cardiopulmonary resuscitation (CPR). Lives are saved when all of these elements are incorporated into emergency action plans. As a safety technician, you are responsible for implementing a plan and making sure everyone on site has been properly trained to follow the plan.

One area often overlooked after an accident is dealing with the news media. When serious accidents happen, the press usually responds. How you react can have a dramatic effect on how the public perceives your company. You should be familiar with your organization's policy on dealing with the news media. If there is no policy, you should request that one be developed.

> **NOTE**
> Emergency action plans are covered in more detail in Volume 2, Module 75207-03, *Emergency-Action Plans.*

5.7.0 Inspections, Employee Observations, and Audits

As the safety technician, you'll have special responsibilities. You will be a resource to the site personnel for safety, health, and in some cases environmental matters. You will also serve as a site consultant and auditor. It will be your job to help find and correct safety hazards. Sometimes you'll be working under the direction of a safety manager. More often, you'll be working on your own. Either way, you'll coordinate and/or perform the following tasks.

- *Safety inspections* – Used to detect unsafe conditions and make plans to correct them
- *Employee observations* – Performed to help detect unsafe work practices and procedures
- *Safety audits* – Designed to monitor the use and effectiveness of the company's safety polices and procedures

Supervisors will also be expected to perform safety inspections and employee observations. You may be asked to train them to do so. You should audit the quality, content, and follow-up of the crew leader's audits and observations and provide feedback as needed.

It is important that you know how to do each of these tasks effectively and efficiently so that problems can be found and fixed as soon as possible.

> **NOTE**
> Inspections, employee observations, and audits are covered in more detail in Volume 1, Module 75204-03, *Inspections, Employee Observations, and Audits* .

5.8.0 Accident Investigation and Analysis

All accidents, injuries, illnesses, and near misses must be reported and promptly investigated. Accident investigations are important because they:

- Determine the causes of accidents
- Help prevent re-occurrences
- Satisfy insurance company requirements and government regulations
- Document facts and preserve evidence for legal purposes
- Detect trends and identify potential problem areas

The front-line supervisor generally performs the primary accident investigation function. This is because it is generally agreed that front-line supervisors are the most familiar with the work area and employees. Depending on the nature of the incident and other conditions, accidents may also be investigated by a safety technician, a safety committee, or management. Safety technicians,

Instructor's Notes:

regardless of their level of participation, should always be available as a resource to the supervisor to provide any guidance or support that is needed. As a safety technician, you may be responsible for maintaining site accident investigation records and completing supporting documents required by the company, OSHA, or the insurance company.

> **NOTE**
> Accident investigation and analysis are covered in more detail in Module 75213-03, *Accident Investigation: Policies and Procedures* and Module 75214-03 *Accident Investigation: Data Analysis*, Volume 4.

5.9.0 Records

Recordkeeping is a critical part of a company's safety and health program. The data that is collected helps keep track of work-related injuries and illnesses. Once this data is gathered and analyzed, it can be used to help identify and correct problem areas. This will help prevent futures illnesses and injuries. Recordkeeping not only provides information to management about illness and injury, it also informs workers about incidents that happen in the work area. When workers are aware of injuries, illnesses, and hazards in the workplace, they are more likely to follow safe work practices and report workplace hazards.

Recordkeeping is also required by OSHA. They use specific illness and injury information that is reported to them as part of the agency's site-specific inspection targeting program. The Bureau of Labor Statistics (BLS) also uses injury and illness records as the source data for the Annual Survey of Occupational Injuries and Illnesses. This report shows safety and health trends nationwide and industry wide.

As a safety technician, you are responsible for making sure recordkeeping is done correctly. You must know all of the OSHA requirements for recording and classifying workplace illnesses and injuries and properly report them as needed. In addition, you may be required to coordinate all of the site safety and health records required by your firm. To do so, you must be familiar with your firm's recordkeeping requirements.

> **NOTE**
> Recordkeeping is covered in more detail in Volume 4, Module 75215-03, *Recordkeeping*.

5.10.0 Program Evaluation and Follow-Up

Program evaluation and follow-up are some of the most important, but often neglected, elements of an effective safety program. To be assured that your program is meeting the company's goals and objectives, it must be evaluated on a regular basis and modified as needed. The safety technician, along with the project manager, has the responsibility of evaluating the job-site safety program on a regular basis and making any needed improvements.

5.11.0 Nine Industry Best Practices

A recent study conducted by the Construction Industry Institute (CII) Committee, titled *Making Zero Accidents a Reality*, identified the following best practices that have the greatest impact on affecting safety performance:

- Demonstrated management commitment
- Staffing for safety
- Safety planning for pre-project and pre-task items
- Safety training and education
- Worker involvement and participation
- Recognition and rewards
- Sub-contractor management
- Accident/incident reporting and investigation
- Drug and alcohol testing

As a safety technician, you should always be aware of these industry best practices and, to the extent possible, integrate them into site-specific safety and health activities.

6.0.0 ◆ REGULATORY REQUIREMENTS

As a safety technician, you have a responsibility to know which local, state, and federal requirements apply to the work being done on your site. Government agencies that provide regulatory requirements applicable to the construction industry follow:

- Occupational Safety and Health Administration (OSHA)
- Mine Safety and Health Administration (MSHA)
- Environmental Protection Agency (EPA)
- Department of Transportation (DOT) – Commercial Drivers License Program

Make sure you are familiar with all of these agencies and their regulations. It could cost your company a great deal of money in fines and lawsuits if you aren't.

Explain the importance of a recordkeeping system.

Describe OSHA's recordkeeping requirements.

Describe the safety technician's role in recordkeeping.

Explain the importance of program evaluation and follow-up.

Discuss the role of the project manager and safety technician during safety program evaluation.

Discuss the construction industry best practices that impact safety performance.

Show Transparency 12, Nine Industry Best Practices.

Discuss the regulatory agencies and requirements that a safety technician must know.

INTRODUCTION TO SAFETY TECHNOLOGY 1.9

Describe the purpose and function of OSHA.

Explain the safety technician's responsibility for knowing and following OSHA requirements. Refer the participants to OSHA's Web site (www.osha.gov) for more information about OSHA.

Discuss how the MSHA Act of 1977 affects work sites.

Discuss the similarities between OSHA and MSHA.

Describe how MSHA inspections are conducted.

6.1.0 Occupational Safety and Health Administration (OSHA)

In 1970, the U.S. Senate and House of Representatives passed an act called the *Occupational Safety and Health Act of 1970*. The passing of this act led to the establishment of the Occupational Safety and Health Administration, also known as OSHA. This act was passed to assure safe and healthful working conditions for all working men and women. This act also:

- Authorizes enforcement of the standards developed under the act
- Assists and encourages states in their efforts to assure safe and healthful working conditions
- Provides for research, information, education, and training in the field of occupational safety and health

As a safety technician, you are required to know and follow the rules and regulations established by OSHA. OSHA offers several sources on standards and policies related to the construction industry. OSHA's Web site (www.osha.gov) provides detailed information such as:

- Standards and policies
- Statistics and data
- Reference documents
- Forms
- Training resources

For additional information about OSHA and the services they provide, contact the OSHA or State Plan Office in your area. It's important to take the time to learn which OSHA regulations apply to the work that is being done on your site. Failure to do so could mean serious injuries, loss of lives, or equipment loss or damage.

> **NOTE**
> OSHA requirements and regulations are covered in more detail throughout the safety technician course.

6.2.0 Mine Safety and Health Administration (MSHA)

On November 9, 1977, President Jimmy Carter signed into law the *U.S. Mine Safety and Health Act of 1977, Public Law 95-464*. The act became effective March 9, 1978. A result of this law, also known as the Mine Act, was the establishment of a regulatory organization called the Mine Safety and Health Administration (MSHA). MSHA's mission is to ensure safe and healthful working conditions for miners and any other workers doing work at a mine, rock quarry, or mine material processing facility. The Mine Act states that mine operators are responsible for preventing unsafe, unhealthful conditions or practices in mines that could endanger the lives and health of miners. It is important for you, as a safety technician, to understand MSHA requirements because any construction or maintenance done at a mine or a facility that handles mine products such as Portland cement, coal, gravel, and limestone falls under MSHA regulations, not OSHA regulations.

Mine operators must comply with the safety and health standards enforced by MSHA, an agency within the Department of Labor (DOL). Like OSHA, MSHA may issue citations and propose penalties for violations. Unlike the employees under the OSHA Act, mine employees are subject to government sanctions for violating safety standards such as smoking in or near mines or mining machinery. Similarly, employers and other supervisory personnel may be held personally liable for civil penalties or may be prosecuted criminally for violations of Mine Act standards.

> **NOTE**
> Mine Act standards are listed in *Title 30 CFR, Parts 70, 71, 74, 75, 77,* and *90.*

MSHA is similar to OSHA in many ways. There are certain areas in which there are marked differences. Some of the more important differences can be seen in:

- Inspections
- Withdrawal orders
- Miner training
- Accident, injury, and illness reporting

6.2.1 *Inspections*

An MSHA representative must inspect underground mines in their entirety at least four times a year. **Surface mines** are to be inspected at least two times a year.

An MSHA inspector will normally begin the inspection at the mine office. The officer will inform the mine operator of the reason for the inspection and request all needed records. The inspector's review of the records will likely focus on the pre-shift or on-shift examination record. Such records help the inspector determine where to concentrate attention during the inspection.

Instructor's Notes:

The mine operator's representative and a representative authorized by the miners must be given the opportunity to accompany the MSHA inspector during the inspection. Similarly, each must be given the opportunity to participate in the post-inspection conference. One miner representative, who is an employee of the operator, must be paid the regular wage for the time spent accompanying the inspector.

Whenever the inspector observes a condition that appears to be a violation of the standards, he or she must issue a citation. If, in the opinion of the mine inspector, an imminent danger condition exists, then the inspector must issue a withdrawal order.

6.2.2 Withdrawal Orders

Describe MSHA's requirements regarding withdrawal orders. Refer the participants to the bulleted list in the Participant Module.

MSHA has the authority, under specified conditions, to order an operator to withdraw miners from all or part of a mine. Miners who must stop work by such an order are entitled to receive compensation at their regular rate of pay for specified periods of time. All miners in the affected area must be withdrawn except:

- Those necessary to eliminate the hazard
- Public officials whose duties require their presence in the area
- Representatives of the miners qualified to make mine examinations
- Consultants

If an imminent danger is found to exist during an inspection, MSHA is required to order the withdrawal of all persons from the affected area, except those referred to in *Section 104* of the act, until the danger no longer exists. The order must describe the conditions or practices both causing and constituting the imminent danger and the area affected. The withdrawal order does not stop issuance of a citation and proposed penalty.

Other situations for which MSHA may issue a withdrawal order include the following:

- During a follow-up inspection, if MSHA finds that a mine operator has failed to abate a cited violation and there is no valid reason to extend the abatement period, MSHA must issue a withdrawal order until the violation is abated.
- If a mine operator fails to abate a breathable dust violation for which a citation has been issued and the abatement period has expired, MSHA must either extend the abatement period or issue a withdrawal order.
- If two violations make up **unwarrantable failures** to comply with the standards that are found during the same inspection, or if the second unwarrantable violation is found within 90 days of the first, a withdrawal order must be issued.
- Miners may be ordered withdrawn from a mine if they have not received the safety training required by the act. Miners who are withdrawn for this reason are protected by the Mine Act from discharge or loss of pay.

6.2.3 Miner/Employee Training

Explain MSHA's miner/employee training requirements.

Show Transparency 13, Required Miner Training.

Mine operators are required to have a safety and health training program approved by MSHA. Miner training must include:

- At least 8 hours of annual refresher training are required for all miners.
- At least 40 hours of instruction for new underground miners including:
  - Rights of miners and their representatives under the act
  - Use of self-rescue and respiratory devices
  - Hazard recognition
  - Escapeways
  - Walk-around training
  - Emergency procedures
  - Basic ventilation
  - Basic roof control
  - Electrical hazards
  - First aid
  - Safety and health aspects of the task assignment
- At least 24 hours of instruction are required for new surface miners. The training must include all of the same items for underground miners except:
  - Escapeways
  - Basic ventilation
  - Basic roof control

> **NOTE**
> The Mine Act requires that the training be conducted during normal working hours and that the miners must be paid at their normal rate during the training period. Regulations concerning training and retraining of miners are stated in *Title 30 CFR, Part 48.*

6.2.4 Accident, Injury, and Illness Reporting

Occupational injury means an injury that results in death, loss of consciousness, medical treatment, temporary assignment to other duties, transfer to another job, or inability to perform all duties on any day after the injury. Occupational illness is an illness or disease that may have

INTRODUCTION TO SAFETY TECHNOLOGY 1.11

Explain MSHA's accident, incident, and illness reporting requirements.

Show Transparency 14, MSHA Defined Accidents.

Explain the purpose and function of the EPA.

Discuss the EPA laws that apply to the construction industry.

Refer the participants to the bulleted list in the Participant Module.

resulted from work at a mine or mine material processing facility for which a workers' compensation award is made.

> **NOTE**
> The definition of occupational injury described here is MSHA specific, not OSHA specific.

All mine operators are required to immediately report accidents to the nearest MSHA district or sub-district office. Also, operators must investigate and submit to MSHA, upon request, an investigation report on accidents and occupational injuries.

MSHA defines the following as accidents for the purpose of reporting accidents, injuries, and illnesses under the Mine Act:

- An unplanned flood of a mine by a liquid or gas
- A fatality at a mine
- An injury to an individual at a mine that has a reasonable potential to result in the worker's death
- An injury that may result in death
- Entrapment for more than 30 minutes
- An unplanned ignition or explosion of gas or dust
- An unplanned fire not extinguished within 30 minutes of its discovery
- An unplanned ignition or explosion of a blasting agent or an explosive
- An unplanned roof fall in active work areas where roof bolts are in use, or a roof fall that impairs ventilation or impedes passage
- Coal or rock outbursts that cause withdrawal of miners or that disrupt mining activity for more than one hour
- An unstable condition at an impoundment, refuse pile, or **culm** bank requiring emergency action
- Damage to hoisting equipment in a shaft or slope that endangers an individual or interferes with use of equipment for more than 30 minutes

> **NOTE**
> For more information about MSHA regulations and procedures, visit their web site at www.msha.gov.

6.3.0 Environmental Protection Agency (EPA)

The U.S. Environmental Protection Agency was established in 1970 to consolidate in one agency a variety of federal research, monitoring, standard-setting and enforcement activities to ensure environmental protection. The EPA's mission is to protect human health and to safeguard the natural environment including the air, water, and land upon which life depends. Therefore, it is important to understand how work being done on a construction site affects the environment. This includes knowing the major environmental regulations covering the release, treatment, storage, and disposal of potentially hazardous materials into air, water, and soil. As a safety technician, you will need to have a working knowledge of the following EPA laws:

- *Comprehensive Environmental Response, Compensation, and Liability Act (CERCLA)* – This law imposes liability on owners or operators of a facility from which there is a release of hazardous substances into the environment.
- *Resource Conservation and Recovery Act (RCRA)* – This law regulates the generation, treatment, storage, and disposal of hazardous waste. RCRA requires an owner/operator of a facility to undertake corrective action to clean up a facility used for the treatment, storage, or disposal of hazardous waste. There is also a complex permit program with which the owner/operator must comply.
- *Underground Storage Tank Regulations (under RCRA and state control)* – This law imposes operating, reporting, financial assurance, and potential cleanup obligations on persons and companies owning and operating underground petroleum storage tanks (USTs). The rules and regulations covering USTs are quite detailed and prescriptive. In addition, many of the states have their own regulations and enforcement policies.
- *Clean Air Act (CAA) of 1990* – This act is a complex, multi-faceted statute that is designed to regulate air emissions from stationary and mobile sources.
- *Clean Water Act (CWA)* – This act prohibits the discharges of pollutants from point sources and storm water into navigable waters of the United States without a permit. The act imposes liability on the person who is responsible for the operation and/or equipment that results in a discharge.

Instructor's Notes:

- *Safe Drinking Water Act (SDWA)* – This act imposes federal drinking water standards on virtually all public water systems. The act requires the establishment of drinking water standards for maximum contaminant levels (MCLs).
- *Toxic Substances Control Act (TSCA)* – This act governs the manufacture and use of chemical products. The importing and exporting of chemicals are regulated under the act. In addition, there are specific regulations regarding the use, management, storage, and disposal of polychlorinated biphenyl (PCB) materials.
- *Oil Pollution Act of 1990* – This act imposes strict liability on responsible parties for removal costs and damages resulting from discharges of oil into navigable waters of the United States. An owner/operator of an onshore facility that resulted in a discharge would be considered a responsible party.

6.4.0 Driver's License Programs

Driving certain commercial motor vehicles (CMVs), such as heavy construction equipment, requires special skills and knowledge. Prior to the implementation of the federal Commercial Driver's License (CDL) Program, in a number of States and the District of Columbia, any person licensed to drive an automobile could also legally drive heavy equipment, a tractor-trailer, or a bus. Even in many of the states that did have a classified licensing system, a person's skills were not tested on the type of vehicle that he or she would be driving. As a result, many drivers were operating motor vehicles that they were not qualified to drive. This led to many serious accidents. Fortunately, because of CDL programs, this is no longer the case.

As a safety technician, you may be responsible for knowing CDL rules and regulations that affect your company. You will need to make sure that any worker operating heavy equipment or other commercial vehicles has a current license. You must also know the regulatory requirements established by the Department of Transportation's Commercial Driver's License Program and the standards set by the *Commercial Motor Vehicle Safety Act of 1986*.

6.4.1 Commercial Motor Vehicle Safety Act of 1986

The *Commercial Motor Vehicle Safety Act of 1986* was signed into law on October 27, 1986. The goal of the act is to improve highway safety by ensuring that drivers of large trucks, heavy equipment, and buses are qualified to operate those vehicles and to remove unsafe and unqualified drivers from the highways. The act retained the state's right to issue a driver's license, but established minimum national standards that states must meet when licensing CMV drivers.

The act makes it illegal to hold more than one license by requiring states to adopt testing and licensing standards for truck and bus drivers. This helps verify a person's ability to operate a specific type of vehicle. It also cuts back on the number of drivers with multiple traffic tickets. Because drivers can only have one license, they cannot register in another state once they've accumulated several citations in order to have a clean record.

It is important to note that the act does not require drivers to obtain a separate federal license. It only requires states to upgrade their existing testing and licensing programs, if necessary, and conform to the federal minimum standards.

One of the outcomes of the act is that drivers have been required to have a CDL in order to drive a CMV since April 1, 1992. The Federal Highway Administration (FHWA) developed and issued standards for testing and licensing CMV drivers. Among other things, the standards require states to issue CDLs to their CMV drivers only after the driver passes knowledge and skills tests administered by the state related to the type of vehicle to be operated. Drivers need CDLs if they are in interstate, intrastate, or foreign commerce and drive a vehicle that meets one of the following classifications of a CMV:

- *Class A* – Any combination of vehicles with a **gross vehicle weight rating** (GVWR) of 26,001 or more pounds, provided the **gross unloaded vehicle weight** (GUVW) of the vehicle(s) being towed is in excess of 10,000 pounds
- *Class B* – Any single vehicle with a GVWR of more than 26,001 pounds, or any vehicle towing another vehicle that is not in excess of a 10,000 pounds GUVW
- *Class C* – Any single vehicle or combination of vehicles that does not meet the definition of Class A or Class B but is either designed to transport 16 or more passengers, including the driver, or is designated for hazardous materials

6.4.2 Endorsements and Restrictions

Drivers who operate special types of CMVs need to pass additional tests in order to be properly licensed. *Table 1* shows the test subjects and the endorsement letter that is assigned to a CDL after the test is passed.

Explain the purpose of commercial driver's license (CDL) programs.

Explain that the safety technician is responsible for knowing the CDL rules and requirements that affect the company.

Describe the requirements of the Commercial Motor Vehicle Safety Act of 1986, including CMV classifications.

Discuss the additional tests a CMV driver must pass in order to operate specific CMVs.

Show Transparency 15 (Table 1).

INTRODUCTION TO SAFETY TECHNOLOGY 1.13

Explain the criteria that allow good drivers to be grandfathered from the skills test for a CDL.

Describe the conditions a driver must verify during the two-year period prior to applying for a CDL.

Explain the state and federal requirements for the CDL document.

Show Transparency 16, CDL Document Requirements.

Explain the state and federal CDL waiver provisions.

Table 1 CMV Endorsement Letter Designations

Test subject	Endorsement
Double/Triple trailers	T
Passenger	P
Tank vehicle	N
Hazardous materials	H
Combination of tank vehicle and hazardous materials	X

201T01.EPS

> **NOTE**
> If a driver either fails the air brake component of the general knowledge test or performs the skills test in a vehicle not equipped with air brakes, the driver is issued an air brake restriction, prohibiting the driver from operating a CMV equipped with air brakes.

6.4.3 Grandfathering Provision

Drivers with good driving records may be **grandfathered** from the skills test according to the following criteria:

- Driver has a current license at time of application
- Driver has a good driving record and previously passed an acceptable skills test
- Driver has a good driving record in combination with certain driving experience

In order for a driver's record to be considered good, a driver must verify that during the two-year period immediately prior to applying for a CDL they:

- Have not had more than one license
- Have not had any license suspended, revoked, or canceled
- Have not had any convictions or any type of offenses relating to motor vehicles
- Have not had more than one conviction for serious traffic violations while operating any type of motor vehicle
- Have not had any violation of state or local law relating to motor vehicle traffic control in connection with any traffic accident
- Have no record of an accident in which they were at fault

6.4.4 Commercial Driver's License Document

A state determines the license fee, license renewal cycle, most renewal procedures, and continues to decide the age, medical, and other driver qualifications of its intrastate commercial drivers. Interstate drivers must meet the longstanding federal driver qualifications (*49 CFR 391*).

All CDLs must contain the following information:

- The words Commercial Driver's License or CDL
- The driver's full name, signature, and address
- The driver's date of birth, sex, and height
- A color photograph or digital image of the driver
- The driver's state license number
- The name of the issuing state
- The date of issuance and the date of the expiration of the license
- The class(es) of vehicle that the driver is authorized to drive
- Notation of the air brake restriction, if issued
- The endorsement(s) for which the driver has qualified

States may issue learner's permits for purposes of behind-the-wheel training on public highways as long as learner's permit holders are required to be accompanied by someone with a valid CDL appropriate for that vehicle and the learner's permit.

6.4.5 Waiver Provisions

All active duty military drivers were waived from the CDL requirements by the Federal Highway Administrator. A state, at its discretion, may waive firefighters, emergency response vehicle drivers, farmers, and drivers removing snow and ice in small communities from the CDL requirements, subject to certain conditions.

Instructor's Notes:

In addition, a state may also waive the CDL knowledge and skills testing requirements for seasonal drivers in farm-related service industries and waive certain knowledge and skills testing requirements for drivers in remote areas of Alaska. The drivers are issued restricted CDLs. A state can also waive the CDL hazardous materials endorsement test requirements for part-time drivers working for the pyrotechnics (fireworks) industry, subject to certain conditions.

6.4.6 Other Requirements

There are several other requirements related to the CDL legislation that affect commercial drivers, their employers, and states with CDL programs. They include:

- *Penalties* – The federal penalty to a driver who violates the CDL requirements is a civil penalty of up to $2,500 or, in aggravated cases, criminal penalties of up to $5,000 in fines and/or up to 90 days in prison. An employer is also subject to a penalty of up to $10,000 if he or she knowingly uses a driver to operate a CMV without a valid CDL.
- *CDLIS clearinghouse* – States must be connected to the Commercial Driver's License Information System (CDLIS) and the National Driver Register (NDR) to exchange information about CMV drivers and traffic convictions and disqualifications. A state must use both the CDLIS and NDR to check a driver's record and the CDLIS to make certain that the applicant does not already have a CDL. Motor carrier employers also have access to the CDLIS.
- *Blood alcohol content standards* – The Federal Highway Administration (FHWA) has established 0.04% as the blood alcohol concentration (BAC) level at or above which a CMV driver is deemed to be driving under the influence of alcohol and subject to the disqualification sanctions in the act. States maintain a BAC level between 0.08% and 0.10% for non-CMV drivers.
- *Employee notifications* – Within 30 days of a conviction for any traffic violation, except parking, a driver must notify his/her employer, regardless of the nature of the violation or the type of vehicle that was driven at the time.

  If a driver's license is suspended, revoked, or canceled, or if the employee is disqualified from driving, FHWA must notify the employer. The notification must be made by the end of the next business day following receipt of the notice of the suspension, revocation, cancellation, lost privilege, or disqualification.

  Employers may not knowingly use a driver who has more than one license or whose license is suspended, revoked, canceled, or is disqualified from driving. Violation of this regulation may result in civil or criminal penalties.
- *Disqualifications* – The following is a list of reasons for disqualifications from the CDL program.
  - For conviction while driving a CMV
  - Two or more serious traffic violations within a 3-year period
  - One or more violations of an out-of-service order within a 10-year period
  - Driving under the influence of a controlled substance or alcohol
  - Leaving the scene of an accident
  - Using a CMV to commit a felony
  - Using a CMV to commit a felony involving manufacturing, distributing, or dispensing controlled substances.

Summary

The role of safety technician on a construction site is an important one. You are responsible for conducting safety inspections; creating job safety plans; organizing and conducting safety training; investigating and maintaining records of accidents, incidents, injuries, and illnesses. You are also responsible for knowing how to prevent accidents and understanding the effects accident costs have on a company. Another important part of your job is understanding government regulatory requirements that must be followed on a work site including the Department of Transportation Commercial Driver's License Program.

Explain the other requirements related to CDL legislation, including penalties, the CDLIS clearinghouse, blood alcohol standards, employee notifications, and disqualifications.

Review the material covered in the module.

Have the participants complete the Review Questions and go over the answers prior to administering the Module Examination.

Administer the Module Examination. Record the results on Craft Training Report Form 200 and submit the results to the Training Program Sponsor.

INTRODUCTION TO SAFETY TECHNOLOGY 1.15

Review Questions

1. Conducting safety meetings is a safety technician's responsibility.
   a. True
   b. False

For Questions 2 through 5, match the safety-related term to the corresponding definition.

Safety Term

2. ____ Safety
3. ____ Risk
4. ____ Hazards
5. ____ Accidents

Description

a. Conditions, changing sets of circumstances, or behaviors that present the potential for injury, illness, or property damage
b. Unplanned and sometimes harmful or damaging events that interrupt the normal progress of a job
c. The chance of injury, damage, or loss
d. A general term that denotes an acceptable level of risk

For Questions 6 through 8, match the level of accident causation to the corresponding description.

Level of Accident Causation

6. ____ Level I
7. ____ Level II
8. ____ Level III

Description

a. The root causes of accidents
b. The direct causes of accidents
c. The indirect causes of accidents

9. All of the following are basic components of a safety program *except* _____.
   a. a management policy statement
   b. a policy on alcohol and drug abuse
   c. emergency action plans
   d. conflict resolution programs

10. Leaving the scene of an accident is *not* a valid reason for disqualification from the Commercial Driver's License (CDL) program.

Instructor's Notes:

PROFILE IN SUCCESS

Harry Conerly, Universal Ensco, Inc.
Corporate Safety Manager

I work for Universal Ensco, Inc., an international engineering, project management, and field services firm serving the pipeline, oil and gas, telecommunications, and electric power industries worldwide.

Being the Corporate Safety Manager, my responsibilities include field work (audits and training) as well as audits and training in the home office. With over 100 survey crews in the field, it becomes challenging to see that everyone is trained and has the proper documentation for their task, as well as the proper training (OQ, Operator Qualification) and documentation for operating our fleet of 4-wheelers and suburbans. The new-hire orientation becomes a challenge, to say the least. Setting up the corporate safety committee was even more challenging. The committee is made up of all levels of staff, divided into five subcommittees, (Evacuation response, Emergency Response, Fleet Safety, Ergonomics, and Safety Awareness) which all make up our team approach to safety.

I came into the safety field at the request of a friend who needed a safety site person on a project in Kansas for a year. I had just sold my funeral home, flower shop, and hair salon, and was looking to get into something different: this was different. My friend told me because of my chemistry background in mortuary school that I would be easy to train. I went through all their training: hazwoper, confined space, trenching, and so forth.

When I came to Universal Ensco, Inc., I came with years of experience in the field in all sorts of projects, such as demolition, landfill construction, Corp of Engineer projects, RCRA, EPA, Superfund sites, wetlands, Air Force, and many more.

What really helped me was becoming involved with NCCER and the structure they gave to the construction industries. For the first time, a person coming into safety or thinking about safety could see a career ladder. Before you just took a course here or there and maybe it would get you the next position.

Now a person who doesn't have a college degree but is gathering experience in the construction industry can go all the way up the career ladder to the Construction Site Safety Program (CSSP) and sit for the national exam. Thanks to NCCER, all the material and courses are of a functional nature and target all employees at all levels. Now that we have the Operators Qualification Program for pipeline it also enables a person just starting out to go through a systematic program on covered tasks with the material supplied by NCCER for a rewarding career that offers job satisfaction and promotions.

INTRODUCTION TO SAFETY TECHNOLOGY 1.17

Instructor's Notes:

GLOSSARY

Trade Terms Introduced in This Module

Construction User's Round Table (CURT): CURT describes itself as an autonomous organization that provides a forum for the exchange of information, views, practices and policies of various owners at the national level. Similar groups, called Local User Councils, function at the local level and seek to address problems of cost, quality, safety and overall cost effectiveness in their respective areas.

Culm: Coal refuse.

Direct labor costs: Costs that can be directly related to an accident such as medical costs, workers' compensation insurance, benefits, and liability and property damage insurance payments.

Experience modification rate (EMR): A numeric factor used in determining workers' compensation costs. It rises for contractors with poor accident experience and falls for those with good accident experience.

Grandfather: To permit continuation of certain actions or conditions that existed before passage of a law or regulation that prohibits of affects the action or condition.

Gross income: Income before expenses.

Gross vehicle weight rating (GVWR): The maximum allowed gross vehicle weight for a vehicle.

Gross unloaded vehicle weight (GUVW): Also known as dry weight. This is the weight of a liquid cargo trailer without liquids.

Profit: Net income over a given period of time.

Release of energy: Events or conditions that release energy from systems, machines, or pieces of equipment.

Surface mines: Mines in which the mining operation is done near the surface of the ground.

Unwarrantable failures: An unwarrantable failure violation (second within 90 days) refers to a situation in which the mine operator knew or should have known that a violation existed and yet failed to take corrective action.

INTRODUCTION TO SAFETY TECHNOLOGY 1.19

Instructor's Notes:

Additional Resources

This module is intended to present thorough resources for task training. The following reference works are suggested for further study. These are optional materials for continued education rather than for task training.

www.osha.gov

www.msha.gov

www.dot.gov

www.epa.gov

www.asse.org

Basic Safety Administration: A Handbook for the New Safety Specialist, 2003. Fred Fanning, CSP. Des Plaines, IL: The American Society of Safety Engineers (ASSE).

Construction Safety Planning, 1995. David V. MacCollum, P.E., CSP. Hoboken, NJ: John Wiley & Sons.

Figure Credits

Professional Safety Associates, Inc.	201F03

INTRODUCTION TO SAFETY TECHNOLOGY 1.21

MODULE 75201-03 — ANSWERS TO REVIEW QUESTIONS

Answer	Section
1. a	2.0.0
2. d	3.0.0
3. c	3.0.0
4. a	3.0.0
5. b	3.0.0
6. b	4.1.1
7. c	4.1.2
8. a	4.1.3
9. d	5.0.0
10. b	6.4.6

CONTREN™ LEARNING SERIES — USER UPDATES

The NCCER makes every effort to keep these textbooks up-to-date and free of technical errors. We appreciate your help in this process. If you have an idea for improving this textbook, or if you find an error, a typographical mistake, or an inaccuracy in NCCER's Contren™ textbooks, please write us, using this form or a photocopy. Be sure to include the exact module number, page number, a detailed description, and the correction, if applicable. Your input will be brought to the attention of the Technical Review Committee. Thank you for your assistance.

Instructors – If you found that additional materials were necessary in order to teach this module effectively, please let us know so that we may include them in the Equipment/Materials list in the Instructor's Guide.

Write: Curriculum Revision and Development Department
National Center for Construction Education and Research
P.O. Box 141104, Gainesville, FL 32614-1104

Fax: 352-334-0932

E-mail: curriculum@nccer.org

Craft ______________________ Module Name ______________________

Copyright Date ____________ Module Number ____________ Page Number(s) ____________

Description

____

____

____

____

(Optional) Correction

____

____

____

(Optional) Your Name and Address

____

____

____

Module 75202-03

Hazard Recognition, Evaluation, and Control

NCCER STANDARDIZED CRAFT TRAINING PROGRAM

The National Center for Construction Education and Research (NCCER) provides a standardized national program of accredited craft training. Key features of the program include instructor certification, competency-based training, and performance testing. The program provides trainees, instructors, and companies with a standard form of recognition through a National Craft Training Registry. The program is described in full in the *Guidelines for Accreditation*, published by the NCCER. For more information on standardized craft training, contact the NCCER by writing us at P.O. Box 141104, Gainesville, FL 32614-1104; calling 352-334-0911; or e-mailing info@nccer.org. More information may be found at our Web site, www.nccer.org.

HOW TO USE THIS ANNOTATED INSTRUCTOR'S GUIDE

Each page presents two sections of information. The larger section displays each page exactly as it appears in the Participant Module. The narrow column ties suggested participant and instructor actions to each page and provides icons (detailed below) to call your attention to material, safety, audiovisual, or testing requirements. The bottom of each page includes space for your notes.

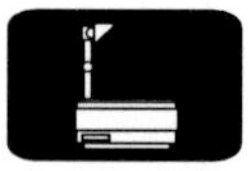

The **Audiovisual** icon indicates an appropriate time to show a transparency or other audiovisual aid.

The **Classroom** icon prompts you to define a term, stress a point, ask participants to explain a concept, or give examples.

The **Demonstration** icon directs you to show participants how to perform tasks.

The **Examination** icon tells you to administer the written module examination.

The **Homework** icon is placed where you may wish to assign reading for the next class, to assign a project, or to advise participants to prepare for an examination.

The **Laboratory** icon is used when participants are to practice performing tasks.

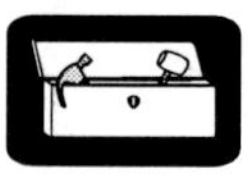

The **Materials** icon is a reminder for you to gather materials needed for classes, labs, and testing.

The **Performance Testing** icon tells you to administer a performance test or a portion thereof.

The **Safety** icon is used to emphasize safety issues. It is often keyed to *Caution* and *Warning* statements in the Participant Module.

The **Teaching Tip** icon indicates additional guidance is available, such as how to conduct an exercise, get the most educational value from a field trip, or encourage class participation. Teaching Tips may expand on a feature (*Think About It*, *Did You Know?*) or provide Quick Quizzes or similar exercises. You will be referred to the Teaching Tips section at the back of the module if there is additional material.

The **Combination** icon indicates that the laboratory listed corresponds with a performance task. If desired, you can note the proficiency of the participants during the laboratory and use it to satisfy performance testing requirements.

PREPARATION

Before teaching this module, you should review the Objectives, Performance Tasks, Materials and Equipment List, and the Module Outline. Be sure to allow ample time to prepare your own training or lesson plan and gather all required materials and equipment.

MODULE OVERVIEW

In this module, participants will learn how to identify hazards, evaluate the risk of hazards, and take the necessary steps to control or correct hazards.

PREREQUISITES

Please refer to the Course Map in the Participant Module. Prior to training with this module, it is recommended that the participant shall have successfully completed the following:

Field Safety; Safety Technology, Module 75201-03

OBJECTIVES

Upon completion of this module, the participant will be able to:

1. Recognize unsafe acts and conditions on a work site.
2. Describe the techniques for recognizing hazards.
3. Evaluate the risk associated with identified hazards.
4. Describe the seven major methods for controlling hazards.

PERFORMANCE TASKS

Under the supervision of the instructor, the participant should be able to:

1. Calculate risk using a written formula.

MATERIALS AND EQUIPMENT LIST

Overhead projector and screen
Transparencies
Whiteboard/chalkboard
Markers/chalk
Blank acetate sheets
Transparency pens
Pencils and scratch paper
Appropriate personal protective equipment
Copy of *29 CFR 1926, OSHA Construction Industry Regulations*
Task safety analysis (TSA) checklist
Horn
Bell
Whistle
Sign
Module Examinations*
Performance Profile Sheets*

* Located in the Test Booklet.

SAFETY CONSIDERATIONS

Ensure that participants are equipped with appropriate personal protective equipment.

ADDITIONAL RESOURCES

This module is intended to present thorough resources for task training. The following reference works are suggested for both instructors and motivated participants interested in further study. These are optional materials for continued education rather than for task training.

Web site: www.osha.gov

Web site: www.asse.org

The Psychology of Safety Handbook, 2001. E. Scott Geller, Ph.D. Boca Raton, FL: CRC/Lewis Publishers.

Root Cause Analysis Handbook: A Guide to Effective Incident Investigation, 1999. JBF Associates Division. Rockville, MD: Government Institutes.

TEACHING TIME FOR THIS MODULE

An outline for use in developing your lesson plan is presented below. Approximately 2½ hours are suggested to cover *Hazard Recognition, Evaluation, and Control*. This includes 10 minutes at the beginning of the session for administrative tasks and one 10-minute break. You may need to adjust the time required for hands-on activities and testing based on your class size and resources. Because laboratories often correspond to Performance Tasks, the proficiency of the participants may be noted during these exercises for Performance Testing purposes.

Topic	Planned Time
A. Introduction	________
B. Hazard Recognition	________
1. Accident/Incident Types and Energy Sources	________
2. Hazard Recognition Techniques	________
C. Hazard Evaluation	________
1. Evaluating the Risk	________
2. Laboratory Provide the participants with a scenario and have them use the formula to calculate risk. This laboratory corresponds with Performance Task 1.	________
D. Identifying the Causes	________
E. Methods of Hazard Control	________
1. Engineering or Substitution	________
2. Reduction of the Hazard Potential	________
3. Safety Devices	________
4. Administrative Controls	________
5. Warning Devices	________
6. Personal Protective Equipment	________
7. Worker Rotation	________
F. Review	________
G. Module Examination 1. Participants must score 70% or higher to receive recognition from the NCCER. 2. Record the testing results on Craft Training Report Form 200 and submit the results to the Training Program Sponsor.	________
H. Performance Testing 1. Participants must perform each task to the satisfaction of the instructor to receive recognition from the NCCER. If applicable, proficiency noted during laboratory exercises can be used to satisfy the Performance Testing requirements. 2. Record the testing results on Craft Training Report Form 200 and submit the results to the Training Program Sponsor.	________

Module 75202-03

Hazard Recognition, Evaluation, and Control

Instructor's Notes:

COURSE MAP

This course map shows all of the modules in Safety Technology. The suggested training order begins at the bottom and proceeds up. The local Training Program Sponsor may adjust the training order.

Assign reading of Module 75202-03.

SAFETY TECHNOLOGY

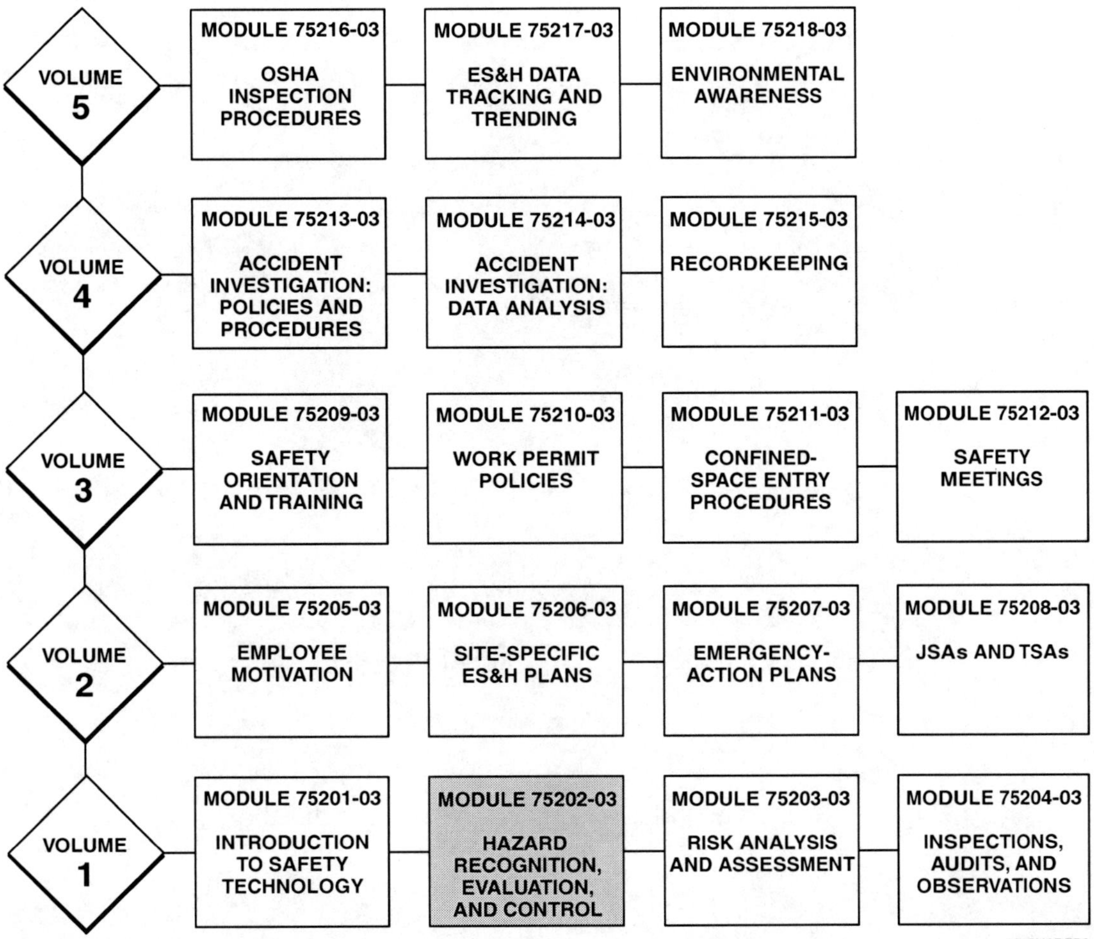

202CMAP.EPS

HAZARD RECOGNITION, EVALUATION, AND CONTROL 2.iii

Instructor's Notes:

MODULE 75202-03 CONTENTS

Figures

HAZARD RECOGNITION, EVALUATION, AND CONTROL 2.v

Instructor's Notes:

MODULE 75202-03

Hazard Recognition, Evaluation, and Control

Objectives

When you have completed this module, you will be able to do the following:

1. Recognize unsafe acts and conditions on a work site.
2. Describe the techniques for recognizing hazards.
3. Evaluate the risk associated with identified hazards.
4. Describe the seven major methods for controlling hazards.

Prerequisites

Before you begin this module, it is recommended that you successfully complete the following: Field Safety; Safety Technology, Module 75201-03.

Required Trainee Materials

1. Pencil and paper
2. Appropriate personal protective equipment
3. Copy of *29 CFR 1926, OSHA Construction Industry Regulations*

1.0.0 ◆ INTRODUCTION

The process of hazard recognition, evaluation, and control is the foundation of an effective safety program. When hazards are identified and assessed, they can be addressed quickly, reducing the hazard potential. Inspections, observations, audits, and new equipment evaluations should be part of a safety program. Each of these techniques helps to evaluate potential hazards and identify the best method of hazard control.

Safety technicians play an important role in identifying and controlling hazards. It is your responsibility to help determine what working conditions are unsafe, inform workers of noted hazards, and provide safe work alternatives. You should also advise workers to tell their supervisor about hazardous conditions they see. If the worker can correct the hazard, he or she should do so. This puts some of the responsibility of hazard recognition and control in the hands of the workers, making them active members of the safety team.

2.0.0 ◆ HAZARD RECOGNITION

There are a number of ways to recognize hazards and potential hazards on a work site. Some techniques are more complicated than others. In order to be effective, they all must answer this question; what could go wrong with this situation or operation? No matter what hazard recognition technique you use, you can be sure that answering that question in advance will save lives and protect equipment from damage.

2.1.0 Accident/Incident Types and Energy Sources

Accident/incident types and energy sources are considered potential hazards indicators. The best approach in determining if a situation or equipment is potentially hazardous is to ask yourself these questions:

- How can this situation or equipment cause harm?
- What types of energy sources are present that can cause an accident?
- What is the magnitude of the energy?

Ensure that you have everything you need to teach the course. Check the Materials and Equipment list in the front of this module.

Explain that terms shown in bold are defined in the Glossary at the back of this module.

Review the goals of the course and explain what will be expected of the participant.

Show Transparency 1, Objectives.

Show Transparency 2, Performance Task.

Explain why hazard recognition, evaluation, and control are the foundation of an effective safety program.

Describe the safety technician's role in hazard recognition, evaluation, and control.

Explain that there are different ways to recognize hazards.

Explain how recognizing accident/incident types and energy sources can identify potential hazards.

Describe the different types of accidents that can cause injury.

Describe the energy sources that are typically released during an uncontrolled release of energy. Refer the participants to the bulleted list in the Participant Module.

Identify the classic techniques of hazard recognition:
- **Job safety analysis (JSA)**
- **Task safety analysis (TSA)**
- **Safety inspections**
- **Pre-job planning checklists**
- **Recognizing accident types and energy sources**
- **Job observations**

Describe how JSAs are used to help identify hazards.

Show Transparency 3 (Figure 1).

Explain the purpose and use of TSAs.

Show Transparency 4 (Figure 2).

- What could go wrong to release the energy?
- How can the energy be eliminated or controlled?
- Will I be exposed to any hazardous materials?

Before you can fully answer these questions, you need to know the different types of accidents that can happen and the energy sources behind the accidents. Some of the different types of accidents that can cause injuries include:

- Falls on the same elevations or falls from elevations
- Being caught in, on, or between equipment
- Coming in contact with acid, electricity, heat, cold, radiation, pressurized liquid, gas, or toxic substances
- Being struck by falling objects
- Being cut by tools or equipment
- Exposure to high noise levels
- Repetitive motion or excessive vibration

When equipment is the cause of an accident, it is usually because there was an **uncontrolled release of energy**. The different types of energy sources that are typically released include:

- Mechanical
- Pneumatic
- Hydraulic
- Electrical
- Chemical
- Thermal
- Radioactive
- Gravitational
- Stored energy

2.2.0 Hazard Recognition Techniques

Several techniques have been developed that can help identify and correct hazards. Some of the classic techniques of hazard recognition are:

- Job safety analysis
- Task safety analysis
- Safety inspections
- Pre-job planning checklists
- Recognizing accident types and energy sources
- Job observations

2.2.1 Job Safety Analysis (JSA)

A job safety analysis (JSA), also known as job hazard analysis (JHA), is one approach to hazard recognition. In a JSA, the task at hand is broken down into its individual parts or steps and then each step is analyzed for its potential hazards. Once a hazard is identified, certain actions or procedures are recommended that will correct that hazard. For example, during a JSA, it is determined that using a come-a-long to install a pump motor in a tight space would be safer than having a worker do it manually. By using the come-a-long, the chance that the worker's hand would get crushed during installation is reduced. Using the JSA saved the worker from injury. *Figure 1* shows an example of a form used to conduct a JSA.

JSAs can also be used as pre-planning tools. This helps to ensure that safety is planned into the job. When JSAs are used as pre-planning, they contain the following information.

- Tools, materials, and equipment needs
- Staffing or manpower requirements
- Duration of the job
- Quality concerns

2.2.2 Task Safety Analysis (TSA)

Another approach to hazard recognition is the task safety analysis (TSA), also called a task hazard analysis (THA). TSAs are similar to JSAs in that both require workers to identify potential hazards and needed safeguards associated with a job they are about to do. The difference is the form that is used to report the hazard. During a TSA, a pre-printed, fill-in-the-blank checklist, like the one shown in *Figure 2,* is often used to document any hazard found during analysis. The conclusions found during a TSA should be discussed with the crew by the first-line supervisor or team leader before work begins. Some companies require workers to sign the completed TSA forms or checklists before they start work. This is so they can document that workers have been told of potential hazards and safety procedures.

> **NOTE**
> Job safety analyses and task safety analyses are explained in more detail in Volume 3, Module 75212-03, *JSAs and TSAs.*

2.2.3 Safety Inspection

Safety inspections should be done on a regular basis. Depending on the size of the company and company policies, they can be done on a daily,

Instructor's Notes:

JOB HAZARD ANALYSIS FORM

Page __ of __

Job Title:
Job Location:
PPE:
Tools, Materials & Equipment:

Date of Analysis:
Conducted By:
Staffing:
Duration:

Step	Hazard	Quality Concern	New Procedure or Protection

202F01.EPS

Figure 1 ◆ Job safety analysis (JSA) form.

Describe how and why safety inspections are done.

Explain how safety checklists can be used during safety inspections.

Show Transparency 5 (Figure 3).

weekly, monthly, semi-annual, or annual basis. For more information about safety inspections, please refer to Volume 1, Module 75204-03, *Inspections, Audits, and Observations.*

During a safety inspection, tools, equipment, and the work area are checked for hazards. It is important to plan in advance what will be inspected. Preparing for safety inspections helps to ensure that the inspection will be thorough and successful in identifying hazards. The following are questions to consider when preparing for a safety inspection:

- Have there been any reported accidents or near miss incidents?
- Who is responsible for inspecting each area?
- What tools, materials, or equipment need regular inspection?
- What did the last inspection reveal?
- Were the control actions adequate and completed?
- How frequently has the work area been inspected?
- Who will correct any noted deficiencies? How? When?

Problem areas and potential hazards can be dealt with faster and more effectively when inspections are approached in this way. Safety inspections checklists *(Figure 3)* are a good way to make sure safety inspections are done properly and that nothing is missed.

HAZARD RECOGNITION, EVALUATION, AND CONTROL 2.3

Provide the participants with a scenario and have them complete a TSA checklist.

TASK SAFETY ANALYSIS CHECKLIST			
Name:	Location:		
Signature:	Date:		
	Yes	**No**	**N/A**
1. Have underground utilities been located prior to excavation?			
2. In areas where there are known or suspected unexploded ordinance, has the area been cleared by qualified explosive ordinance disposal (EOD) personnel?			
3. Are excavations, the adjacent areas, and protective systems inspected and documented daily?			
4. Are excavations over 5 feet in depth adequately protected by shoring, trench box or sloping?			
5. When excavations are undercut, is the overhanging material safely supported?			
6. Have methods been taken to control the accumulation of water in excavations?			
7. Are employees protected from falling materials (loose rock or soil)?			
8. Are substantial stop logs or barricades installed where vehicles or equipment are used or allowed adjacent to an excavation?			
9. Have steps been taken to prevent the public, workers or equipment from falling into excavations?			
10. Are all wells, calyx holes, pits, shafts, etc. barricaded or covered?			
11. Are walkways provided where employees or equipment are required or permitted to cross over excavations?			
12. Where employees are required to enter excavations is access/egress provided every 25 feet laterally?			
13. For excavations less than 20 feet, is the maximum slope 1-1/2 horizontal to 1 vertical?			
14. Are support systems drawn from manufacturer's tabulated data in accordance with all manufacturer's specifications?			
15. Are copies of the tabulated data maintained at the job site?			
16. Are members of support systems securely connected together?			
17. Are shields installed in a manner to restrict lateral or other hazardous movement?			
Comments:			

202F02.EPS

Figure 2 ◆ Task safety analysis (TSA) checklist.

Instructor's Notes:

SAFETY INSPECTION CHECKLIST FOR SCAFFOLDING

1. Are scaffolds and scaffold components inspected before each work shift by a competent person?

   _______ YES _______ NO

2. Have employees who erect, disassemble, move, operate, repair, maintain, or inspect the scaffold been trained by a *competent person* to recognize the hazards associated with this type of scaffold and with the performance of their duties related to this scaffold?

   _______ YES _______ NO

3. Have employees who use the scaffold been trained by a qualified person to recognize the hazards associated with this scaffold and know the performance of their duties related to it?

   _______ YES _______ NO

4. Is the maximum load capacity of this scaffold known and has it been communicated to all employees?

   _______ YES _______ NO

5. Is the load on the scaffold (including point loading) within the maximum load capacity of this particular scaffold?

   _______ YES _______ NO

6. Is the scaffold plumb, square, and level?

   _______ YES _______ NO

7. Is the scaffold on base plates and are mudsills level, sound, and rigid?

   _______ YES _______ NO

8. Is there safe access to all scaffold platforms?

   _______ YES _______ NO

9. Are all working platforms fully planked?

   _______ YES _______ NO

10. Do planks extend at least 6" and no more than 12" over the supports?

    _______ YES _______ NO

11. Are the planks in good condition and free of visible defects?

    _______ YES _______ NO

12. Does the scaffold have all required guardrails and toe-boards?

    _______ YES _______ NO

13. Are 4:1 (height to width) scaffolds secure to a building or structure as required?

    _______ YES _______ NO

202F03.EPS

Figure 3 ◆ Safety inspection checklist for scaffolding.

Take the participants on a field trip where they can observe a safety inspection being conducted with a checklist.

Discuss the pre-planning questions that should be considered when preparing for a safety inspection. Refer the participants to the bulleted list in the Participant Module.

Explain the importance of using job observations to help eliminate unsafe behavior. Remind the participants that 85% of all accidents happen because of unsafe behavior.

Discuss the questions that should be considered when observing workers. Describe the actions that should be taken once unsafe behavior has been observed.

Describe how risk factors into hazard evaluation. Explain the terms probability and consequences.

Explain how categorizing and prioritizing hazards is an important part of evaluating risk.

Discuss the three categories used to determine the seriousness of a hazard.

Describe the formula that helps prioritize risk.

Show Transparency 6, Risk Formula.

2.2.4 Pre-Job Planning Safety Checklist

Before each job starts, an assessment of the hazards, risks, and associated safety needs for that job should be completed. The following safety pre-planning questions should be considered during the assessment process.

- Are there any special laws or regulations that will affect job-site operations?
- Are there any hazardous processes, materials, or equipment that are going to be on site at the same time?
- What personal protective equipment or safety gear is needed for the job?
- Have workers been trained on any special tools and equipment involved in the job?
- Are all of the necessary work permits in place?

Knowing the answers to these questions, and addressing any potential problems before a job starts, will help prevent accidents and injuries.

2.2.5 Job Observations

Approximately 85% of all accidents happen because of the unsafe behavior of workers. Because of that, there should be a program or procedure in place to make sure workers are doing their jobs safely. These programs are typically called job observation programs. With a job observation program, poor performance and at-risk behavior can be identified by actually watching people as they work. When observing workers on a job, consider the following questions to make sure workers are doing their jobs safely.

- Is the worker following the right procedure for the job?
- Is the worker using the right tools and equipment properly?
- Is the worker using personal protective equipment properly?
- Is the worker over-reaching or lifting improperly?
- Are others at risk because of this operation?
- Is the area clear of trash and debris?

If you find that workers are in imminent danger, immediately ask them to stop, express concern for their safety, let them know specifically what they are doing wrong, and ask them what steps need to be taken to do their jobs safely. In your discussions, try to find out why they are doing what they are doing. Try to determine if it is a training issue or a motivational issue. Be sure to discuss your findings with the workers' supervisor(s).

3.0.0 ◆ HAZARD EVALUATION

Risk is a measure of both the **probability** and **consequences** of an event. A safe operation is one in which there is an **acceptable level of risk**. For example, climbing a ladder has risk that is considered to be acceptable if the proper ladder is being used on the job, if it is erected correctly, and if it is in good condition. If any one of these conditions were different, climbing the ladder would not have an acceptable level of risk.

Once hazards have been identified, they must be evaluated. Typically this is done by determining an acceptable level of risk based on the classification and prioritization of the hazard. These tasks require a combination of good judgment and knowledge about the work process. For example, if you see every hazard as life threatening, or you shut down a job for a very insignificant hazard, you can cost the company a great deal of money and lose your credibility. On the other hand, if you are able to evaluate the risk and put the identified hazards in proper perspective, you will likely prevent the loss of both lives and money.

3.1.0 Evaluating the Risk

An important part of evaluating a hazard is categorizing and prioritizing it. This is helpful because there may be limited resources to handle all safety problems. Workers and work sites are safer when the most hazardous conditions are fixed first.

That's why it is important to learn how to classify and prioritize hazards. Determining the seriousness of the hazard will depend on these three categories:

- *Probability* – The chance that a given event will occur
- *Consequences* – The results of an action, condition, or event
- *Exposure* – The amount of time and the degree to which someone or something is exposed to an unsafe condition, material, or environment

This formula helps prioritize risks:

Probability + Consequences + Exposure = Risk

The formula works by assigning a numerical value to each hazard category. This is done by taking the following steps:

Step 1 Assign a value of 1 to 4 from the following probability categories:

1. Unlikely to happen
2. Possibly will happen in time
3. Probably will happen in time
4. Likely to happen very soon

Instructor's Notes:

> **NOTE**
> This step helps to determine the likelihood of an undesired event occurring.

Step 2 Assign a value of 1 to 4 from this list of consequences.

1. *Negligible* – An injury is not likely.
2. *Marginal* – Minor illness, injury, or property damage are likely.
3. *Critical* – Severe illness, injury, or property damage are likely.
4. *Catastrophic* – Death or permanent disability are likely.

Step 3 Assign a value of 1 to 4 based on the exposure to the hazard.

1. There is little exposure to the hazardous condition or task. Controls are in place to limit exposures and are deemed effective.
2. Exposure is small. A limited number of workers have been exposed to the hazard. Some controls are in place to limit or control exposure.
3. Exposure is still significant. At least 4 workers have been exposed to the hazard. There are no controls in place to limit or control exposures.
4. Total number of workers exposed to the task or hazardous condition in a day or shift is high, the frequency of exposure or duration of over-exposure to contaminants is high. Or, those exposed, though fewer in number, are not within your control, such as members of the general public, or the employees of another contractor.

Once a numerical value has been determined for each category, the numbers should be added together to get a total. They are then compared to a specific range of numbers designed to prioritize hazards. The range can vary from 3 to 12. A high number in the range means there is a greater risk of illness, injury, or death. A low number means there is a lower risk of illness, injury, or death. When the risk is high, the hazard should be placed at the top of the priority list and fixed immediately. *Figure 4* shows an example of how risk is calculated.

> **NOTE**
> These risk values should be considered as guidelines or a management tool, and are not intended to be used as a definitive measurement system.

EVENT: TRIPPING OVER DEBRIS IN THE WORK AREA

P = PROBABILITY
C = CONSEQUENCES
E = EXPOSURE

P = 4 (LIKELY TO HAPPEN VERY SOON)
C = 2 (MINOR ILLNESS, INJURY OR PROPERTY DAMAGE)
E = 1 (THERE IS LITTLE EXPOSURE TO THE HAZARDOUS CONDITION OR TASK)

P + C + E = 7

RISK: ON A SCALE OF 1 TO 12, THE RISK OF TRIPPING OVER DEBRIS IN THE WORK AREA IS MODERATE TO HIGH.

202F04.EPS

Figure 4 ◆ Calculating risk.

4.0.0 ◆ IDENTIFYING THE CAUSES

Once a hazard has been identified, the next step is to determine the causes so that corrective actions can be taken. In most instances, there are two types of fixes, the quick fix and the permanent fix. The quick fix, often referred to as an Immediate Temporary Control (ITC), includes marking the hazard, or repairing or replacing a damaged piece of equipment such as a broken ladder. The permanent or long-range fix addresses the root cause(s) of a hazard. Root causes are management system failures that failed to detect, correct, or anticipate the unsafe act or condition *(Figure 5)*. Root causes, when properly addressed will affect permanent change and reduce the likelihood of similar incidents.

Hazard analysis flow charts *(Figure 6)* can be helpful in identifying root causes and the associated root actions. The flow chart suggests steps that should be taken to fix this problem and ones similar to it. You will notice questions such as "Was the condition discovered during the last inspection? Should it have been?" and "Did anyone report the hazard? Determine why the hazard was not reported." These questions represent the recommended corrective actions or system fixes for the problem. They are designed to make you think of ways to improve hazard control. See the *Appendix* for more examples of hazard analysis flow charts.

Once the recommendations have been made and approved, deadlines and responsibilities for completing the correction must be assigned and communicated to both the management and employees. Recommendations should be tracked until the recommended actions have been

Explain each step of the risk formula in detail. Refer the participants to the Participant Module.

Show the participants how to calculate risk, using examples from personal experience.

Provide the participants with a scenario and have them use the formula to calculate risk. Note the proficiency of each participant. This laboratory corresponds with Performance Task 1.

Explain that identifying the causes of hazards helps with identifying corrective actions.

Describe the different methods for identifying hazard causes, specifically root cause analysis and hazard analysis flow charts.

Show Transparencies 7 and 8 (Figures 5 and 6).

HAZARD RECOGNITION, EVALUATION, AND CONTROL 2.7

![icon]

Identify the seven major methods of controlling hazards.

Show Transparency 9, Seven Major Methods of Controlling Hazards.

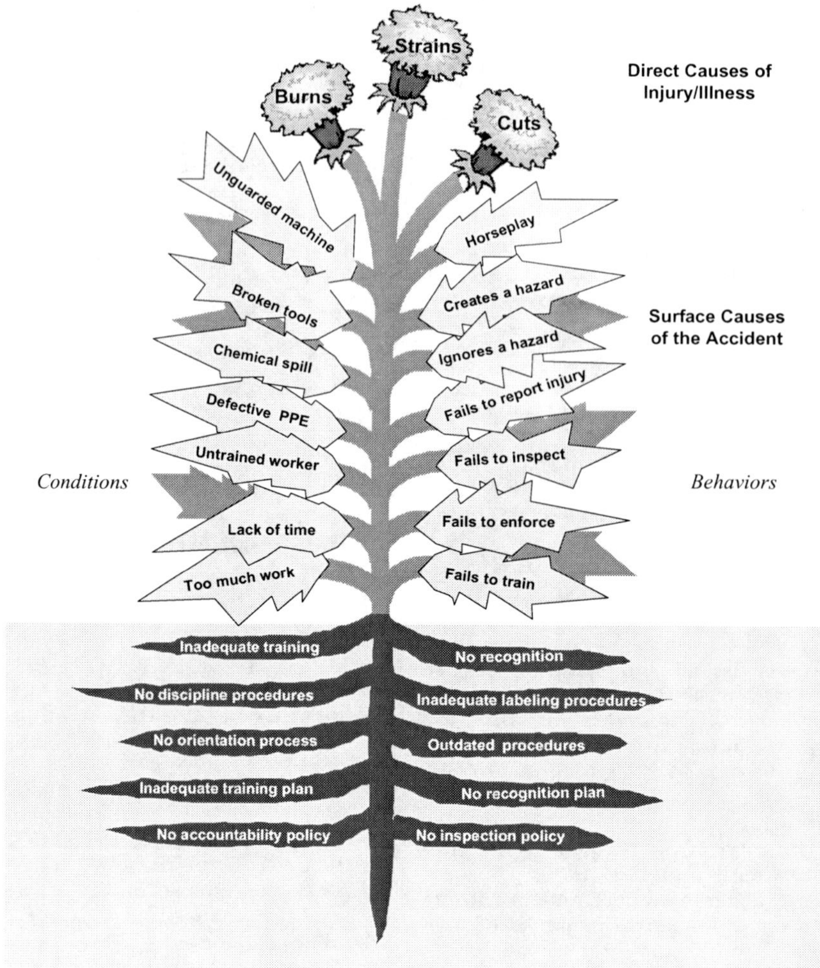

Figure 5 ◆ Root causes of accidents.

completed. Make sure there is a way to document the corrective actions that have been made and that they are effective. This will be helpful during the recordkeeping process. Temporary fixes must also have a deadline by which they must be completed; otherwise, quick fixes may become permanent.

> **NOTE**
> For more information about recordkeeping, refer to Volume 4, Module 75216-03, *Recordkeeping*.

5.0.0 ◆ METHODS OF HAZARD CONTROL

In addition to determining the root causes of hazards and the appropriate corrective action, it is important to understand how to control potential hazards. The following are the seven major methods of controlling hazards.

1. Engineering or substitution
2. Reduction of the hazard potential
3. Safety devices
4. Administrative controls

Instructor's Notes:

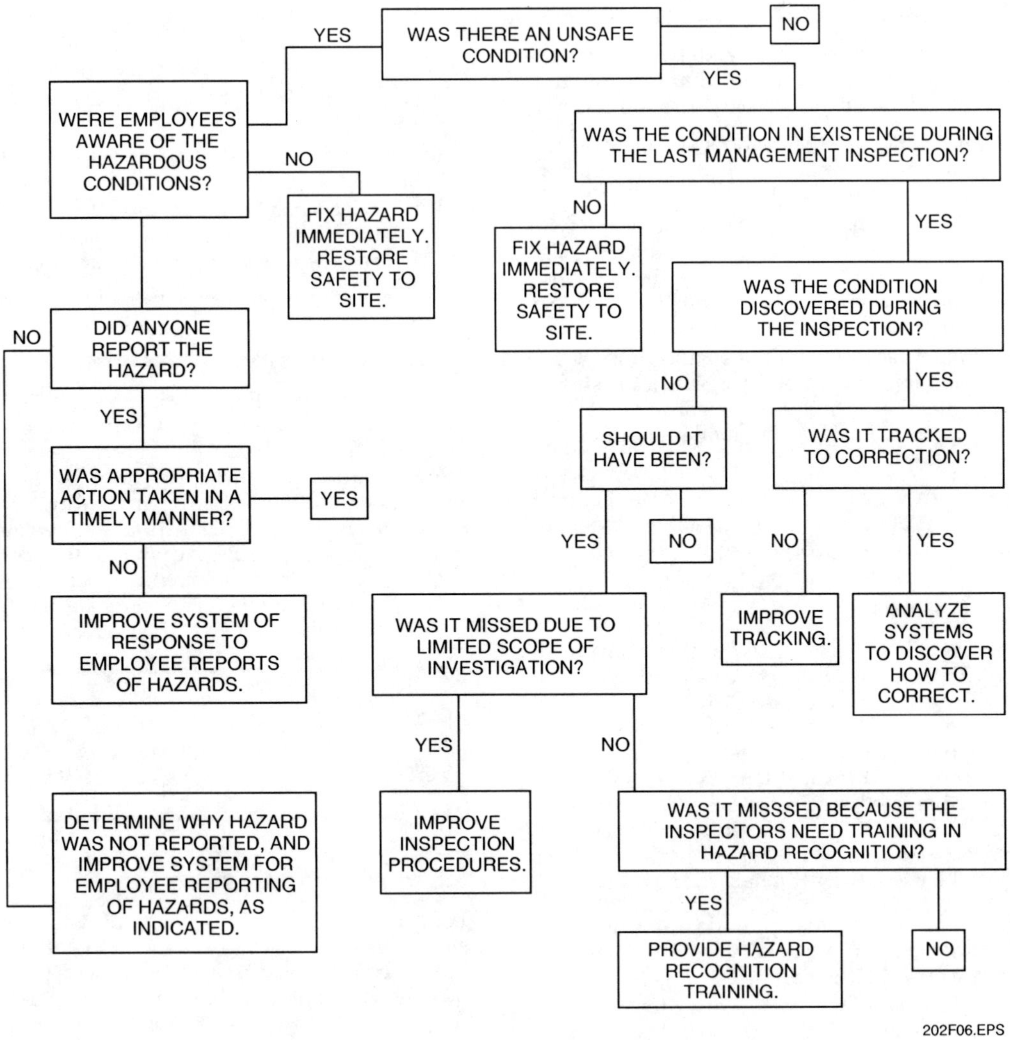

Figure 6 ◆ Root causes flow chart.

5. Warning devices
6. Personal protective equipment
7. Worker rotation

5.1.0 Engineering or Substitution

Sometimes equipment is hazardous because of a **flaw** in the design. For example, during an inspection, you may find that three machine guards have been removed to make it easier to lubricate pieces of equipment. You've found this problem several other times. One way to fix the problem is by redesigning the equipment. This can be done by extending grease fittings outside the guard, which will eliminate the need to remove the guard. Another possible engineering solution is to develop engineering design standards for lubricating equipment without removing the guards. Either way, you are controlling potential hazards by using engineering or substitution.

Explain why using engineering or substitutions can control hazards. Include personal experiences in your discussion.

Take the participants on a field trip and show them real examples of engineering or substitutions that help control hazards. Provide the participants with photographs if taking them on a field trip is not an option.

HAZARD RECOGNITION, EVALUATION, AND CONTROL 2.9

Explain the elements of a hazard potential reducing program.

Show Transparency 10, Elements of a Hazard Potential Reduction Program.

5.2.0 Reduction of the Hazard Potential

Hazards can be controlled by reducing the potential for hazards. Elements of a hazard potential reducing program usually have considerations for:

- *Distance* – Reduce the chances of a massive fire by relocating refueling areas 50' away from flammable materials.
- *Quantity* – Replace one 20-ton storage tank with five one-ton cylinders. This will reduce the size of a leak in the event of container failure because the amount of hazardous materials has been reduced.
- *Limiting personnel* – Train a small group of five individuals, instead of all workers on site, to perform all of the required hazardous work on a site. This will reduce the total number of persons exposed to the hazard.

By incorporating each of these elements into a hazard control program, you are taking the steps necessary to protect workers and equipment.

Describe the difference between a passive and active safety device. Show the participants an example of each.

5.3.0 Safety Devices

Safety devices can be classified as passive or active. An example of a passive safety device is an air bag in your car. You don't have any control over when the safety device is used. An example of an active safety device is the seat belt on a fork truck. The operator must decide to use the belt. When active safety devices are used, the company must develop and enforce policies and procedures mandating use. The types of safety devices that can be used to control hazards include:

- *Barricades* – Barricades provide a physical barrier that prevents workers from entering an unsafe area. *(PASSIVE)*
- *Power transmission guards* – Power transmission guards cover chains, belts, sprockets, or pulleys. *(ACTIVE)*
- *Point of operation guards* – Point of operation guards protect workers from the cutting action or movement of the tool. The bottom blade guard on a radial saw is an example of a point of operation guard. *(ACTIVE)*
- *Interlocks on safety gates* – Interlocks stop equipment when the safety gate is opened. *(ACTIVE)*
- *Hand switches* – Switches are located outside the danger zone and must be touched and/or held down for the machine to operate or cycle. An example of a hand switch is the safety bar on a power lawn mower. *(ACTIVE)*

Workers must also be trained on the hazards, safeguards, and limitations of the guards or safety devices that are used in their work. This ensures the effectiveness of using safety devices as a hazard-control method.

> **WARNING**
> Workers must be told that safety devices should never be destroyed, removed, or modified, no matter how inconvenient they might seem.

Explain the importance of administrative controls when trying to control hazards. Ask the participants to share some rules that they have been required to follow on site.

5.4.0 Administrative Controls

Safety rules, operating procedures, and maintenance procedures are examples of administrative controls. They are based on procedures that have been documented and formalized by management. Lockout/tagout, confined-space entry procedures, and work permits are examples of administrative controls. Administrative controls require training and enforcement to be effective. When administrative controls are specified, they should be audited to verify their effectiveness.

> **NOTE**
> For more information on audits, please refer to Volume 1, Module 75204-03, *Inspections, Audits, and Observations.*

Describe how warning devices are used to control hazards. Show the participants an example of a warning device such as a horn, bell, whistle, or sign.

5.5.0 Warning Devices

Warning devices can be horns, bells, whistles, and signs. They are used to alert workers about hazardous conditions. Their effectiveness as warning devices is sometimes limited, however, because they are not always distinguishable from other sounds on a work site. In some cases, workers hear alarms or signals so often, they unconsciously ignore them. In order for these warning signs to be effective, they must be distinctive and **audible** over **ambient noise levels**. Detailed messages on signs must be simple and easy to understand. Symbols, instead of words, may be used to help any workers with limited reading skills or language barriers. Warning devices are more effective as a hazard control method when they are clear and easy to understand. Make sure everyone on site knows what warning devices look like and/or how they sound.

Instructor's Notes:

5.6.0 Personal Protective Equipment

Personal protective equipment is the most basic method of hazard control. It should be considered the last line of defense as a method of hazard control because it protects individuals, but it does not necessarily create a safe work environment. Hazard controls should do both at all times. Make personal protective equipment an important part of your safety program, but understand that it must be used in combination with other hazard controls to be effective.

> **NOTE**
> For more information on personal protective equipment, please refer to Volume 1, Module 75103-03, *Personal Protective Equipment* of Field Safety.

5.7.0 Worker Rotation

Worker rotation is last method of controlling hazards and is actually an administrative control. Worker rotation simply involves limiting worker exposure in certain situations. Workers should be rotated if their work involves hot or cold work environments, constant or repetitive motion, or if they are bored. This can be accomplished by **cross-training** workers for different jobs on the site, adjusting the work schedule, or providing frequent breaks. It's important to rotate workers to ensure that everyone is alert and paying attention to safety.

Summary

Hazard recognition, evaluation, and control are critical components of an effective safety and health program. Most importantly, they play a big role in accident prevention. When accidents are prevented, the overall health and well being of everyone on site is improved, and operating costs remain affordable. As a safety technician, you are responsible for being actively involved in the process.

In this module, you learned the tools used to recognize hazards, as well as how to determine an acceptable level of risk. You also learned the methods used to control the hazards that cause accidents, injury, and death.

Review Questions

1. All of the following are hazard recognition techniques *except* _____.
   a. job observations
   b. task safety analysis
   c. pre-job planning checklists
   d. engineering and substitution

2. Safety inspection checklists are a good way to make sure JSAs are done correctly.
   a. True
   b. False

3. Risk is a measure of both probability and _____.
   a. an acceptable level or risk
   b. chance
   c. poor planning
   d. consequences

4. Determining the seriousness of a hazard will depend on probability, consequences, and _____.
   a. hazard control measures
   b. the acceptable level of risk
   c. exposure
   d. root causes

5. ITCs address the root causes of accidents.
   a. True
   b. False

6. Safety devices can be classified as active or passive.
   a. True
   b. False

7. Administrative controls are informal procedures created by front-line supervisors.
   a. True
   b. False

8. All of the following are considered warning devices *except* _____.
   a. horns
   b. bells
   c. hand signals
   d. signs

Explain that personal protective equipment is the most basic method of hazard control. Ask the participants to give examples of the types of personal protective equipment that is commonly used on work sites.

Describe how worker rotation can help control hazards.

Review the material covered in the module.

Have the participants complete the Review Questions and go over the answers prior to administering the Module Examination.

Administer the Module Examination. Record the results on Craft Training Report Form 200 and submit the results to the Training Program Sponsor.

Administer the Performance Test and fill out Performance Profile Sheets for each participant. If desired, participant proficiency noted during laboratory exercises may be used to complete the Performance Test. Record the testing results on Craft Training Report Form 200 and submit the results to the Training Program Sponsor.

HAZARD RECOGNITION, EVALUATION, AND CONTROL 2.11

9. The last line of defense for controlling hazards is _____.
   a. administrative controls
   b. warning devices
   c. worker rotation
   d. personal protective equipment

10. Workers should be rotated if there is a potential to become bored.
   a. True
   b. False

Instructor's Notes:

GLOSSARY

Trade Terms Introduced in This Module

Acceptable level of risk: The level of risk that is reasonable when working in hazardous conditions.

Ambient noise levels: Background noise that is related to the jobs done on a work site.

Audible: When a noise or sound is heard or capable of being heard.

Consequences: Something that happens as a result of a set of conditions or actions.

Cross-training: Training workers to do multiple jobs.

Flaw: A part of the design of equipment, parts, or a process that creates a hazard or operational or maintenance difficulties.

Probability: The chance that something will happen.

Uncontrolled release of energy: Energy that is released as result of an energy source that is uncontrolled. Energy sources can include tools, equipment, machinery, temperature, pressure, gravity, or radiation.

HAZARD RECOGNITION, EVALUATION, AND CONTROL 2.13

Instructor's Notes:

Hazard Analysis Flow Charts

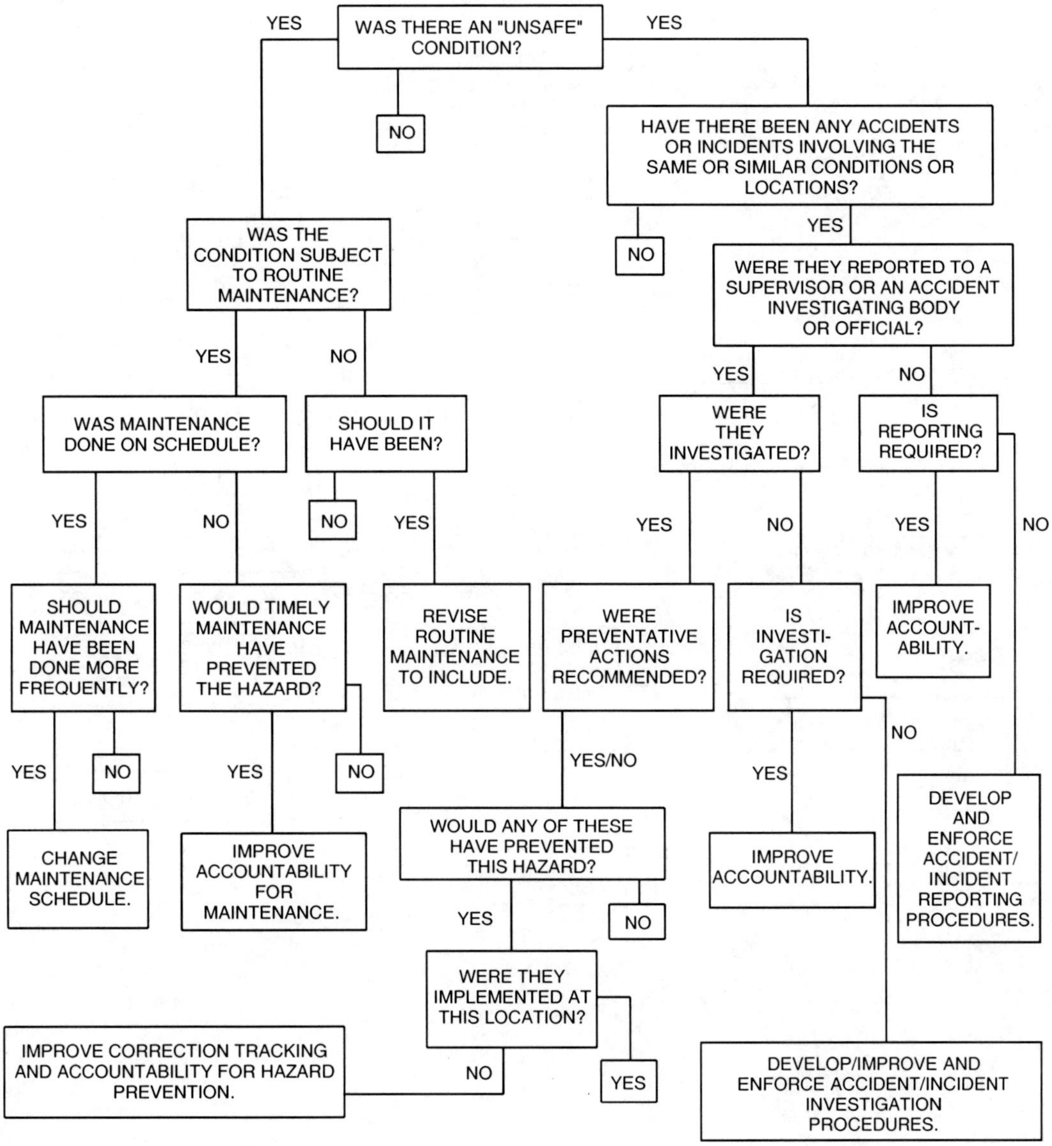

202A01.EPS

HAZARD RECOGNITION, EVALUATION, AND CONTROL 2.15

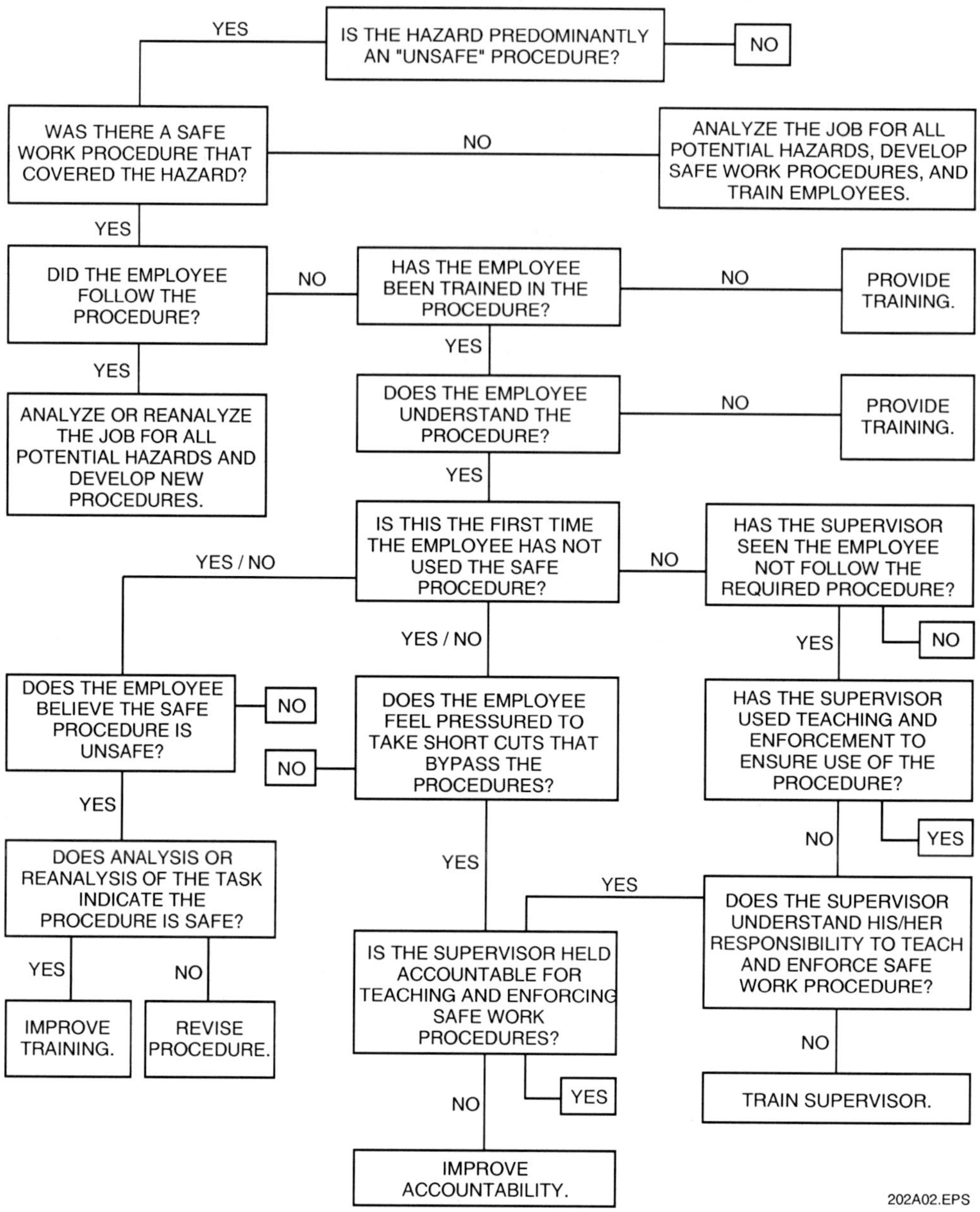

Instructor's Notes:

Additional Resources

This module is intended to present thorough resources for task training. The following reference works are suggested for further study. These are optional materials for continued education rather than for task training.

www.osha.gov

www.asse.org

The Psychology of Safety Handbook, 2001. E. Scott Geller,Ph.D. Boca Raton, FL: CRC/Lewis Publishers.

Root Cause Analysis Handbook: A Guide to Effective Incident Investigation, 1999. JBF Associates Division. Rockville, MD: Government Institutes.

Figure Credits

Professional Safety Associates, Inc.	202F01, 202F06, Appendix
US Army Corp of Engineers	202F02
OSHA Job Safety and Health Quarterly Spring 1999	202F03

HAZARD RECOGNITION, EVALUATION, AND CONTROL 2.17

MODULE 75202-03 — ANSWERS TO REVIEW QUESTIONS

Answer		Section
1.	d	2.2.0
2.	b	2.2.2
3.	d	3.0.0
4.	c	3.1.0
5.	b	4.0.0
6.	a	5.3.0
7.	b	5.4.0
8.	c	5.5.0
9.	d	5.6.0
10.	a	5.7.0

CONTREN™ LEARNING SERIES — USER UPDATES

The NCCER makes every effort to keep these textbooks up-to-date and free of technical errors. We appreciate your help in this process. If you have an idea for improving this textbook, or if you find an error, a typographical mistake, or an inaccuracy in NCCER's Contren™ textbooks, please write us, using this form or a photocopy. Be sure to include the exact module number, page number, a detailed description, and the correction, if applicable. Your input will be brought to the attention of the Technical Review Committee. Thank you for your assistance.

Instructors – If you found that additional materials were necessary in order to teach this module effectively, please let us know so that we may include them in the Equipment/Materials list in the Instructor's Guide.

Write: Curriculum Revision and Development Department
National Center for Construction Education and Research
P.O. Box 141104, Gainesville, FL 32614-1104

Fax: 352-334-0932

E-mail: curriculum@nccer.org

Craft ______ Module Name ______

Copyright Date ______ Module Number ______ Page Number(s) ______

Description ______

______

______

______

(Optional) Correction ______

______

______

(Optional) Your Name and Address ______

______

______

Module 75203-03

Risk Analysis and Assessment

NCCER STANDARDIZED CRAFT TRAINING PROGRAM

The National Center for Construction Education and Research (NCCER) provides a standardized national program of accredited craft training. Key features of the program include instructor certification, competency-based training, and performance testing. The program provides trainees, instructors, and companies with a standard form of recognition through a National Craft Training Registry. The program is described in full in the *Guidelines for Accreditation*, published by the NCCER. For more information on standardized craft training, contact the NCCER by writing us at P.O. Box 141104, Gainesville, FL 32614-1104; calling 352-334-0911; or e-mailing info@nccer.org. More information may be found at our Web site, www.nccer.org.

HOW TO USE THIS ANNOTATED INSTRUCTOR'S GUIDE

Each page presents two sections of information. The larger section displays each page exactly as it appears in the Participant Module. The narrow column ties suggested participant and instructor actions to each page and provides icons (detailed below) to call your attention to material, safety, audiovisual, or testing requirements. The bottom of each page includes space for your notes.

The **Audiovisual** icon indicates an appropriate time to show a transparency or other audiovisual aid.

The **Classroom** icon prompts you to define a term, stress a point, ask participants to explain a concept, or give examples.

The **Demonstration** icon directs you to show participants how to perform tasks.

The **Examination** icon tells you to administer the written module examination.

The **Homework** icon is placed where you may wish to assign reading for the next class, to assign a project, or to advise participants to prepare for an examination.

The **Laboratory** icon is used when participants are to practice performing tasks.

The **Materials** icon is a reminder for you to gather materials needed for classes, labs, and testing.

The **Performance Testing** icon tells you to administer a performance test or a portion thereof.

The **Safety** icon is used to emphasize safety issues. It is often keyed to *Caution* and *Warning* statements in the Participant Module.

The **Teaching Tip** icon indicates additional guidance is available, such as how to conduct an exercise, get the most educational value from a field trip, or encourage class participation. Teaching Tips may expand on a feature (*Think About It, Did You Know?*) or provide Quick Quizzes or similar exercises. You will be referred to the Teaching Tips section at the back of the module if there is additional material.

The **Combination** icon indicates that the laboratory listed corresponds with a performance task. If desired, you can note the proficiency of the participants during the laboratory and use it to satisfy performance testing requirements.

PREPARATION

Before teaching this module, you should review the Objectives, Performance Tasks, Materials and Equipment List, and the Module Outline. Be sure to allow ample time to prepare your own training or lesson plan and gather all required materials and equipment.

MODULE OVERVIEW

In this module, participants will learn how to identify, analyze, and correct at-risk worker behavior.

PREREQUISITES

Please refer to the Course Map in the Participant Module. Prior to training with this module, it is recommended that the participant shall have successfully completed the following:

Field Safety; Safety Technology, Modules 75201-03 and 75202-03

OBJECTIVES

Upon completion of this module, the participant will be able to:

1. Explain the factors involved in analyzing performance.
2. Discuss the relationship between human behavior and work-site safety.
3. Explain the techniques used to coach and counsel workers with performance problems.
4. Explain the ABC model.

PERFORMANCE TASKS

There are no performance tasks for this module.

MATERIALS AND EQUIPMENT LIST

Overhead projector and screen

Transparencies

Whiteboard/chalkboard

Markers/chalk

Blank acetate sheets

Transparency pens

Pencils and scratch paper

Appropriate personal protective equipment

Copy of *29 CFR 1926, OSHA Construction Industry Regulations*

Module Examinations*

* Located in the Test Booklet.

SAFETY CONSIDERATIONS

Ensure that participants are equipped with appropriate personal protective equipment.

ADDITIONAL RESOURCES

This module is intended to present thorough resources for task training. The following reference works are suggested for both instructors and motivated participants interested in further study. These are optional materials for continued education rather than for task training.

Web site: www.osha.gov

Web site: www.asse.org

The Psychology of Safety Handbook, 2001. E. Scott Geller, Ph.D. Boca Raton, FL: CRC/Lewis Publishers.

The Participation Factor—How to Increase Involvement in Occupational Safety, 2002. E. Scott Geller, Ph.D. Des Plaines, IL: The American Society of Safety Engineers (ASSE).

Root Cause Analysis Handbook: A Guide to Effective Incident Investigation, 1999. JBF Associates Division. Rockville, MD: Government Institutes.

TEACHING TIME FOR THIS MODULE

An outline for use in developing your lesson plan is presented below. Approximately 2½ hours are suggested to cover *Risk Analysis and Assessment*. This includes 10 minutes at the beginning of the session for administrative tasks and one 10-minute break. You may need to adjust the time required for hands-on activities and testing based on your class size and resources. Because laboratories often correspond to Performance Tasks, the proficiency of the participants may be noted during these exercises for Performance Testing purposes.

Topic	Planned Time
A. Introduction	________
B. Analyzing Performance Problems	________
1. The Behavioral Law of Effect	________
2. Observation and Reinforcement	________
3. Performance Barriers	________
4. Impairment Factors	________
5. Unpreventable Employee Misconduct	________
6. Chances and Risks	________
7. Loss of Focus	________
C. Coaching and Counseling	________
1. Dealing with At-Risk Behavior	________
2. Dealing with Safe Behavior	________
D. The ABC Model	________
1. Activator	________
2. Behavior	________
3. Consequences	________
E. Review	________
F. Module Examination	________

1. Participants must score 70% or higher to receive recognition from the NCCER.
2. Record the testing results on Craft Training Report Form 200 and submit the results to the Training Program Sponsor.

Module 75203-03

Risk Analysis and Assessment

Instructor's Notes:

COURSE MAP

This course map shows all of the modules in Safety Technology. The suggested training order begins at the bottom and proceeds up. The local Training Program Sponsor may adjust the training order.

Assign reading of Module 75203-03.

SAFETY TECHNOLOGY

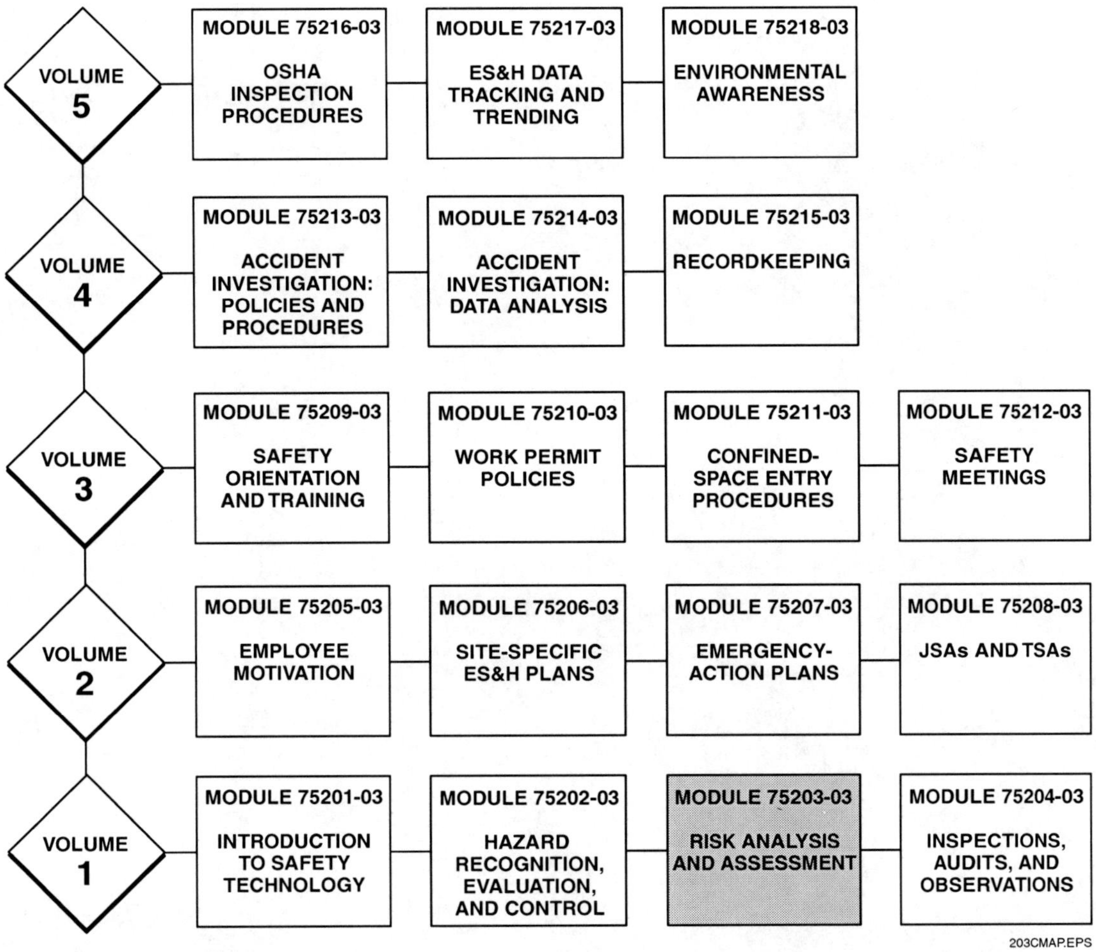

RISK ANALYSIS AND ASSESSMENT 3.iii

Instructor's Notes:

MODULE 75203-03 CONTENTS

Figures

RISK ANALYSIS AND ASSESSMENT 3.v

Instructor's Notes:

MODULE 75203-03

Risk Analysis and Assessment

Objectives

When you have completed this module, you will be able to do the following:

1. Explain the factors involved in analyzing performance.
2. Discuss the relationship between human behavior and work-site safety.
3. Explain the techniques used to coach and counsel workers with performance problems.
4. Explain the ABC model.

Prerequisites

Before you begin this module, it is recommended that you successfully complete the following: Field Safety; Safety Technology, Modules 75201-03 and 75202-03.

Required Materials

1. Pencil and paper
2. Appropriate personal protective equipment
3. Copy of *29 CFR 1926, OSHA Construction Industry Regulations*

1.0.0 ◆ INTRODUCTION

Understanding human behavior is a key element in analyzing and assessing why people take risks. It is important to know why workers behave the way they do and the factors that influence their ability to work safely. This is the first step in preventing workplace accidents.

As a safety technician, you are responsible for identifying at-risk behavior. In order to do this, you must learn to assess worker behavior. This means you must be able to accurately observe workers, understand barriers and impairment issues, and provide **consequences** for at-risk behavior. Once you learn these skills, you will be a valuable member of the company because you are working to create a safe and productive environment.

> **NOTE**
> For more information on risk analysis, please refer to Volume 1, Module 75202-03, *Hazard Recognition, Evaluation, and Control.*

2.0.0 ◆ ANALYZING PERFORMANCE PROBLEMS

The majority of all accidents are the result of at-risk behavior *(Figure 1)*. It is important that you understand human behavior and the human factors that lead to accidents. Some of the concepts you need to understand when analyzing performance problems include:

- The behavioral law of effect
- Observation and **reinforcement**
- Performance barriers
- Impairment factors
- Unpreventable employee misconduct
- Chances and risks
- Loss of focus

2.1.0 The Behavioral Law of Effect

The Behavioral Law of Effect states: "Behavior that brings **reward** is repeated, whereas behavior that does not bring reward or is punished is not

Ensure that you have everything you need to teach the course. Check the Materials and Equipment list in the front of this module.

Explain that terms shown in bold are defined in the Glossary at the back of this module.

Review the goals of the course and explain what will be expected of the participant.

Show Transparency 1, Objectives.

Describe the safety technician's role in analyzing and assessing risk.

Explain the importance of understanding human behavior when analyzing and assessing risk.

Show Transparency 2 (Figure 1).

Identify the behavioral concepts a safety technician should know and use when analyzing performance problems.

Explain the Behavioral Law of Effect using practical, work-related examples.

Describe how observing worker behavior can reinforce both safe and/or at-risk behavior.

Explain how the safety technician must be consistent and firm when observing and correcting at-risk behavior.

Explain that there are three types of barriers that can prevent workers from performing a job safely.

Describe the physical barriers that can hinder a worker's performance. Explain how to overcome the barriers.

Explain the knowledge barriers that can hinder a worker's performance. Discuss possible solutions.

203F01.EPS

Figure 1 ◆ Horseplay is at-risk behavior.

repeated." This law of behavior does not, however, say whether the behavior is good or bad. It simply states that if the behavior is rewarded, it will be repeated.

This law also does not indicate the source of the reward. This can be a problem because the source of the reward could be positive or negative. For example, a crew experienced a mechanical problem with some heavy equipment. The operator quickly diagnosed the problem and fixed it. A little while later, the supervisor told the operator that he did a great job keeping the unit on line. The operator felt good that his efforts were recognized and he will likely fix the same problem the same way in the future. What the supervisor did not know was that the operator did not follow the correct procedures in making the repair and, as a result, unnecessarily exposed himself to a significant hazard. The reward here should be considered negative because the operator's behavior was not safe but it was still rewarded.

Additionally, if every at-risk or hazardous act resulted in injury, the incident itself would reinforce the need to follow the procedure. But, since literally thousands of at-risk or hazardous acts are committed each day with no adverse consequences, the absence of adverse consequences tends to reward at-risk behavior. Make sure you know all of the details of incidents and accidents before offering praise or rewards. This will help to ensure that good behavior is rewarded properly and that the proper consequences result from at-risk behavior.

2.2.0 Observation and Reinforcement

Observing a worker confirms and reinforces their behavior, whether it's safe or at-risk. This is because the worker being observed assumes that their behavior is acceptable if it is done openly without any adverse consequences. Overlooking or ignoring a safety violation essentially gives approval to the violation and also gives permission for the condition to exist.

There will be times when you will be asked to overlook at-risk behavior. Asking you to overlook simple safety violations is asking you to compromise your entire attitude towards safety. Never do this. Always let workers know that you mean what you say and that if you see an at-risk act or condition, you will ask them to correct the situation. This will help reinforce safe behavior.

2.3.0 Performance Barriers

Often, there are obstacles that prevent workers from performing their jobs safely. These obstacles are called performance barriers. Common performance barriers include:

- Physical barriers
- Knowledge barriers
- Execution barriers

2.3.1 Physical Barriers

Physical barriers are barriers that cause a worker to be physically unable to perform a job safely. Some examples of physical barriers include:

- Poor eyesight
- Loss of hearing
- Lack of individual strength or agility
- Illness

In order to overcome physical barriers, the worker should be assigned a different job or the job task should be restructured if possible.

2.3.2 Knowledge Barriers

Knowledge barriers present a problem when the worker does not know or remember how, when, or why to do an assigned task. The following factors should be considered when trying to determine if a knowledge barrier does exist.

- Education
- Training
- Skill level
- Ability to understand written or oral instructions
- Language barriers

Training and practice are possible solutions to correct knowledge barriers.

Instructor's Notes:

2.3.3 Execution Barriers

An execution barrier exists when a worker is physically and mentally capable of completing a job task but does not do it. This type of behavior may be the result of either a task barrier or an incentive barrier.

Task barriers are caused when the worker is not provided with a proper work environment, the appropriate number of people for the task, or the right tools or equipment to do the job. Task barriers are most easily fixed by providing workers with what they need to get the job done safely.

Incentive barriers exist when the worker has been rewarded for doing the job incorrectly, has been punished even though the job was done correctly, or has never received feedback for both positive and negative behavior. For example, even though a worker may be using the improper equipment to complete a task, he or she is still completing the task and therefore gets praise from his or her supervisor for finishing the task on time. It is likely that this worker will continue this behavior until there is an accident. The worker will do so because he or she continually receives positive feedback for negative behavior, thus encouraging the behavior.

Positive feedback can be in the form of rewards. Negative feedback can be in the form of disciplinary actions or lost privileges. Incentive barriers can be fixed by noting the performance problem and providing positive or negative feedback as needed.

2.4.0 Impairment Factors

Workers can sometimes get confused, distracted, or disoriented and do something they otherwise would not have done (*Figure 2*). This behavior can be the result of fatigue, illness, intense or prolonged stress, or substance abuse. These are considered impairment factors. Impairment factors are different from performance barriers because the impairment factor affects the worker in such a way that it would be difficult for them to do any other type of work; in contrast, performance barriers only affect one task.

It is very easy to say an individual just forgot to follow the correct procedure and leave it at that. But, if an individual just forgot to follow the procedure because of prescription drug use or substance abuse, then the behavior cannot be ignored or overlooked. Pay attention to the worker's behavior so that you can properly judge the nature of the impairment and take the proper actions to correct the situation.

Impairment factors such as fatigue, stress, or illness are less clearly identifiable. These factors are sometimes only found during an accident investigation. If an accident investigation is being done, interview those directly involved in the accident to determine if any of these factors may have contributed to the accident. It may also be helpful to explore these issues with witnesses or with people close to those directly involved in the accident. It may even be necessary to contact a doctor to establish and/or confirm the effects of the particular impairment factor on behavior.

Define the term execution barriers. Explain that there are two types of execution barriers: task barriers and incentive barriers. Explain appropriate corrective actions.

Explain the factors that can impair worker behavior and performance.

Show Transparency 3 (Figure 2).

Describe the safety technician's role in and responsibility for identifying these factors.

203F02.EPS

Figure 2 ◆ Example of an impaired worker.

RISK ANALYSIS AND ASSESSMENT 3.3

Define the term unpreventable employee misconduct.

Explain the importance of confirming unpreventable employee misconduct before classifying a worker's behavior.

Explain the correlation between chance and risk.

Discuss the two factors that influence personal risk acceptance.

Explain the difference between real and perceived risk.

Discuss the factors that influence perceived risk. Refer the participants to the bulleted list in the Participant Module.

2.5.0 Unpreventable Employee Misconduct

Unpreventable employee misconduct is behavior that happens when workers clearly disregard rules, procedures, and regulations. It is important to be aware of behavior that is inappropriate and correct it as needed. Before classifying a worker's behavior as unpreventable employee misconduct, first confirm that other factors did not contribute to the behavior.

2.6.0 Chances and Risks

Some people are willing to accept a higher level of risk than others. For example, if there were two aquariums with rattlesnakes and one had $100.00 inside and the other had a $1,000.00 inside, some people would take the chance of a snake bite for the $1,000.00, others would do it for the $100.00, and some wouldn't take the chance at all. Why do you think people take these kinds of risks? The answer is that there are consequences that influence both real and perceived risk. People generally move towards those consequences that are soon, certain, and positive.

2.6.1 Personal Risk Acceptance

Personal risk acceptance is the level of risk at which a worker is willing to accept the consequences of their actions. This type of risk is categorized by two factors of personal acceptance:

- *It won't happen to me* – This is a common belief among workers who take risks. Past experience has led the worker to believe there is a relatively low level of risk. They say to themselves, "I've done it this way a thousand times and never had an incident." And they are right. They haven't been hurt and therefore will continue to ignore the risk.
- *Real versus perceived risk* – The actual risk associated with a specific hazard or behavior is determined by the likelihood and severity of a possible incident, not a worker's previous experience with the same situation. Very often, it's easy to overlook this and take a risk based on personal acceptance rather than a real risk factor. For example, the risk of dying in a car wreck while traveling to and from work is much higher than dying from radiation exposure from a nearby nuclear power plant. Yet, if asked, it is likely that workers wouldn't volunteer to go near a nuclear power plant, but they would still travel to and from work every day.

2.6.2 Factors That Influence Perceived Risk

Everyone's perception of risk is different. Keep this in mind when you are having performance problems with a worker. This knowledge will be helpful in determining a solution. The following are factors and situations that influence perceived risk:

- *Voluntary vs. mandatory exposure* – Hazards we accept voluntarily can include motorcycling, snowmobiling, or jet skiing. These activities are generally perceived as low risk because they are enjoyable. However, if we were required to work from a 24" × 24" unprotected platform 25′ off the ground, we would perceive that risk as much greater. It's important to recognize that different people will always perceive risks differently. This factor will influence a worker's willingness to complete a task.
- *Familiar hazards are considered to be lower risk* – The more we know about a hazard, the more risk we are willing to take. For example, when you first learned to drive, you probably drove slowly and carefully. As you became more familiar with the vehicle and its hazards, you were willing to drive faster and with less caution. This is also true with tools and equipment. Over-confidence leads to accidents. That's why it is important to pay as much attention to workers with a lot of experience as you do to those with little work experience.
- *Controllable vs. uncontrollable* – If a worker feels a hazard can be controlled, the hazard is often perceived as less risky. If the hazard is considered uncontrollable, as in the case of severe weather, the hazard is considered dangerous. Workers can become complacent when it comes to controllable hazards because the more we know how to control a hazard, the less respect we have for it. For example, when workers first learn to work with sulfuric acid, they are extremely cautious and wear the prescribed personal protective equipment. After working with the material for several years without major incidents, however, they may become complacent. They know that if they get to water quickly, the burn can be minimized significantly. As a result, they often don't treat acid with the respect it deserves. Monitoring and observing can help eliminate this problem.
- *Risk compensation* – When people are provided with personal protective equipment or safety devices, they often become careless. In one example, a group of workers was asked to drive forklifts through a practice course. Their speed was clocked each time they completed the course. When seat belts were required, the

Instructor's Notes:

drivers increased their speed considerably. It was concluded that personal protective equipment sometimes gives users a false sense of security. It also concluded that knowledge of risks and proper safety training can prevent this from happening.

- *No one is watching* – Some workers only follow the rules when they are being watched. They are often only willing to take risks when they know they can get away with it. Make sure workers know the consequences of their actions, even if no one is watching.
- *Peer pressure and social conformity* – Peer pressure is an extremely powerful motivator. It can be positive or negative. Very few workers want to be different. Most want to be part of a group and will change their behavior to fit into that group. For example, many people in the workplace recognize the need for and value of hard hats and other forms of personal protective equipment, but often don't use them because their co-workers don't. Make sure everyone is following all of the rules at all times and always wear your hard hat as a good example. This will help eliminate peer pressure.
- *Acceptable consequences* – If the consequences of an action are soon, certain, sizable, and acceptable, they are likely to be repeated. For example, a worker is required to grind a piece of metal using a bench grinder approximately 10 times a day. The task of grinding takes only 30 to 45 seconds. The next time the worker uses the grinder, he or she doesn't use the face shield because it is dirty and needs to be readjusted. The worker feels confident an accident will not happen because the task takes less than a minute and he or she is wearing safety glasses. The worker has done it this way for six years without an incident; therefore, to the worker, the consequences have been acceptable. The worker has a false sense of safety because he or she has gotten the job done without an incident. This doesn't mean, however, that an accident couldn't happen the next time the grinder is used. One way to stop this behavior is to make workers accountable for their actions. This particular worker should be appropriately disciplined and made aware of the consequences of his or her actions.

2.7.0 Loss of Focus

Loss of focus is a situation in which a worker lets his or her mind wander and he or she has difficulty doing a job. He or she temporarily loses focus, and commits a hazardous act. This act puts the worker at risk and sometimes causes an accident. This loss of focus may be due to daydreaming, concern about off-the-job problems, or a sudden distraction. If you suspect a worker isn't focusing, talk to him or her. Find out what may be causing the distraction and figure out a solution such as rotating the worker to a different job task.

3.0.0 ◆ COACHING AND COUNSELING

Now that you know what causes performance problems, you need to know how to coach or counsel those who are placing themselves or others in a hazardous or at-risk situation. Coaching and counseling are important tools to help communicate with workers. These skills make correcting at-risk behavior a much easier task. When a worker feels that he or she can communicate openly, it builds trust and respect between both parties. Once you have the confidence of a worker and he or she feels secure enough to communicate, he or she will be more open to suggestions, ideas, and actions that will reduce at-risk behavior.

3.1.0 Dealing with At-Risk Behavior

At-risk behavior affects everyone on the site, not just the worker taking risks. When you find someone's behavior is too risky, you must:

- Stop the worker without startling him or her.
- Take the worker aside and speak to him or her privately.
- Express concern for his or her safety and well being.
- State what action or behavior was incorrect or at risk.
- Explain why the action or behavior was incorrect or at risk.
- Try to find out if there are any barriers or impairment factors.
- Ask the worker what the proper way to perform the job is, but be prepared to show or explain the proper way to do the job.
- Leave on a positive note.

Use the same technique when dealing with coworkers, visitors, or supervisors. One important thing to remember is that everyone deserves to be treated with dignity and respect. They don't want to be publicly embarrassed. Do your coaching and counseling privately.

Describe how workers can lose focus.

Discuss ways to help workers regain focus.

Discuss coaching and counseling techniques that can influence both at-risk and safe worker behavior.

Provide the participants with several scenarios in which coaching and counseling are needed. Have the participants use role playing to practice these skills.

RISK ANALYSIS AND ASSESSMENT 3.5

Explain each component of the ABC model in detail: activator, behavior, and consequences. Include practical, real-life examples of each component in the explanation.

Show Transparency 4 (Figure 3).

3.2.0 Dealing with Safe Behavior

People want and need compliments. When you see someone performing a job safely, provide positive feedback. Remember, behavior that is rewarded is repeated. That which is not rewarded or is punished is not repeated. When you observe a worker performing a job safely, you should:

- Speak to the worker privately.
- Specifically state the correct behavior observed.
- Compliment the worker on a job well done.

4.0.0 ◆ THE ABC MODEL

The ABC model is a tool that helps you understand human behavior. Its main objective is to find positive ways to change behavior through positive rewards. It also provides an understanding about why people behave a certain way and it helps determine the patterns of consequences associated with the behavior. The ABC model assigns the following terms to A, B, and C:

- A – activator
- B – behavior
- C – consequences

By understanding the ABC model *(Figure 3)*, including the relationship between the activators and behavior, behavior and consequences, and consequence and activator, you can work to change the behavior.

4.1.0 Activator

An activator is any condition that prompts behavior. For example, if the phone rings, we answer it. The ringing phone is considered the activator.

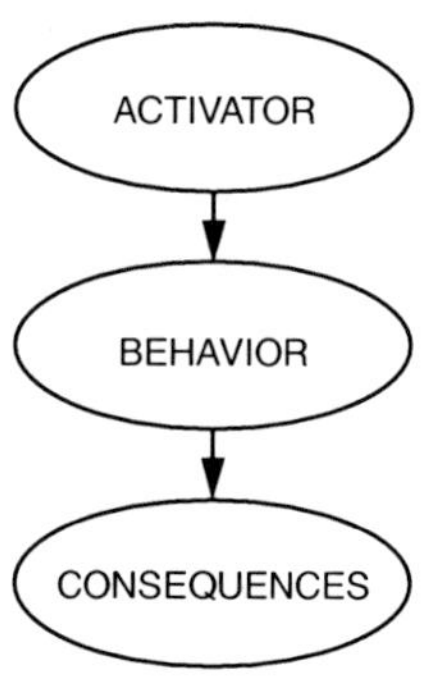

Figure 3 ◆ ABC model.

Sometimes activators are clear and other times they are not. Some examples of activators that are easy to point out and correct include:

- Incentive programs
- Policies
- Signs

When activators are not clear, it is difficult to find an immediate cause for at-risk behavior. If you find that the activator is not clear and an investigation must be done, use the following questions to determine what prompted the at-risk behavior:

- Is there a rule, policy, or procedure that addresses the behavior?
- Is the rule, policy, or procedure specific to the behavior?
- Is the rule, policy, or procedure known, understood, and accepted as valid?
- Is the behavior activated by practice, habit, and peer influence or job requirements?

Once you answer these questions, you will be able to determine the cause of the behavior and immediately correct the problem.

4.2.0 Behavior

Behavior is the second part of the ABC model and probably the most easily explained. In the context of this model, behavior refers to the specific behavioral response to the activator. For example, a safety rule exists stating that safety glasses must be worn. The rule is the activator. The wearing of safety glasses is the behavior. Behavior is both observable and measurable.

4.3.0 Consequences

Consequences are the results of the behavior, positive or negative. It's important that workers are aware of both the positive and negative consequences of their behavior. For example, certain pieces of personal protective equipment can be uncomfortable and therefore may not be worn. The consequences for this action are immediate comfort for the worker, which is perceived as positive by the worker, but is actually negative because of the injuries the worker can receive. To encourage workers to wear safety glasses, you must have consequences that are soon, certain, and positive. Otherwise, we can actually discourage safe actions.

Instructor's Notes:

In the context of the ABC model, consider the following factors when examining the consequences of behavior.

- Is the reward pleasurable or positive?
- Is the reward certain?
- Is the reward soon?
- Is the reward sizeable?

If the answer to any of these questions is yes, then it is likely that at-risk behavior will continue unless corrected. Training and strong leadership can help workers understand the consequences of their behavior.

Summary

Understanding human behavior is a key element to successful risk analysis and assessment. It also significantly affects the number of accidents and incidents on a site. Once you know why workers behave the way they do and the factors that influence their behavior, you will be better able to prevent accidents and injuries.

In this module, you learned the factors involved in human behavior, as well as techniques to improve your observation skills. You also learned how to communicate the consequences of human behavior to workers and in turn make everyone on site accountable for safety.

Review Questions

1. The majority of accidents are the result of at-risk behavior.

a. True
b. False

2. The Behavioral Law of Effect states that _____.

a. if something bad is going to happen, it will happen regardless of one's behavior
b. you can provide good behavioral examples but a worker's will is the ultimate tool for making decisions
c. behavior that brings rewards is repeated, whereas behavior that does not bring rewards or is punished is not repeated
d. it is often necessary to punish workers in order for them to work harder

For Questions 3 through 5, match the performance barrier to the corresponding description.

Performance Barrier

3. ____ Physical

4. ____ Knowledge

5. ____ Execution

Description

a. Barriers that are a result of workers not knowing or remembering how to complete a task
b. Barriers that cause a worker to not complete a task even though he or she is physically capable of doing so
c. Barriers that prevent a worker from physically performing a job

6. The phrase "it won't happen to me" is a common belief among workers who take risks.

a. True
b. False

7. All of the following are ways to deal with at-risk behavior *except* _____.

a. taking workers aside and speaking to them privately
b. explaining why the action or behavior was incorrect or at risk
c. trying to find out if there are any performance barriers involved
d. speaking in a raised tone of voice to the worker performing at-risk behavior

8. When you observe a worker performing a job safely, you should speak to the worker in front of co-workers so that other workers who aren't doing their jobs safely will get a hint.

a. True
b. False

9. The A in the ABC model stands for activator.

a. True
b. False

10. Consequences only result from bad behavior.

a. True
b. False

Review the material covered in the module.

Have the participants complete the Review Questions and go over the answers prior to administering the Module Examination.

Administer the Module Examination. Record the results on Craft Training Report Form 200 and submit the results to the Training Program Sponsor.

RISK ANALYSIS AND ASSESSMENT 3.7

Instructor's Notes:

GLOSSARY

Trade Terms Introduced in This Module

Consequences: The final outcome of actions and behaviors.

Reinforcement: Supporting or strengthening of a behavior or action.

Reward: Something that is given or returned verbally, through body language, or in writing for good or bad behavior.

RISK ANALYSIS AND ASSESSMENT 3.9

Instructor's Notes:

REFERENCES

Additional Resources

This module is intended to present thorough resources for task training. The following reference works are suggested for further study. These are optional materials for continued education rather than for task training.

www.osha.gov

www.asse.org

The Psychology of Safety Handbook, 2001. E. Scott Geller,Ph.D. Boca Raton, FL: CRC/Lewis Publishers.

The Participation Factor—How to Increase Involvement in Occupational Safety, 2002. E. Scott Geller, Ph.D. Des Plaines, IL: The American Society of Safety Engineers (ASSE).

Root Cause Analysis Handbook: A Guide to Effective Incident Investigation, 1999. JBF Associates Division. Rockville, MD: Government Institutes.

RISK ANALYSIS AND ASSESSMENT 3.11

MODULE 75203-03 — ANSWERS TO REVIEW QUESTIONS

Answer		Section
1.	a	2.0.0
2.	c	2.1.0
3.	c	2.3.1
4.	a	2.3.2
5.	b	2.3.3
6.	a	2.6.1
7.	d	3.1.0
8.	b	3.2.0
9.	a	4.0.0
10.	b	4.3.0

CONTREN™ LEARNING SERIES — USER UPDATES

The NCCER makes every effort to keep these textbooks up-to-date and free of technical errors. We appreciate your help in this process. If you have an idea for improving this textbook, or if you find an error, a typographical mistake, or an inaccuracy in NCCER's Contren™ textbooks, please write us, using this form or a photocopy. Be sure to include the exact module number, page number, a detailed description, and the correction, if applicable. Your input will be brought to the attention of the Technical Review Committee. Thank you for your assistance.

Instructors – If you found that additional materials were necessary in order to teach this module effectively, please let us know so that we may include them in the Equipment/Materials list in the Instructor's Guide.

Write: Curriculum Revision and Development Department
National Center for Construction Education and Research
P.O. Box 141104, Gainesville, FL 32614-1104

Fax: 352-334-0932

E-mail: curriculum@nccer.org

Craft __________ Module Name __________

Copyright Date __________ Module Number __________ Page Number(s) __________

Description

____

____

____

(Optional) Correction

____

____

(Optional) Your Name and Address

____

____

Module 75204-03

Inspections, Audits, and Observations

NCCER STANDARDIZED CRAFT TRAINING PROGRAM

The National Center for Construction Education and Research (NCCER) provides a standardized national program of accredited craft training. Key features of the program include instructor certification, competency-based training, and performance testing. The program provides trainees, instructors, and companies with a standard form of recognition through a National Craft Training Registry. The program is described in full in the *Guidelines for Accreditation*, published by the NCCER. For more information on standardized craft training, contact the NCCER by writing us at P.O. Box 141104, Gainesville, FL 32614-1104; calling 352-334-0911; or e-mailing info@nccer.org. More information may be found at our Web site, www.nccer.org.

HOW TO USE THIS ANNOTATED INSTRUCTOR'S GUIDE

Each page presents two sections of information. The larger section displays each page exactly as it appears in the Participant Module. The narrow column ties suggested participant and instructor actions to each page and provides icons (detailed below) to call your attention to material, safety, audiovisual, or testing requirements. The bottom of each page includes space for your notes.

The **Audiovisual** icon indicates an appropriate time to show a transparency or other audiovisual aid.

The **Classroom** icon prompts you to define a term, stress a point, ask participants to explain a concept, or give examples.

The **Demonstration** icon directs you to show participants how to perform tasks.

The **Examination** icon tells you to administer the written module examination.

The **Homework** icon is placed where you may wish to assign reading for the next class, to assign a project, or to advise participants to prepare for an examination.

The **Laboratory** icon is used when participants are to practice performing tasks.

The **Materials** icon is a reminder for you to gather materials needed for classes, labs, and testing.

The **Performance Testing** icon tells you to administer a performance test or a portion thereof.

The **Safety** icon is used to emphasize safety issues. It is often keyed to *Caution* and *Warning* statements in the Participant Module.

The **Teaching Tip** icon indicates additional guidance is available, such as how to conduct an exercise, get the most educational value from a field trip, or encourage class participation. Teaching Tips may expand on a feature (*Think About It*, *Did You Know?*) or provide Quick Quizzes or similar exercises. You will be referred to the Teaching Tips section at the back of the module if there is additional material.

The **Combination** icon indicates that the laboratory listed corresponds with a performance task. If desired, you can note the proficiency of the participants during the laboratory and use it to satisfy performance testing requirements.

PREPARATION

Before teaching this module, you should review the Objectives, Performance Tasks, Materials and Equipment List, and the Module Outline. Be sure to allow ample time to prepare your own training or lesson plan and gather all required materials and equipment.

MODULE OVERVIEW

In this module, participants will learn the safety technician's role in inspections, audits, and observations and to perform these functions as required.

PREREQUISITES

Please refer to the Course Map in the Participant Module. Prior to training with this module, it is recommended that the participant shall have successfully completed the following:

Field Safety; Safety Technology, Modules 75201-03 through 75203-03

OBJECTIVES

Upon completion of this module, the participant will be able to:

1. Describe the role and responsibility of the safety technician in on-site inspections, audits, and observations.
2. State the purpose of a safety inspection.
3. Explain how to conduct a safety audit.
4. Describe how to conduct an employee observation.

PERFORMANCE TASKS

Under the supervision of the instructor, the participant should be able to:

1. Conduct a safety audit.
2. Perform a safety observation.

MATERIALS AND EQUIPMENT LIST

Overhead projector and screen
Transparencies
Whiteboard/chalkboard
Markers/chalk
Blank acetate sheets
Transparency pens
Pencils and scratch paper
Appropriate personal protective equipment
Copy of *29 CFR 1926, OSHA Construction Industry Regulations*
Module Examinations*
Performance Profile Sheets*

* Located in the Test Booklet.

SAFETY CONSIDERATIONS

Ensure that participants are equipped with appropriate personal protective equipment.

ADDITIONAL RESOURCES

This module is intended to present thorough resources for task training. The following reference works are suggested for both instructors and motivated participants interested in further study. These are optional materials for continued education rather than for task training.

Web site: www.osha.gov

Web site: www.asse.org

TEACHING TIME FOR THIS MODULE

An outline for use in developing your lesson plan is presented below. Approximately 2½ hours are suggested to cover *Inspections, Audits, and Observations*. This includes 10 minutes at the beginning of the session for administrative tasks and one 10-minute break. You may need to adjust the time required for hands-on activities and testing based on your class size and resources. Because laboratories often correspond to Performance Tasks, the proficiency of the participants may be noted during these exercises for Performance Testing purposes.

Topic	Planned Time
A. Introduction	________
B. Safety Inspections	________
1. Different Types of Safety Inspections	________
2. Items Included in a Safety Inspection	________
3. Safety Inspection Time Frame	________
4. Safety Inspection Findings	________
C. Safety Audits	________
1. Safety Audit Time Frame	________
2. Laboratory	________
Provide the participants with a scenario and have them conduct a safety audit. This laboratory corresponds with Performance Task 1.	
D. Safety Observations	________
1. Performing a Safety Observation	________
2. Safety Observations Time Frame	________
3. Safety Observation Findings	________
4. Laboratory	________
Provide the participants with a scenario in which a worker has been observed performing an unsafe act. Have the participants use role playing to appropriately correct the noted behavior. This laboratory corresponds with Performance Task 2.	
E. Government Regulations	________
1. The Occupational Safety and Health Act	________
2. Other Federal Agencies	________
3. OSHA Inspections	________
F. Review	
G. Module Examination	________
1. Participants must score 70% or higher to receive recognition from the NCCER.	
2. Record the testing results on Craft Training Report Form 200 and submit the results to the Training Program Sponsor.	
H. Performance Testing	________
1. Participants must perform each task to the satisfaction of the instructor to receive recognition from the NCCER. If applicable, proficiency noted during laboratory exercises can be used to satisfy the Performance Testing requirements.	
2. Record the testing results on Craft Training Report Form 200 and submit the results to the Training Program Sponsor.	

Module 75204-03

Inspections, Audits, and Observations

Instructor's Notes:

COURSE MAP

This course map shows all of the modules in Safety Technology. The suggested training order begins at the bottom and proceeds up. The local Training Program Sponsor may adjust the training order.

Assign reading of Module 75204-03.

SAFETY TECHNOLOGY

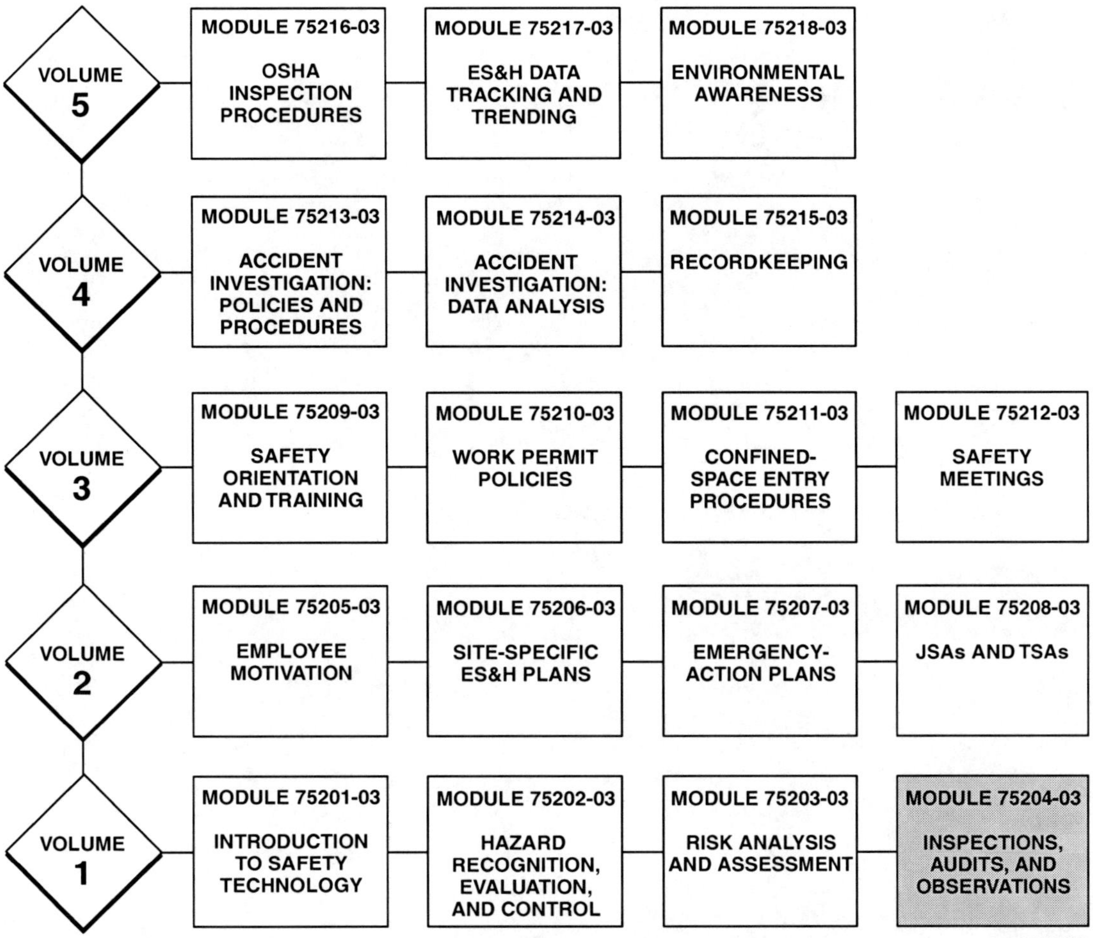

204CMAP.EPS

INSPECTIONS, AUDITS, AND OBSERVATIONS 4.iii

Instructor's Notes:

MODULE 75204-03 CONTENTS

Figures

INSPECTIONS, AUDITS, AND OBSERVATIONS 4.v

Instructor's Notes:

MODULE 75204-03

Inspections, Audits, and Observations

Ensure that you have everything you need to teach the course. Check the Materials and Equipment list in the front of this module.

Objectives

When you have completed this module, you will be able to do the following:

1. Describe the role and responsibility of the safety technician in on-site inspections, audits, and observations.
2. State the purpose of a safety inspection.
3. Explain how to conduct a safety audit.
4. Describe how to conduct an employee observation.

Prerequisites

Before you begin this module, it is recommended that you successfully complete the following: Field Safety; Safety Technology, Modules 75201-03 through 75203-03.

Required Materials

1. Pencil and paper
2. Appropriate personal protective equipment
3. Copy of *29 CFR 1926, OSHA Construction Industry Regulations*

1.0.0 ◆ INTRODUCTION

In 2001, the construction industry reported 1,264 work-related deaths. That's 20% of all workplace deaths in the country. On any given day, four to five construction workers die from on-the-job injuries. On that same day, nine hundred construction workers are seriously injured.

These four causes make up 90% of construction fatalities:

- Falling
- Transportation incidents
- Contact with objects and equipment
- Exposure to harmful substances or toxic environments

Figure 1 shows a breakdown of the causes of deaths in the construction industry for 2001.

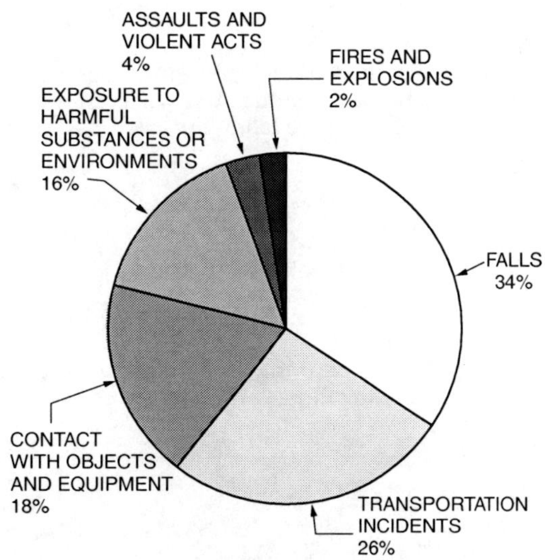

Figure 1 ◆ Causes of deaths in construction for 2001.

Explain that terms shown in bold are defined in the Glossary at the back of this module.

Review the goals of the course and explain what will be expected of the participant.

Show Transparency 1, Objectives.

Show Transparency 2, Performance Tasks.

Explain the impact accidents have on the construction industry.

Show Transparency 3 (Figure 1).

INSPECTIONS, AUDITS, AND OBSERVATIONS 4.1

Explain why safety is part of everyone's job.

Show Transparency 4 (Figure 2).

Explain the reasons for conducting safety inspections.

Explain the hazard control process.

Show Transparency 5 (Figure 3).

Explain what is typically done during a safety inspection.

In addition to deaths and immediate injuries to workers, unsafe conditions and practices can cause other potential problems. They include:

- Damage or destruction of materials and equipment
- Scheduling delays or work shutdowns
- Future health problems for the workers
- Damage to the environment

There are other, less obvious areas that are affected by unsafe conditions and practices, and accidents and injuries. These things affect the financial well being of the company and can determine whether your company wins contract bids. They include:

- High cost of workers' compensation insurance
- Occupational Safety and Health Administration (OSHA) fines
- Legal action

What does all this mean to you and your co-workers? The obvious answer is possible death or physical disability. But take a look at *Figure 2*. Injuries and accidents cost money. Your company needs to pass the increased cost on to their customers. That can mean losing bids, which means less work and lower pay for you.

Safety is part of everyone's job. As the safety technician, you'll have special responsibilities. You'll be responsible for coordinating site safety activities. It will be your job to help find and correct safety hazards. Sometimes you'll be working under the direction of the safety manager. More often, you'll be working on your own. Either way, you'll often perform the following safety tasks:

- **Inspections**
- **Audits**
- **Observations**

To do these tasks well, you'll need to:

- Serve as a resource to site personnel for safety and health matters.
- Review your company's past safety performance including illness and injury statistics.
- Understand the hazards associated with the type of work that is performed on site.
- Learn your company's policies, procedures, and work practices for the work performed.
- Make intelligent decisions to develop solutions to hazards and unsafe practices.
- Develop people-skills so that you can talk to personnel on the site, including workers, supervisors, and superintendents.

The main reason for conducting safety inspections, audits, and observations is to identify and fix unsafe conditions and practices before an accident or injury can happen. To do this, several steps must be taken. *Figure 3* shows the hazard control process.

Step 1 Identify the unsafe condition or practice.

Step 2 If the condition or practice is an immediate danger to workers or property, provide an immediate temporary solution.

Step 3 Determine the cause of the unsafe condition or practice.

Step 4 Take steps to correct the condition or practice.

Step 5 Follow up to be sure that the corrective action really fixed the problem.

Step 6 Document the problem, the solution, and how and when you followed up to be sure that the solution really fixed the problem.

A good safety program, including safety inspections, audits, and observations, saves the company more money than it costs them.

2.0.0 ◆ SAFETY INSPECTIONS

Safety inspections involve checking work areas to identify, report, and correct hazards to workers, materials, and equipment. Every day, you and other workers on construction sites perform safety inspections. You change a burned-out light bulb in a storage area, clean oil spots off the floor, or return an abandoned tool to storage. You see an unsafe condition and you fix it. This type of inspection is called continuous or ongoing and it

WHY WORRY ABOUT SAFETY?

ACCIDENTS AND INJURIES	=	• INCREASED INSURANCE PREMIUMS • LOSS OF EQUIPMENT • SCHEDULE DELAYS • LEGAL COSTS • WORK SHUTDOWNS • FINES	=	• HIGH BIDS • POOR REPUTATION • SCHEDULE OVERRUNS	=	• COMPANY WINS FEWER JOB CONTRACTS • LESS WORK • LOWER PAY

204F02.EPS

Figure 2 ◆ Why worry about safety?

Instructor's Notes:

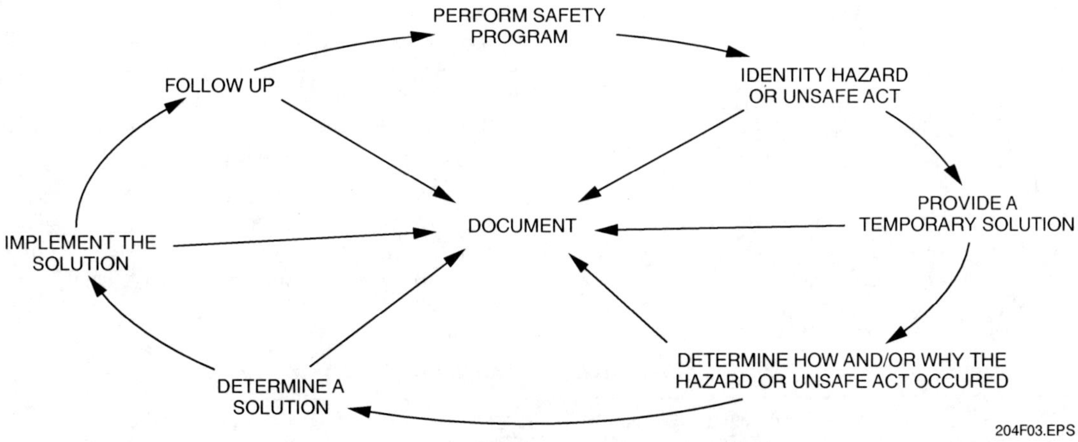

Figure 3 ◆ Hazard control process.

is informal. Safety technicians also perform formal inspections.

For an inspection to be formal, it needs to be written. Usually people use a checklist of things to be reviewed before a job can be safely done. The checklist must be written so that nothing is forgotten or missed. A written inspection is also needed to make sure your company has proof that it has an ongoing safety program. A properly prepared inspection checklist serves a dual purpose. It provides a listing of items or areas to be inspected and it serves as a permanent record of the inspection.

Part of your job as the safety technician is to continuously evaluate the checklist. The checklist needs to do its job, too. Safety inspections help to determine whether your job site is safe; if your checklist is out of date, it's not doing its job. Update the checklist as new equipment, materials, and work procedures are added to your work site.

> **NOTE**
> Check the OSHA Web site (www.osha.gov) or check with your local OSHA office for examples of checklists.

On most sites, you'll need more than one checklist because some areas or equipment need to be inspected more or less frequently than others. For example, fire extinguishers may need to be inspected monthly, but a ladder or scaffold should be inspected before each use. Keep safety checklists close to where they're needed. However, be sure that all completed safety checklists are stored in a safe location.

2.1.0 Different Types of Safety Inspections

In addition to the continuous inspections, there are three types of scheduled inspections. They are as follows:

- *Periodic inspections* – These inspections are made weekly, monthly, semi-annually, or at other set intervals. They are for specific items and equipment such as storage areas or heavy machinery.
- *Intermittent inspections* – These inspections are not performed at a set interval. They are sometimes called special inspections. You might need to do a special inspection when you notice that a large number of near misses have been occurring at your work site.
- *General inspections* – These inspections include areas not covered by periodic inspections. They might include walkways, tool rooms, or maintenance shops.

2.2.0 Items Included in a Safety Inspection

The items included in a safety inspection vary. It depends on the type of the work, how much hazard and risk are involved, and the number of people at the site. Some of the things you may need to include are as follows:

- Environmental factors such as lighting and ventilation
- Hazardous supplies and materials
- Production and related equipment

Explain the difference between informal and formal inspections.

Discuss the safety technician's role during both types of inspections.

Explain that checklists are often used during safety inspections.

Show Transparency 6 (Figure 4).

Describe the different types of safety inspections that are typically done at a work site.

Describe the items and areas that are generally inspected on a work site. Refer the participants to the bulleted list in the Participant Module.

Download a variety of safety inspection checklists from OSHA's Web site (www.osha.gov). Plan a field trip to a site where these checklists may be used. Under your supervision, have the participants perform an inspection using a checklist or have the participants observe a worker conducting an inspection. Ensure that participants use any necessary personal protective equipment.

- Power source equipment
- Electrical equipment
- Hand and power-tools
- Personal protective equipment (PPE)
- First-aid supplies and equipment
- Fire protection equipment
- Walkways, sidewalks, and roadways
- Elevators, stairways, and manlifts
- Working surfaces
- Transportation equipment
- Warning and signaling devices
- Containers
- Storage facilities
- Structural openings such as doorways or windows
- Building and structures
- Grounds and storage areas
- Loading and shipping platforms

Figure 4 shows a sample checklist for lockout/tagout procedures. You can find this checklist and many others on the OSHA Web site (www.osha.gov).

LOCKOUT/TAGOUT PROCEDURES

	IS ALL MACHINERY OR EQUIPMENT CAPABLE OF MOVEMENT REQUIRED TO BE DE-ENERGIZED OR DISENGAGED AND LOCKED-OUT DURING CLEANING, SERVICING, ADJUSTING, OR SETTING UP OPERATIONS, WHENEVER REQUIRED?
	WHERE THE POWER DISCONNECTING MEANS FOR EQUIPMENT DOES NOT ALSO DISCONNECT THE ELECTRICAL CONTROL CIRCUIT:
	ARE THE APPROPRIATE ELECTRICAL ENCLOSURES IDENTIFIED?
	ARE MEANS PROVIDED TO ASSURE THE CONTROL CIRCUIT CAN ALSO BE DISCONNECTED AND LOCKED-OUT?
	IS THE LOCKING-OUT OF CONTROL CIRCUITS IN LIEU OF LOCKING-OUT MAIN POWER DISCONNECTS PROHIBITED?
	ARE ALL EQUIPMENT CONTROL VALVE HANDLES PROVIDED WITH A MEANS FOR LOCKING-OUT?
	DOES THE LOCK-OUT PROCEDURE REQUIRE THAT STORED ENERGY (MECHANICAL, HYDRAULIC, AIR, ETC.) BE RELEASED OR BLOCKED BEFORE EQUIPMENT IS LOCKED-OUT FOR REPAIRS?
	ARE APPROPRIATE EMPLOYEES PROVIDED WITH INDIVIDUALLY KEYED PERSONAL SAFETY LOCKS?
	ARE EMPLOYEES REQUIRED TO KEEP PERSONAL CONTROL OF THEIR KEY(S) WHILE THEY HAVE SAFETY LOCKS IN USE?
	IS IT REQUIRED THAT ONLY THE EMPLOYEE EXPOSED TO THE HAZARD, PLACE OR REMOVE THE SAFETY LOCK?
	IS IT REQUIRED THAT EMPLOYEES CHECK THE SAFETY OF THE LOCK-OUT BY ATTEMPTING A STARTUP AFTER MAKING SURE NO ONE IS EXPOSED?
	ARE EMPLOYEES INSTRUCTED TO ALWAYS PUSH THE CONTROL CIRCUIT STOP BUTTON IMMEDIATELY AFTER CHECKING THE SAFETY OF THE LOCK-OUT?
	IS THERE A MEANS PROVIDED TO IDENTIFY ANY OR ALL EMPLOYEES WHO ARE WORKING ON LOCKED-OUT EQUIPMENT BY THEIR LOCKS OR ACCOMPANYING TAGS?
	ARE A SUFFICIENT NUMBER OF ACCIDENT PREVENTIVE SIGNS OR TAGS AND SAFETY PADLOCKS PROVIDED FOR ANY REASONABLY FORESEEABLE REPAIR EMERGENCY?
	WHEN MACHINE OPERATIONS, CONFIGURATION, OR SIZE REQUIRES THE OPERATOR TO LEAVE HIS OR HER CONTROL STATION TO INSTALL TOOLS OR PERFORM OTHER OPERATIONS, AND WHEN PART OF THE MACHINE COULD MOVE IF ACCIDENTALLY ACTIVATED, IS SUCH ELEMENT REQUIRED TO BE SEPARATELY LOCKED OR BLOCKED OUT?
	IN THE EVENT THAT EQUIPMENT OR LINES CANNOT BE SHUT DOWN, LOCKED-OUT, AND TAGGED, IS A SAFE JOB PROCEDURE ESTABLISHED AND RIGIDLY FOLLOWED?

204F04.EPS

Figure 4 ◆ Sample lockout/tagout procedure checklist.

Instructor's Notes:

2.3.0 Safety Inspection Time Frame

The frequency of safety inspections depends on your work site. In some cases, inspections are done on a regularly scheduled basis. In other cases, they're done randomly and without notice. Sometimes inspections are done only when needed. For example, cranes and rigging equipment may need to be checked before each use. Most of the time, this will be done during normal working hours.

More detailed inspections and tests are usually needed at monthly and yearly intervals. However, you need to constantly evaluate the frequency of your inspections. If your work site starts having more minor accidents or near misses, you may need to conduct frequent, unannounced inspections. Ask yourself the following questions.

- How severe is the loss potential?
- What is the potential for injury?
- How fast does a part or equipment wear out?
- What is my site's past safety history?
- Does OSHA or another agency require inspections?

The answers to these questions will help you decide how often you need to perform inspections.

2.4.0 Safety Inspection Findings

When you find an unsafe condition or work practice that presents an immediate danger to workers or property, have it fixed right away. Waiting increases the risk that someone will get hurt or that equipment will be damaged.

For example, you come upon a piece of equipment that is leaking oil onto the floor. What are some of the things you could do? Your first actions will probably be immediate and temporary, but enough to prevent a serious fall:

- If the equipment is powered on, turn it off.
- Attach a DANGER – DO NOT OPERATE tag to the equipment.
- Set up safety cones to alert workers that the floor may be slippery.
- Arrange for the floor to be cleaned.

After the immediate hazard has been fixed, you'll need to follow up with the worker's supervisor to make sure that the equipment has been properly repaired. You should also check to see if this is an isolated problem or a symptom of a bigger issue.

Of course, not all safety issues will be as simple as this one. Some issues will be complicated and will need in-depth analysis to determine the cause and to find a solution. For more information about accident investigation, please refer to the accident investigation modules in Volume 4.

When you find an unsafe condition or work practice that is outside of your area of responsibility, notify the responsible person. This is often the job superintendent.

Remember, always document your findings and what was done to fix the problem. Record the date and time you did the inspection. Note any problems that were found and what was done to fix them. Write clearly. Record who was notified, including the date and time this person was notified. Then, record that the fix was verified to be working properly. If the problem has been successfully fixed, record that there is no longer a hazard.

After you perform a safety inspection, you will need to tell management, job-site supervisors, and co-workers of the results. Let them know both good and bad results. This will help provide management with quality information that can be used to make informed decisions in the future.

3.0.0 ◆ SAFETY AUDITS

Safety audits are reviews of safety policies and procedures to see if they are accurate and being used. You'll need to use your experience during safety audits. You'll also need to review safety policies and procedures to determine whether they are adequate, realistic, and up to date.

3.1.0 Safety Audit Time Frame

Safety audits are usually performed on a periodic basis. Depending on who is doing the audit, however, they can be done more or less frequently. For example, in the case of a site audit, where the audit is done by the corporate office, it could be done on a quarterly basis. On the other hand, if the audit is done by the site safety technician, it could be done on a weekly basis. The size of a project and the company's safety requirements have an impact on how often safety audits are done.

During an audit, you'll need to review applicable safety policies and procedures as well as safety and health regulations. You will also need to determine whether the policies and procedures are accurate and up to date. Ask yourself:

- Does this policy or procedure still apply to your work site?
- Is the procedure accurate?
- Do the policies and procedures address the hazards and required safety precautions?
- Is the procedure being followed?

Explain the factors that are considered when determining the length of time between inspections.

Show Transparency 7, Inspection Consideration Questions.

Explain the steps that must be taken when an unsafe condition or at-risk practice is found during an inspection.

Explain that the response to the finding will vary based on the finding's severity.

Define the term safety audit. Explain the safety technician's role during safety audits.

Explain the factors that are considered when determining the length of time between audits.

Describe a typical safety audit procedure.

Show Transparency 8 (Figure 5).

INSPECTIONS, AUDITS, AND OBSERVATIONS 4.5

Provide the participants with a scenario and have them conduct a safety audit. Note the proficiency of each participant. This laboratory corresponds with Performance Task 1.

Review the safety policies and procedures that apply to your areas of responsibility. Discuss any ideas you have to improve the safety policies and procedures with site personnel and your safety manager.

When your work site starts to use new equipment or materials, don't wait for the next scheduled safety audit. Review the applicable policy and procedure to be sure that it's still accurate.

Be sure to document that you performed a safety audit. Your findings and the control actions taken should also be recorded. *Figure 5* shows an example of a documented safety audit.

Define the term safety observation. Explain the safety technician's role during safety observations.

Describe a typical safety observation procedure.

Explain the factors that are considered when determining the length of time between safety observations.

Discuss unsafe acts and at-risk behavior commonly observed during a safety observation.

4.0.0 ◆ SAFETY OBSERVATIONS

A safety observation is the act of watching a worker or group of workers during the performance of their jobs. The purpose is to determine whether the workers are working safely or committing an unsafe act.

Your company will provide you and your co-workers with written policies and procedures to ensure a safe working environment. Safety observation on your part as the safety technician not only ensures that the policies and procedures are carried out, but also ensures that the policies and procedures themselves are safe and practical. Stay up to date; make sure you know your company's safety policies and procedures as well as the applicable OSHA regulations.

4.1.0 Performing a Safety Observation

To perform a job safety observation, you will need to do the following:

- Observe the worker in the performance of his or her job.
- Validate training and job instructions.
- Note safe and at-risk behavior.
- Provide immediate feedback to the worker.

4.2.0 Safety Observations Time Frame

The frequency of safety observations depends on the job site but may be done continuously. As the safety technician, you'll need to stay alert and continuously monitor your co-workers' activity for unsafe practices. You'll also need to monitor the work site for unsafe acts and conditions.

The following is a list of some common unsafe acts:

- Failing to use personal protective equipment
- Failing to warn co-workers of possible hazards or at-risk behaviors
- Failing to follow instructions, procedures, and/or sound advice
- Using defective tools or equipment
- Improper lifting
- Taking an improper working position
- Making safety devices inoperable
- Operating equipment at improper speeds
- Operating equipment without authority
- Servicing, repairing, or adjusting equipment while it is in motion or energized
- Loading or placing equipment or supplies improperly or dangerously
- Using equipment improperly
- Working under a suspended load or in an obviously dangerous area
- Working while impaired by alcohol or drugs, legal or illegal
- Engaging in horseplay

SAFETY AUDIT REPORT LOG

April 4, 2003 –
Performed safety audit on using the paint booth safety policy and procedure. Found that the recommended respirator filter does not filter particle size for new metal primer paint. Recommended to Ralph Brown, Safety Manger, that we change filters. He agreed. Ordered new filters.

H. White, Site Safety Tech

April 5, 2003 –
Received filters. Inked change into current procedure and sent it to the office for word processing. Notified all workers at site to use the new filters. Hung sign on paint booth supply cabinet as a reminder.

H. White, Site Safety Tech

204F05.EPS

Figure 5 ◆ Sample documentation of safety audit.

Instructor's Notes:

> **NOTE**
> The terms *unsafe behavior* and *at-risk* behavior are sometimes used interchangeably.

The following is a list of the most commonly noted unsafe conditions:

- Congested workplaces
- Defective tools, equipment, or supplies
- Excessive noise
- Fire and explosive hazards
- Hazardous atmospheric conditions
  - Gases
  - Dusts
  - Fumes
  - Vapors
- Inadequate supports or guards
- Inadequate warning systems
- Poor housekeeping
- Poor illumination
- Poor ventilation

4.3.0 Safety Observation Findings

When you see a worker performing an unsafe act or practice that represents an immediate risk of accident or injury, stop him or her immediately. You'll need to talk to the employee about the unsafe practice. Try to keep the following in mind:

- Stay calm. Try not to startle the worker.
- Tell the worker that you are concerned for his or her well being and safety.
- State the nature of the unsafe act or practice; be specific.
- Inform the worker of the appropriate safety rule, practice, or procedure.
- Explain the need for following the safety rule, practice, or procedure. Cite federal and state governmental regulations. Explain that these rules require employers to train their employees on the hazards associated with their work and the safeguards to be taken to prevent accidents or injury.
- Demonstrate the correct procedure if necessary.
- Let the worker practice the procedure with your guidance.
- Close on a positive note.

> **NOTE**
> Recurring problems should be documented and handled according to your company's disciplinary policy.

Don't blame workers for unsafe acts when other factors are involved. Help the worker become aware of his or her surroundings. If workers can recognize unsafe conditions on their own, it will make your job easier. Safety-conscious workers recognize hazards and take steps to do something about them. Being a safety-conscious worker requires that you stop and talk to a co-worker when at-risk or hazardous behavior is observed. This must be done in a non-threatening way to be effective.

In addition to telling workers that they have done something wrong, you need to tell them when they're doing a good job. You don't need to pat them on the back all of the time, but positive feedback is needed. Remember: behavior that is rewarded is repeated. Behavior that is punished is not repeated.

Document the results of safety observations *(Figure 6)*. You may need these records to identify workers with poor safety habits.

EMPLOYEE OBSERVATION REPORT LOG

April 3, 2003 -
During a routine employee observation, I noticed that John Smith didn't have eye protection on while using an angle grinder. All other guards were in place. I stopped Smith and explained the importance of eye protection while using the grinder. Smith said he understood and would wear eye protection from now on.

H. White, Site Safety Tech

204F06.EPS

Figure 6 ◆ Employee observation documentation.

Explain what the safety technician should to do when an unsafe act or at-risk behavior is observed.

Show Transparency 9, Tips for Correcting Unsafe Acts or At-Risk Behavior.

Explain how to provide positive feedback to workers.

Emphasize that behavior that is rewarded is repeated and that behavior that is punished is not repeated.

Discuss the importance of documenting the results of safety observations.

Show Transparency 10 (Figure 6).

Provide the participants with a scenario in which a worker has been observed performing an unsafe act. Have the participants use role playing to appropriately correct the noted behavior. Note the proficiency of each participant. This laboratory corresponds with Performance Task 2.

Identify the regulatory bodies that regulate the local, state, and federal requirements for inspections, audits, and observations.

Describe the safety technician's role during an OSHA inspection.

Explain OSHA's requirements regarding OSHA inspections.

5.0.0 ◆ GOVERNMENT REGULATIONS

There are city, county, state, and federal regulations that affect your own company policies and procedures. As a safety technician, you'll need to keep current on the regulatory agencies that govern the work that you do. Part of your job is to make sure that your company is following all of the safety and reporting rules of the different agencies.

Because the structure of these regulatory bodies is different in each area, you should familiarize yourself with the regulatory requirements for each job. Then make sure that your company policies reflect reporting and regulatory requirements.

5.1.0 The Occupational Safety and Health Act

Almost all industries, including construction, are subject to federally prescribed mandatory rules for workers' safety and health under the Occupational Safety and Health Act (OSHA). OSHA sets the rules and regulations that form the basis for inspections, citations, penalties, and appeals.

5.2.0 Other Federal Agencies

OSHA is not the only federal agency affecting worker safety and industry standards. Other administrations involved in safety efforts include the Mine Safety and Health Administration (MSHA), the Environmental Protection Agency (EPA), and the Consumer Product Safety Commission (CPCS). Whether you need to comply with rules of these agencies depends on the location of work, type of work, and specific hazards at your job site.

5.3.0 OSHA Inspections

In your role as a safety technician, you may need to escort OSHA officials on an inspection tour of your work site. Be aware that your company does not need to grant OSHA official access to your work site except as noted in this section. By law, OSHA can be required to obtain search warrants to conduct inspections. However, few employers feel the need to exercise this right. To ensure compliance with OSHA standards, inspections are made where and when considered advisable by the agency. Inspections are made without prior notice to the company being inspected. It's important to know your own company's policy and to adhere to it when dealing with OSHA or any other state, city, local, and federal agencies.

Some exceptions that allow OSHA inspectors immediate access to a work site are as follows:

- Accidental death or a catastrophic incident that resulted in the hospitalization of three or more people
- Verbal or written complaints received by OSHA concerning an immediate hazard
- Routine inspections of sites that are very hazardous
- Routine inspections of other industries
- Follow-up inspections to see if corrections for hazards found during an initial inspection have been put in place
- Referrals from local law enforcement agencies or other federal agencies.

If a safety inspector unexpectedly appears at your work site, keep in mind the following:

- Be polite. Remember, the inspector is just doing his or her job.
- If your company's policy is to have you or a supervisor escort the inspector on the tour or if you are unsure of your company's policy about outside inspections, call your safety manger or construction superintendent.
- Don't leave the inspector standing in the rain or hot sun while you are contacting your supervisor.
- Show the inspector where to get water. Offer coffee if it's available at your site.

> **NOTE**
> For more information about OSHA inspections, please refer to Volume 5, Module 75216-03, *OSHA Inspection Procedures.*

Summary

As the safety technician, you'll be responsible for coordinating site safety and health activities. As such, your job is to help find and correct safety hazards before an accident or injury can occur. You'll be performing safety inspections, audits, and employee observations. It will be your job to see that inspections, audits, and observations are performed and properly documented, including corrective actions and follow-ups. The frequency with which inspections, audits, and observations are performed will depend on your work site and company.

During a safety inspection, you'll check an area to identify, report, and correct hazards to workers, materials, and equipment. You'll be doing these inspections continuously, as well as on a sched-

Instructor's Notes:

uled periodic basis. Some inspections will be done on an intermittent basis; like when you notice an increase in reported near misses at your work site. Other inspections are general; they might include tool rooms, lay down areas, or maintenance shops. The type of work being done at your site will govern the frequency of all inspections and the items that are covered in the inspections. You'll need to notify your supervisor and co-workers of the results of your inspections.

During safety audits, you'll review the safety policies and procedures that apply to your areas of responsibility to see if they are adequate and are being used.

During a safety observation, you'll watch a worker during the performance of his or her job. The purpose is to determine whether the worker is working safely or committing an unsafe act. You'll need to provide the worker with immediate feedback (positive and negative) about his or her work.

As a safety technician, you'll need to keep current on the regulatory agencies that govern the work you do, especially OSHA regulations. Part of your job is to make sure that your company is following the safety and reporting regulations from the different regulatory agencies.

All of these tasks are designed to decrease the number and severity of accidents. A good safety record at your job site will help to ensure your well being and that of your co-workers. It also helps to ensure the financial health of your company.

Review Questions

1. A safety technician performs _____.
   a. OSHA inspections
   b. safety inspections, audits, and observations
   c. accident cost evaluations
   d. inspections only when directed to do so

2. A safety technician is primarily responsible for _____.
   a. coordinating site safety activities
   b. contacting OSHA when any accident occurs
   c. investigating all accidents at the work site
   d. computing the cost of accidents at the work site

3. The main reason for conducting safety inspections, audits, and observations is to _____.
   a. identify and fix unsafe conditions before an accident occurs
   b. evaluate whether a worker performs safely while doing his or her job
   c. determine the cause of an accident
   d. determine whether all safety policies and procedures are being followed

4. Safety inspections _____.
   a. are not required on most construction sites
   b. check a particular area for hazards and unsafe conditions
   c. verify the accuracy of safety policies and procedures
   d. are the responsibility of the safety manager

5. Safety inspections are performed only when it is convenient so that work will not be interrupted.
   a. True
   b. False

6. Safety technicians need to notify site management, supervisors, and co-workers of the results of a safety inspection.
   a. True
   b. False

7. During a safety audit, the safety technician _____.
   a. reviews safety policies and procedures for accuracy
   b. checks the work area for hazards and unsafe conditions
   c. observes whether a co-worker performs unsafe acts
   d. escorts OSHA inspectors around the work site

8. Safety audits are usually conducted _____.
   a. only after an accident at the site
   b. at the same time as the safety inspection
   c. annually
   d. periodically

9. The main reason for performing safety observations is to _____.
   a. identify and correct unsafe acts and conditions
   b. discipline employees who violate safety practices
   c. have observations on record of OSHA
   d. make employees feel accountable

10. What should the safety technician do if he or she sees a co-worker performing an unsafe act that could cause an immediate accident?
    a. Write down all of the information about the unsafe act and talk to the foreman about it.
    b. Wait until a quiet time and then talk to the co-worker about what he or she did wrong.
    c. Stop the worker immediately and tell him or her what they did wrong and why.
    d. Ask the foreman or safety manager to correct the worker.

Review the material covered in the module.

Have the participants complete the Review Questions and go over the answers prior to administering the Module Examination.

Administer the Module Examination. Record the results on Craft Training Report Form 200 and submit the results to the Training Program Sponsor.

Administer the Performance Test and fill out Performance Profile Sheets for each participant. If desired, participant proficiency noted during laboratory exercises may be used to complete the Performance Test. Record the testing results on Craft Training Report Form 200 and submit the results to the Training Program Sponsor.

INSPECTIONS, AUDITS, AND OBSERVATIONS 4.9

Instructor's Notes:

GLOSSARY

Trade Terms Introduced in This Module

Audit: To review safety policies and procedures to see if they are adequate and being used.

Inspection: The act of checking an area to identify, report, and correct hazards to workers, materials, and equipment.

Observation: Watching a worker during the performance of his or her job for the purpose of determining whether the worker is working safely or committing an unsafe act.

INSPECTIONS, AUDITS, AND OBSERVATIONS 4.11

Instructor's Notes:

REFERENCES

Additional Resources

This module is intended to present thorough resources for task training. The following reference works are suggested for further study. These are optional materials for continued education rather than for task training.

www.osha.gov

www.asse.org

INSPECTIONS, AUDITS, AND OBSERVATIONS 4.13

MODULE 75204-03 — ANSWERS TO REVIEW QUESTIONS

Answer		Section
1.	b	1.0.0
2.	a	1.0.0
3.	a	1.0.0
4.	b	2.0.0
5.	b	2.3.0
6.	a	2.4.0
7.	a	3.0.0, 3.1.0
8.	d	3.1.0
9.	a	4.0.0
10.	c	4.3.0

CONTREN™ LEARNING SERIES — USER UPDATES

The NCCER makes every effort to keep these textbooks up-to-date and free of technical errors. We appreciate your help in this process. If you have an idea for improving this textbook, or if you find an error, a typographical mistake, or an inaccuracy in NCCER's Contren™ textbooks, please write us, using this form or a photocopy. Be sure to include the exact module number, page number, a detailed description, and the correction, if applicable. Your input will be brought to the attention of the Technical Review Committee. Thank you for your assistance.

Instructors – If you found that additional materials were necessary in order to teach this module effectively, please let us know so that we may include them in the Equipment/Materials list in the Instructor's Guide.

Write: Curriculum Revision and Development Department
National Center for Construction Education and Research
P.O. Box 141104, Gainesville, FL 32614-1104

Fax: 352-334-0932

E-mail: curriculum@nccer.org

Craft ____________________ Module Name ____________________

Copyright Date ____________ Module Number ____________ Page Number(s) ____________

Description ____________________

____

____

____

(Optional) Correction ____________________

____

____

(Optional) Your Name and Address ____________________

____

____

Module 75205-03

Employee Motivation

NCCER STANDARDIZED CRAFT TRAINING PROGRAM

The National Center for Construction Education and Research (NCCER) provides a standardized national program of accredited craft training. Key features of the program include instructor certification, competency-based training, and performance testing. The program provides trainees, instructors, and companies with a standard form of recognition through a National Craft Training Registry. The program is described in full in the *Guidelines for Accreditation*, published by the NCCER. For more information on standardized craft training, contact the NCCER by writing us at P.O. Box 141104, Gainesville, FL 32614-1104; calling 352-334-0911; or e-mailing info@nccer.org. More information may be found at our Web site, www.nccer.org.

HOW TO USE THIS ANNOTATED INSTRUCTOR'S GUIDE

Each page presents two sections of information. The larger section displays each page exactly as it appears in the Participant Module. The narrow column ties suggested participant and instructor actions to each page and provides icons (detailed below) to call your attention to material, safety, audiovisual, or testing requirements. The bottom of each page includes space for your notes.

The **Audiovisual** icon indicates an appropriate time to show a transparency or other audiovisual aid.

The **Classroom** icon prompts you to define a term, stress a point, ask participants to explain a concept, or give examples.

The **Demonstration** icon directs you to show participants how to perform tasks.

The **Examination** icon tells you to administer the written module examination.

The **Homework** icon is placed where you may wish to assign reading for the next class, to assign a project, or to advise participants to prepare for an examination.

The **Laboratory** icon is used when participants are to practice performing tasks.

The **Materials** icon is a reminder for you to gather materials needed for classes, labs, and testing.

The **Performance Testing** icon tells you to administer a performance test or a portion thereof.

The **Safety** icon is used to emphasize safety issues. It is often keyed to *Caution* and *Warning* statements in the Participant Module.

The **Teaching Tip** icon indicates additional guidance is available, such as how to conduct an exercise, get the most educational value from a field trip, or encourage class participation. Teaching Tips may expand on a feature (*Think About It*, *Did You Know?*) or provide Quick Quizzes or similar exercises. You will be referred to the Teaching Tips section at the back of the module if there is additional material.

The **Combination** icon indicates that the laboratory listed corresponds with a performance task. If desired, you can note the proficiency of the participants during the laboratory and use it to satisfy performance testing requirements.

PREPARATION

Before teaching this module, you should review the Objectives, Performance Tasks, Materials and Equipment List, and the Module Outline. Be sure to allow ample time to prepare your own training or lesson plan and gather all required materials and equipment.

MODULE OVERVIEW

In this module, participants will learn the how to effectively communicate with all work-site employees.

PREREQUISITES

Please refer to the Course Map in the Participant Module. Prior to training with this module, it is recommended that the participant shall have successfully completed the following:

Field Safety; Safety Technology, Modules 75201-03 through 75204-03

OBJECTIVES

Upon completion of this module, the participant will be able to:

1. Effectively communicate safety policies and procedures to all employees on a job site.
2. Describe the correct way to provide motivation, recognition, and discipline as needed.

PERFORMANCE TASKS

Under the supervision of the instructor, the participant should be able to:

1. Demonstrate how to effectively communicate with a co-worker.

MATERIALS AND EQUIPMENT LIST

Overhead projector and screen
Transparencies
Whiteboard/chalkboard
Markers/chalk
Blank acetate sheets
Transparency pens
Pencils and scratch paper
Appropriate personal protective equipment
Safety sign or posting
Employee award plaque
Module Examinations*
Performance Profile Sheets*

* Located in the Test Booklet.

SAFETY CONSIDERATIONS

Ensure that participants are equipped with appropriate personal protective equipment.

TEACHING TIME FOR THIS MODULE

An outline for use in developing your lesson plan is presented below. Approximately 2½ hours are suggested to cover *Employee Motivation*. This includes 10 minutes at the beginning of the session for administrative tasks and one 10-minute break. You may need to adjust the time required for hands-on activities and testing based on your class size and resources. Because laboratories often correspond to Performance Tasks, the proficiency of the participants may be noted during these exercises for Performance Testing purposes.

Topic	**Planned Time**
A. Introduction	________
B. Communication	________
1. Verbal Communication	________
2. Non-Verbal Communication	________
3. Written or Visual Communication	
4. Communication Issues	________
5. Laboratory Provide the participants with a scenario and have them demonstrate how to communicate with another worker. This laboratory corresponds with Performance Task 1.	________
C. Motivation	________
1. Gaining Attention	________
2. Building Relevance	________
3. Instilling Confidence	________
4. Fostering Satisfaction	________
D. Discipline	________
1. Poor Attitude Towards the Workplace	________
2. Inability to Work with Others	________
3. Absenteeism and Turnover	________
E. Review	________
F. Module Examination	________
1. Participants must score 70% or higher to receive recognition from the NCCER.	
2. Record the testing results on Craft Training Report Form 200 and submit the results to the Training Program Sponsor.	
G. Performance Testing	________
1. Participants must perform each task to the satisfaction of the instructor to receive recognition from the NCCER. If applicable, proficiency noted during laboratory exercises can be used to satisfy the Performance Testing requirements.	
2. Record the testing results on Craft Training Report Form 200 and submit the results to the Training Program Sponsor.	

Module 75205-03

Employee Motivation

Instructor's Notes:

COURSE MAP

This course map shows all of the modules in Safety Technology. The suggested training order begins at the bottom and proceeds up. The local Training Program Sponsor may adjust the training order.

Assign reading of Module 75205-03.

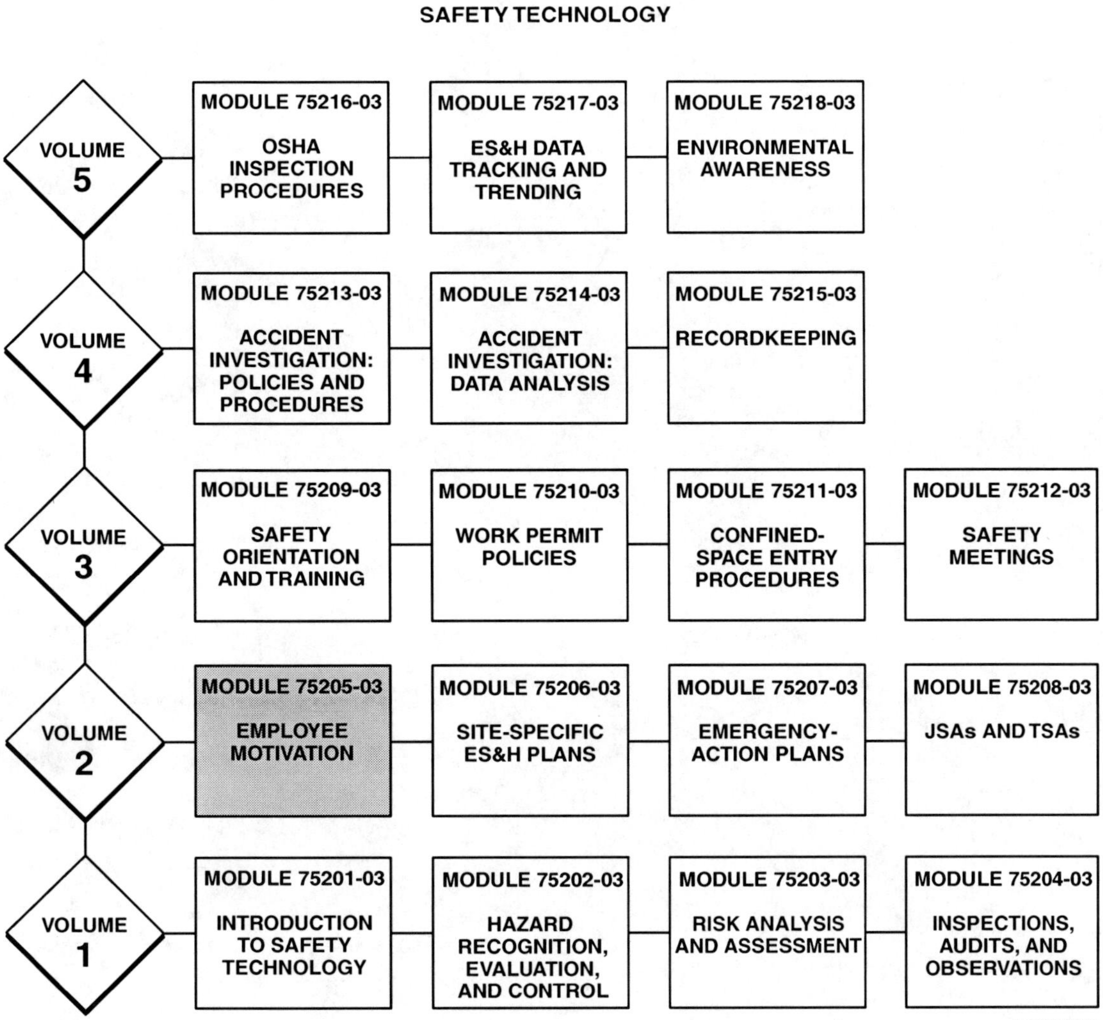

205CMAP.EPS

EMPLOYEE MOTIVATION 5.iii

Instructor's Notes:

MODULE 75205-03 CONTENTS

Figures

EMPLOYEE MOTIVATION 5.v

Instructor's Notes:

MODULE 75205-03

Employee Motivation

Ensure that you have everything you need to teach the course. Check the Materials and Equipment list in the front of this module.

Objectives

When you have completed this module, you will be able to do the following:

1. Effectively communicate safety policies and procedures to all employees on a job site.
2. Describe the correct way to provide motivation, recognition, and discipline as needed.

Prerequisites

Before you begin this module, it is recommended that you successfully complete the following: Field Safety; Safety Technology, Modules 75201-03 through 75204-03.

Required Materials

1. Pencil and paper
2. Appropriate personal protective equipment

1.0.0 ◆ INTRODUCTION

Workers are more likely to work safely if they are motivated, recognized for their good behavior and accomplishments, and disciplined fairly, consistently, and properly when necessary. In your role as safety technician, you must communicate with workers to ensure that they know what is expected of them.

You will have a leadership role on a site. This means that workers will be looking to you to provide information and guidance about how to get the job done safely. In this role, you will need to have the respect of the workers. When an individual is respected and approachable, discipline problems are reduced. This makes it easier for you to concentrate on keeping the work site and workers safe.

2.0.0 ◆ COMMUNICATION

Effective **communication** with people at all levels on a job site is important. In order to be successful, safety technicians must develop an understanding of human behavior and have communication skills that enable them to understand and influence others.

There are many definitions for communication. One is that communication is the act of accurately and effectively conveying or transmitting facts, feelings, and opinions to another person. Another definition is that communication is the method of exchanging information and ideas.

Just as there are many definitions of communication, it also comes in more than one form. The different types of communication are:

- **Verbal**
- **Non-verbal**
- **Written**

A typical person spends about 80 percent of his or her day communicating through writing, speaking, listening, or using non-verbal communication such as body language. Of that time, studies suggest that approximately 20 percent of communication is written, and 80 percent involves speaking or listening.

2.1.0 Verbal Communication

Verbal communication refers to the spoken words exchanged between two or more people. It can be done face-to-face or by other means such as the telephone or two-way radios.

Figure 1 shows the relationship of the four parts of verbal communication:

1. **Sender**
2. **Message**
3. **Receiver**
4. **Feedback**

Explain that terms shown in bold are defined in the Glossary at the back of this module.

Review the goals of the course and explain what will be expected of the participant.

Show Transparency 1, Objectives.

Show Transparency 2, Performance Task.

Explain the purpose and benefits of communication.

Explain the three types of communication: verbal, non-verbal, and written.

Define verbal communication.

Explain the four parts of verbal communication.

Show Transparency 3 (Figure 1).

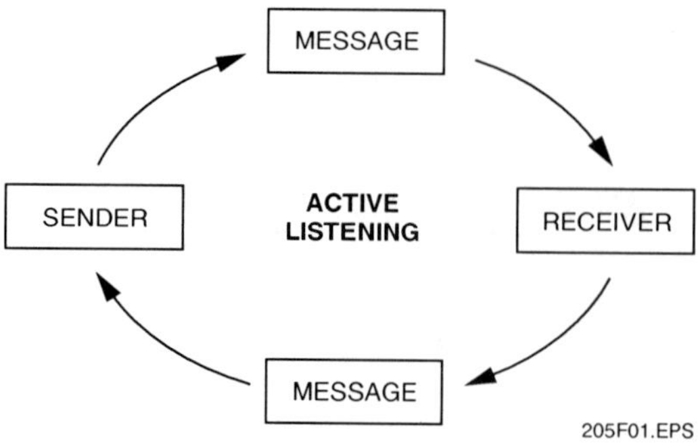

Figure 1 ◆ Communication process.

Explain the role of the sender in communication.

2.1.1 The Sender

The sender is the person who creates the message to be communicated. In verbal communication, the sender actually says the message aloud to the person or persons for whom it is intended.

The sender must be sure to speak in a clear and concise manner that can be easily understood by others. This is not an easy task, but with practice, you will be a very effective sender.

Show Transparency 4, Tips For Becoming an Effective Sender.

Some basic tips to consider for becoming an effective sender include:

- Avoid talking with anything in your mouth, such as food or gum.
- Find an appropriate rate of speech. Don't talk too quickly or too slowly. This is important because in some situations, people tend to focus on your rate of speech instead of on what you are saying.
- Be aware of your tone. Remember, it's not so much what you say, but how you say it.
- Enunciate to prevent misunderstandings, as many letters, such as T, D, B, and E sound similar.
- Don't talk in a monotone. Put some enthusiasm and feeling in your voice.

Explain the process in which messages are organized and delivered.

Describe the role of the receiver in communication.

Explain the barriers to effective listening and how to overcome these barriers.

2.1.2 The Message

The message is what the sender is attempting to communicate to the receiver. A message can be a set of directions, an opinion, or a feeling. Whatever its function, a message is an idea or fact that the sender wants the receiver to know.

Before speaking, the sender should determine what it is he or she wants to communicate. The sender should then organize what to say, ensuring that the message is logical and complete. Taking the time to clarify your thoughts prevents rambling, not getting the message across effectively, or confusing the audience. It also permits the sender to get to the point quickly.

In delivering the message, the sender should consider the audience. It is important not to talk down to the audience. Remember that everyone, whether in a senior or junior position, deserves respect and courtesy. Therefore, the sender should use words and phrases that the audience can understand. Try to avoid technical language or slang when possible. In addition, the sender should use short sentences, which give the audience time to understand and digest one point or fact at a time.

2.1.3 The Receiver

The receiver is the person to whom the message is communicated. For the communication process to be successful, it is important that the receiver understands the message as the sender intended. Therefore, the receiver must listen to what is being said.

The first step to becoming a good listener involves realizing that there are many barriers that get in the way of your listening, particularly on a busy construction job site.

Show Transparency 5, Barriers to Effective Listening.

The following are barriers to effective listening.

- Noise, visitors, telephone, or other distractions
- Preoccupation, being under pressure, or daydreaming
- Reacting emotionally to what is being communicated
- Thinking about how to respond instead of listening
- Giving an answer before the message is complete
- Personal biases
- Finishing the sender's sentence

Show Transparency 6, Tips For Overcoming Barriers to Effective Listening.

The following tips will help you to overcome these barriers.

- Take steps to minimize or remove distractions; learn to tune-out your surroundings.
- Listen for key points.
- Take notes.
- Try not to take things personally.
- Allow yourself time to process your thoughts before responding.
- Let the sender communicate the message without interruption.
- Be aware of your personal biases, and try to stay open-minded.

Instructor's Notes:

There are many ways for a receiver to show that he or she is actively listening to what is being said. This can even be accomplished by maintaining eye contact, nodding your head, and taking notes. It may also be accomplished through feedback.

2.1.4 Feedback

Feedback refers to the communication that occurs after the message has been sent and received. It involves the receiver responding to the message. Feedback is a very important part of the communication process because it allows the receiver to communicate how he or she interpreted the message. It also lets the sender know whether the message was understood as intended. In other words, feedback is a way to make sure that the receiver and the sender understand each other.

The receiver can use the opportunity of providing feedback to **paraphrase** back what was heard. When paraphrasing what you heard, it is best to use your own words. That way, you can show the sender that you interpreted the message correctly and could explain it to others if needed.

In addition, feedback gives the receiver the opportunity to clarify the meaning of the message and request additional information if needed. This is generally done by asking questions.

2.2.0 Non-Verbal Communication

Non-verbal communication refers to things that you can actually see when communicating with others. Examples include facial expressions, body movements, hand gestures, and eye contact.

Non-verbal communication can provide an external signal of an individual's inner emotions. It occurs simultaneously with verbal communication and the sender of the non-verbal communication is often not even aware of it.

Because it can be physically observed, non-verbal communication is just as important as the words used in conveying the message. This is true because people are influenced more by non-verbal signals than by spoken words. Therefore, it is important that you be conscious of the non-verbal cues that you send because you don't want the receiver to interpret your message incorrectly based on your posture or an expression on your face. If you do, it can affect the current communication exchange. When communicating, try not to carry over issues and incidents from earlier in the day.

> **NOTE**
> Standing with your arms crossed over your chest or with your hands on your hips is considered an aggressive or threatening stance. Consider standing with your arms at your side when speaking to another individual. This helps to ensure that the receiver does not feel threatened or intimidated.

2.3.0 Written or Visual Communication

Some communication will have to be written or **visual**. Written or visual communication refers to communication that is documented on paper or transmitted electronically via the computer using words or visuals. Written and visual communication can also include signs, hand signals, and postings.

Many messages on a job have to be communicated in text form. Examples include weekly reports, requests for changes, purchase orders, and correspondence on specific subjects. These items are written because they must be recorded for contractual and historical purposes. In addition, some communication on the job will have to be visual because items that are difficult to explain verbally or by the written word can best be explained through diagrams or graphics. Examples include the plans or drawings used on a job.

It is important to understand your audience before writing or creating a visual message. This helps to ensure that the receiver is able to read the message and understand the content. If he or she is unable to do this, the communication process will be unsuccessful. That's why the writer should consider the actual meaning of words or diagrams and how others might interpret them. It's also important for all handwriting to be legible if the message is to be understood.

Some basic tips for writing include:

- Avoid emotion-packed words or phrases.
- Be positive whenever possible.
- Avoid using technical language or **jargon**.
- Stick to the facts.
- Provide an adequate level of detail.
- Present the information in a logical manner.
- Avoid making judgments, unless asked to do so.
- Check the message for spelling and grammatical errors.
- Make sure the document is legible.

Explain the purpose of providing feedback.

Describe techniques to provide accurate and effective feedback.

Define non-verbal communication.

Describe the circumstances in which non-verbal communication is used.

Define written or visual communication.

Explain how written or visual communication is used by showing participants a sign, hand signal, or posting.

Discuss the basic tips for written communication. Refer the participants to the bulleted list in the Participant Module.

EMPLOYEE MOTIVATION 5.3

Discuss the issues that may affect communication: language, ethnic diversity, reading ability, or physical disabilities.

Show Transparency 7 (Figure 2).

Explain how workplace stress affects the communication process.

Describe OSHA's suggestions for relieving stress.

Show Transparency 8, Tips for Relieving Stress.

Provide the participants with scenarios involving the use of communication. Use role playing techniques to guide the participants through the use of each type of communication. Note the proficiency of each participant. This laboratory corresponds with Performance Task 1.

- Be prepared to provide a verbal or visual explanation, if needed.

Some basic tips for creating visuals include:

- Provide an adequate level of detail.
- Ensure that the diagram is large enough to be seen.
- Avoid creating complex visuals.
- Present the information in a logical order.
- Be prepared to provide a written or verbal explanation of the visual, if needed.

2.4.0 Communication Issues

It is important to understand that each person communicates a little differently because we are all unique individuals. The **diversity** of the workforce can also make communication challenging. You must consider that your audience consists of individuals from different **ethnic groups** and cultural backgrounds, with varying educational levels, and economic status. Therefore, you must learn to understand your audience to determine how to effectively communicate with each individual.

The key to effective communication is to acknowledge that people are different. It is important to be able to adjust your communication style to meet the needs of those on the receiving end of your message. This involves relaying your message in the simplest way possible and avoiding the use of words that people may find confusing. Be aware of how you use technical language, slang, jargon, and words that have multiple meanings. Present the information in a clear, concise manner. Avoid rambling and always speak clearly, using good grammar.

In addition, you may have to communicate your message in multiple ways or adjust your level of detail or terminology to ensure that everyone understands your meaning as intended. For instance, you may have to draw a map for a visual person who cannot comprehend directions in a verbal or written form, or you may have to overcome language barriers on the job site by using graphics or visual aids to relay your message.

Figure 2 shows how to tailor your message to your audience.

2.4.1 Stress

Workplace stress can affect communication. It has also been associated with loss of appetite, ulcers, mental disorders, migraines, difficulty in sleeping, emotional instability, disruption of social and family life, and the increased use of cigarettes, alcohol, and drugs. Stress also affects worker attitudes and behavior. Some frequently reported consequences of stress are difficulties in communicating, maintaining pleasant relations with co-workers, and judging the seriousness of a potential emergency.

OSHA suggests the following to help relieve worker stress.

- Educate employees about job stress.
- Address work-related stresses, such as unreasonable workload, lack of readily available resources, or inadequate and unsafe equipment.
- Have regular staff meetings and discussions.
- Establish stress-management programs.
- Provide flexibility and innovation by supervisors to create alternative job arrangements.
- Provide an organized and efficient work environment.

As a safety technician, it is important to make sure workers are taking positive steps to relieve stress. Safety is one of the most important things

VERBAL INSTRUCTIONS Experienced Crew	VERBAL INSTRUCTIONS Newer Crew	WRITTEN INSTRUCTIONS
"Please drive to the supply shop to pick up our order."	"Please drive to the supply shop. Turn right here and left at Route 1. It's at 75th Street and Route 1. Tell them the company name and that you're there to pick up our order."	1. Turn right at exit. 2. Drive 2 miles to Route 1. Turn LEFT. 3. Drive 1 mile (pass the tire shop) to 75th Street. 4. Look for supply store on right.

DIAGRAM/MAP

Different people learn in different ways. Be sure to communicate so you can be understood.

205F02.EPS

Figure 2 ◆ Tailor your message.

Instructor's Notes:

to be affected if stress is not addressed. Quality and productivity can also be impacted.

3.0.0 ◆ MOTIVATION

Motivating means getting people to act or perform at a high level consistently. The ability to motivate others is a key leadership skill that effective safety technicians must possess. Because safe practices often add steps to a task, time to a job, or paperwork to a process, employees must be motivated to perform their job duties safely. Occasionally, the decision to work safely comes down to a choice between doing a task the quick and easy way, or doing the same task the correct and safe way. If properly instructed and motivated, an employee will choose the correct and safe way every time.

We often consider motivational techniques to be limited to praising an employee, giving bonuses and pay raises, scolding someone, or firing an employee. While those all may be effective pieces of a motivational strategy, much more should be considered. According to Dr. John Keller of Florida State University, there are four major parts to a successful motivational strategy: gaining attention, building relevance, instilling confidence, and fostering satisfaction. *Figure 3* illustrates why Dr. Keller's model is referred to as the **ARCS Model of Motivation.**

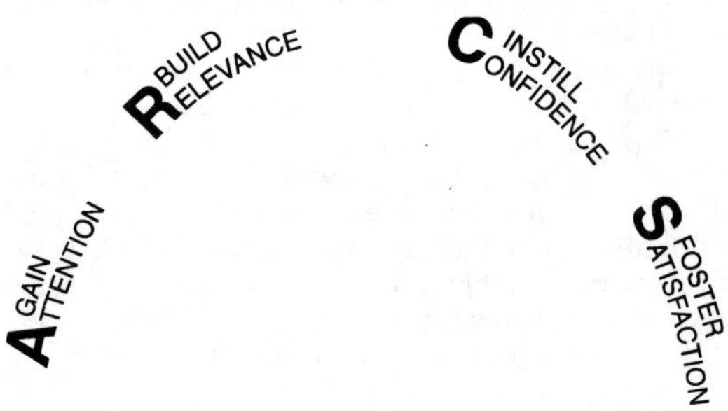

205F03.EPS

Figure 3 ◆ The ARCS Model of Motivation.

As a safety technician, it is very unlikely that you will have the authority to promote someone, give bonuses or raises, or fire someone, but you can still be an effective and persuasive motivator by using the ARCS Model.

3.1.0 Gaining Attention

For communication to be effective, the receiver must be open and receptive to the message. To motivate employees, the safety technician must first gain their attention. This can be done in a number of ways. Some ways to get the employees' attention include having employees talk about hazardous situations or near misses they have experienced, citing accurate and remarkable statistics, and showing or describing shocking but not excessively graphic images.

3.1.1 Sharing Experiences

When working to get someone's attention, it is very useful to have employees share actual work-related experiences involving hazardous situations. These accounts are usually of great interest and, because they come from co-workers, are credible. These experiences can be brought up in a conversation or safety meeting by asking for stories about past mishaps relevant to the topic of importance. For example, if the issue of wearing hard hats needs to be addressed, ask if anyone has seen an accident or near miss where a hard hat saved a life or prevented a serious injury. It is likely that if there are some experienced employees in the group, someone will have a story to tell. When asking for stories, tell workers not to use the specific names of companies involved. You should also have an example or two of your own ready in case no one volunteers.

3.1.2 Accurate and Remarkable Statistics

Numbers often help get someone's attention. You could say, for example, that according to the Bureau of Labor Statistics, construction work is the seventh most lethal occupation a person can have. That will likely grab the attention of a construction worker and make them ready to listen to the information coming next.

Another way might be to ask a challenging question. You might begin to talk about eye protection with a question like, "What is your vision worth to you?" Then answer your own question with some statistics. For example, "According to this recent news article, one person was awarded $250,000 when they lost their sight due to a work-related injury. Is your sight worth a quarter of a million dollars to you, or would you rather find ways to prevent damaging or losing your vision?"

3.1.3 Show or Describe Shocking Images

If done using good judgment and care, showing, describing, or illustrating an accident scenario can

Discuss the importance of employee motivation as it relates to on-site safety.

Explain the ARCS model of motivation.

Show Transparency 9 (Figure 3).

Discuss the importance of gaining the attention of your audience.

Explain the three techniques most commonly used to gain attention: sharing experiences, using accurate and remarkable statistics, and showing or describing shocking images.

Use one of the three techniques discussed to show the participants how to gain attention from employees. Have a participant demonstrate the same techniques. Remind the participants to use good judgment and care when showing or describing shocking images. Ask the participants if they have any questions.

EMPLOYEE MOTIVATION 5.5

Explain that the best way to build relevance into communication is to show employees how the information that is presented directly relates to them and their work habits. Provide an example.

Explain that building confidence in employees helps prevent accidents, injuries, and deaths.

Show Transparency 10, The Best Ways to Build Confidence.

Discuss the importance of fostering job satisfaction.

Explain the techniques that help foster job satisfaction. Show the participants an example of an employee award.

Explain the safety technician's role in disciplining employees.

Discuss common employee discipline problems on a work site.

be a very effective way to get someone's attention. Pictures that show the nonhuman impact of an accident are often enough for the employees to personalize the image so that you gain their attention. Images of tipped vehicles may be a good way to initialize a conversation about speed limits. Images of a burned out building would be a very effective way to begin a conversation on fire prevention.

You can also use your experience in the safety field to describe accidents or near misses that you have seen. For example, describe how removing the guards on a saw caused a worker to lose his or her fingers.

You can also illustrate how accidents happen by using models, props, or videos. Many equipment manufacturers, for example, can supply you with a safety video for their equipment. These will often have animations showing how accidents can happen. You can also get creative and use props or models to show how events may turn dangerous. To illustrate the importance of chocking wheels for example, you might have someone sit in a wheeled office chair, and then toss them a rubber ball. They will see that the shift in weight and momentum caused by catching the ball causes the chair to move. This will pique their interest and get their attention for a conversation about how wheel chocking can prevent dangerous situations.

3.2.0 Building Relevance

Stories, statistics, images, and illustrations used to gain someone's attention will have no long-lasting impact on their own. To effectively motivate someone, they must see how the information is relevant to them. This is done by finding out what is important to the person you're talking to and then helping them to make the link between what you have to say and what is already important to them. For example, to begin a safety meeting on hearing protection, you might ask a question like, "What is one of your favorite sounds?" This question will almost certainly bring many personal responses. To make hearing protection relevant to the employee, follow those responses with a statement like: "Wearing proper hearing protection will make the difference between losing your hearing and being able to enjoy those sounds for the rest of your life." By making the topic of hearing protection relevant and important on a personal level, it becomes much more likely that they will take the time to obtain and use appropriate hearing protection on the job.

3.3.0 Instilling Confidence

Giving employees the sense that they can be successful in preventing accidents, injuries, and fatalities is a very important key to motivation. If a person expects to fail, he or she will likely not put much effort into a given job. The opposite is true of workers who are confident that they can achieve their goals. They will put in the effort needed to get the job done. Confidence comes from knowing what is expected and having the tools and skills to do it.

The best ways to build confidence is by:

- Communicating clear and realistic goals
- Providing employees with the proper tools and equipment
- Providing adequate training
- Reinforcing safe behavior

3.4.0 Fostering Satisfaction

For someone to continue to act or perform at a high level, they must get some satisfaction from their actions. For many employees, there is an internal sense of satisfaction that comes from performing their job well and staying safe on the job. However, as a safety technician you must find ways to create additional reinforcing opportunities for employees by developing recognition and reward programs.

Many organizations keep track of and set goals for days in operation without a lost-time injury. These provide the whole organization with a goal to attain or a record to break. Once that record is broken, the organization can provide further satisfaction by hosting an event or a party to celebrate.

It's important to note that these types of incentive programs can be counter-productive. They can inadvertently discourage reporting and drive accidents underground where the problems that caused the accident are not resolved. Make sure that there are positive incentives for reporting. This will encourage workers to report accidents and incidents and help management make the needed repairs or improvements.

To provide satisfaction for individual employees, most organizations provide employee awards for safe practices. These awards may include plaques or trophies. *Figure 4* shows typical employee award plaques. Monetary awards can also be offered. Those might include bonuses, gift certificates, or pay raises. Long-term and consistently high performance is sometimes rewarded with job opportunities or promotions.

Instructor's Notes:

205F04.EPS

Figure 4 ◆ Safety plaques.

4.0.0 ◆ DISCIPLINE

Discipline is a large part of everyday work. There will always be problems to be resolved and decisions to be made, especially in a fast-paced, deadline-oriented industry such as construction.

In the role as safety technician, it is likely that you will encounter both simple and complex behavioral problems and be forced to make decisions about how to respond.

A few of the most common employee problems include:

- Poor attitude towards the workplace
- Inability to work with others
- Absenteeism and turnover

4.1.0 Poor Attitude Towards the Workplace

Sometimes, employees have poor attitudes towards the workplace. An employee with a poor attitude may create safety hazards in two ways. First, an employee with a poor attitude may unintentionally create unsafe working conditions through sloppy or shoddy work, unprofessional behavior, and carelessness. Second, an employee with a poor attitude may intentionally create unsafe working conditions through tool and equipment sabotage, material damage, and fighting.

A poor attitude on the job may be caused by bad relationships with fellow employees, personal problems, negative perceptions of supervision, or dislike for the job in general. The first step in changing an employee's poor attitude is to determine the cause of the problem.

The best way to determine the cause of a poor attitude is to talk with that employee one-on-one, listen to what the employee has to say, and ask questions to uncover information. Once this conversation has occurred and the facts have been assessed, you can determine how to correct the situation and turn the negative attitude into a positive one.

If it is discovered that a problem stems from factors in the workplace or the surrounding environment, there are several choices to remedy the situation. First, the worker can be moved from the situation to a more acceptable work environment. Next, the part of the work environment found to be causing the poor attitude can be changed. Finally, steps can be taken to change the employee's attitude so that the work environment is no longer a negative factor.

4.2.0 Inability to Work with Others

There will be situations in which an employee has a difficult time working with others. This could be the result of personality differences, an inability to communicate, or some other cause. Whatever the reason, the issues must be addressed and the crew must be able to work together as a team. Teamwork is one of the best ways to prevent accidents and to create a safe working environment.

The best way to determine the reason that individuals don't get along or don't work well together is to talk to the parties involved. Speak with employees and any other individuals on site in order to find out the source of conflict.

Once the reason for the conflict is found, it is easier to determine how to respond. There may be a way to resolve the problem and get the workers communicating and working as a team again. On the other hand, there may be nothing that can be done that will lead to a harmonious situation.

Explain how a poor attitude towards the work place can affect safety.

Discuss the ways to find out why an employee has a poor attitude towards the work place and what can be done to help change it.

Discuss how to determine if an employee has an inability to work with others.

Explain possible resolutions.

Ask the participants if they have ever worked with someone with these problems and then ask them to share their experiences.

Discuss the impact absenteeism and turnover have on safety.

Explain how absenteeism can be prevented.

Explain that knowing the major causes of turnover can help eliminate the problem.

Show Transparency 11, Major Causes of Turnover.

Since this can lead to unsafe working conditions, the safety technician should report the situation to the employee's supervisor or Human Resources Department. In this case, the employee would need to be transferred to another crew or terminated to maintain safe working conditions.

4.3.0 Absenteeism and Turnover

Absenteeism and turnover are big problems on construction jobs. Workers may feel pressured to do jobs for which they are not trained or certified in order to fill in for someone who has left. Absenteeism and turnover may also force an understaffed crew to take on tasks that are too large. Two workers, for example, may find themselves working on a task that requires a crew of three to be done safely. To maintain a safe work environment, therefore, it is important to take steps to reduce absenteeism and turnover.

4.3.1 Absenteeism

Absenteeism has many causes, some of which are inevitable. For instance, people get sick, they have to take time off for family emergencies, and they have to attend family events such as funerals. However, some causes of absenteeism can be prevented.

The most effective way to control absenteeism is to make the company's policy clear to all employees. Companies that do this find that chronic absenteeism becomes less of a problem. New employees should have the policy explained to them. This explanation should include the number of absences allowed and the reasons for which sick or personal days can be taken. In addition, all workers should know how to inform their supervisors when they miss work and understand the consequences of exceeding the number of sick or personal days allowed. The safety technician should be sure to reinforce this policy and describe the effect absenteeism has on the safety of co-workers.

Once the policy on absenteeism is explained to employees, it must be implemented consistently and fairly. If the policy is administered equally, employees will likely follow it. However, if the policy is not administered equally and some employees are given exceptions, then it will not be effective. Consequently, the rate of absenteeism will increase.

Despite having a policy on absenteeism, there will always be employees who are chronically late or miss work. As a safety technician, if you notice cases in which an employee abuses the absenteeism policy and causes unsafe working conditions, discuss the situation directly with the employee's supervisor. Confirm that the employee understands the company's policy and that the employee complies with it. If the employee's behavior continues, disciplinary action may be in order.

4.3.2 Turnover

Turnover refers to the loss of an employee when an employee is fired or leaves the company to work elsewhere. Like absenteeism, there are some causes of turnover that cannot be prevented. For instance, it is unlikely that an employee who finds a job elsewhere earning twice as much money will stay at his or her current job. However, some employee turnover situations can be prevented. This can be done by ensuring safe working conditions for the crew, treating workers fairly and consistently, and helping promote good working conditions. As a safety technician, be alert for any signs of discontent, especially those due to unsafe or perceived unsafe working conditions. Then work with the employees and management to provide resolutions to minimize turnover. The key to doing so is communication. Problems need to be known if they are going to be successfully resolved.

Some of the major causes of turnover include:

- Non-competitive wages and benefits may lead workers to leave one company or industry for another that pays higher wages and/or offers better benefits.
- A lack of job security may cause workers to leave to find more permanent employment.
- Unsafe project sites will cause workers to leave for safer projects.
- Workers will leave to find a less stressful working environment if they perceive unfair and/or inconsistent treatment.
- Workers will move on if they feel that the working conditions are inadequate.

Essentially, the same actions described for absenteeism are effective for reducing turnover.

Instructor's Notes:

Past studies have shown that maintaining harmonious relationships on the job site will go a long way toward reducing both turnover and absenteeism. This will take effective leadership on the part of the supervisor and all other management-level employees.

Summary

The three types of communication are verbal, non-verbal, and written. You must understand how to utilize all of these forms of communication in order to properly interact with others. Communication with site personnel is essential. If you are able to communicate with them, you will be able to provide the recognition, motivation, and feedback they need to keep the workplace safe. You will also be able to motivate workers and impart fair and proper discipline when needed. Establishing open communication is key. Open communication allows you to speak honestly and respectfully to all workers at all times, making even short conversations about safety meaningful and easy to understand.

Review Questions

1. Facial expressions and eye contact are forms of non-verbal communication.
   a. True
   b. False

For Questions 2 through 4, match the correct form of communication with its corresponding description.

Form of Communication

2. ____ Verbal
3. ____ Non-Verbal
4. ____ Written

Description

a. Communication that is documented on paper or transmitted electronically via a computer using words or visuals
b. Spoken words exchanged between two or more people when communicating
c. Communications that you can actually see when communicating with others

For Questions 5 through 8, match the correct component of verbal communication with its corresponding description.

Component of verbal communication

5. ____ The sender
6. ____ The message
7. ____ The receiver
8. ____ Feedback

Description

a. Communication that occurs after the message has been sent and received
b. The person to whom the message is communicated
c. The person who creates the message to be communicated
d. What the sender is attempting to communicate to the audience

9. All of the following are components of the ARCS Model of Motivation *except* _____.
   a. attention
   b. rewards
   c. confidence
   d. satisfaction

Review the material covered in the module.

Have the participants complete the Review Questions and go over the answers prior to administering the Module Examination.

Administer the Module Examination. Record the results on Craft Training Report Form 200 and submit the results to the Training Program Sponsor.

Administer the Performance Test and fill out Performance Profile Sheets for each participant. If desired, participant proficiency noted during laboratory sessions may be used to complete the Performance Test. Record the results on Craft Training Report Form 200 and submit the results to the Training Program Sponsor.

EMPLOYEE MOTIVATION 5.9

PROFILE IN SUCCESS

Oscar Paredes, Qualified Contractors, Inc.
Corporate Environmental Safety & Health Manager

What is your proudest career-related accomplishment?
One of my biggest accomplishments was completing the 100-Hour Construction Site Safety Technician Course (CSST). Also, I used to be a scaffold builder by trade, and I helped to achieve the OSHA VPP Star on a very large project with my former employer, Black & Veatch.

How did you choose a career in the Safety Industry?
I always have liked safety, and one day a Brown & Root safety manager gave me the opportunity to work in safety. Later, I graduated in the very first 100-Hour CSST course, and I have been in safety since then.

What types of training have you been through?
I've taken the 100-Hour CSST course, OSHA 500, OHST, CHST, Scaffold master instructor, NCCER master safety instructor, and other miscellaneous training.

What kinds of work have you done in your career?
I was a laborer and a scaffold builder for 6 years. I've been in safety since 1990. On my very last job I was the project loss-control manager for a large project (about $800,000,000).

Tell us about your present job.
Currently I'm the corporate environmental safety & health manager for Qualified Contractors, Inc., a General Electrical affiliate. I'm responsible for the safety and health of our US-based and international projects.

What factors have contributed most to your success?
I seriously believe that if I had not taken the 100-Hour Safety Technician Course (CSST) through NCCER, I would not be where I am today. It helped me not only to learn but to be motivated to do more.

What advice would you give to those new to Safety industry?
First, I encourage anyone that has not taken the CSST course to make the time for it. I would advise them to continue to stay on top of all the new changes and to participate with their local American Society of Safety Engineers (ASSE) chapters.

Instructor's Notes:

GLOSSARY

Trade Terms Introduced in This Module

ARCS Model of Motivation: A strategy for motivating people developed by Dr. John Keller. The strategy states that to motivate someone you must gain their attention, make the issue relevant to them, help them to feel confident that they can be successful, and make them feel a sense of satisfaction once they have achieved their goal.

Communication: A process by which information is exchanged between individuals through a common system of symbols, signs, or behavior.

Diversity: Differences between individuals, particularly with regard to race, religion, ethnicity, and gender.

Ethnic groups: Large groups of people classed according to common racial, national, tribal, religious, linguistic, or cultural origin or background.

Feedback: The communication that occurs after a message has been sent and received. This communication enables the sender to determine whether his or her message has been accurately received.

Jargon: Technical terminology known only by people who work directly with the technology being discussed.

Message: The information that the sender is attempting to communicate to the receiver.

Non-verbal communication: Communication achieved through non-spoken means such as body language, facial expressions, hand gestures, and eye contact.

Paraphrase: A restatement of a text, passage, conversation, or work process that is explained without changing its meaning.

Receiver: The person to whom the sender is communicating a message.

Sender: The person who creates the message to be communicated.

Verbal communication: Transfer of information through spoken word. This process involves a sender, receiver, message, and feedback.

Visual communication: Communication through visual aids such as signs, postings, and hand signals.

Written communication: Transfer of information through the written word.

EMPLOYEE MOTIVATION 5.11

Additional Resources

This module is intended to present thorough resources for task training. The following reference works are suggested for further study. These are optional materials for continued education rather than for task training.

www.osha.gov

www.asse.org

The Psychology of Safety Handbook, 2001. E. Scott Geller,Ph.D. Boca Raton, FL: CRC/Lewis Publishers.

The Participation Factor—How to Increase Involvement in Occupational Safety, 2002. E. Scott Geller, Ph.D. Des Plaines, IL: The American Society of Safety Engineers (ASSE).

Figure Credit

plaquemaster.com 205F04

Instructor's Notes:

MODULE 75205-03 — ANSWERS TO REVIEW QUESTIONS

Answer		Section
1.	a	2.2.0
2.	b	2.1.0
3.	c	2.2.0
4.	a	2.3.0
5.	c	2.1.1
6.	d	2.1.2
7.	b	2.1.3
8.	a	2.1.4
9.	b	3.0.0
10.	b	4.3.0

CONTREN™ LEARNING SERIES — USER UPDATES

The NCCER makes every effort to keep these textbooks up-to-date and free of technical errors. We appreciate your help in this process. If you have an idea for improving this textbook, or if you find an error, a typographical mistake, or an inaccuracy in NCCER's Contren™ textbooks, please write us, using this form or a photocopy. Be sure to include the exact module number, page number, a detailed description, and the correction, if applicable. Your input will be brought to the attention of the Technical Review Committee. Thank you for your assistance.

Instructors – If you found that additional materials were necessary in order to teach this module effectively, please let us know so that we may include them in the Equipment/Materials list in the Instructor's Guide.

Write: Curriculum Revision and Development Department
National Center for Construction Education and Research
P.O. Box 141104, Gainesville, FL 32614-1104

Fax: 352-334-0932

E-mail: curriculum@nccer.org

Craft _______________ Module Name _______________

Copyright Date _______________ Module Number _______________ Page Number(s) _______________

Description _______________

____

____

____

(Optional) Correction _______________

____

____

(Optional) Your Name and Address _______________

____

____

Module 75206-03

Site-Specific ES&H Plans

NCCER STANDARDIZED CRAFT TRAINING PROGRAM

The National Center for Construction Education and Research (NCCER) provides a standardized national program of accredited craft training. Key features of the program include instructor certification, competency-based training, and performance testing. The program provides trainees, instructors, and companies with a standard form of recognition through a National Craft Training Registry. The program is described in full in the *Guidelines for Accreditation*, published by the NCCER. For more information on standardized craft training, contact the NCCER by writing us at P.O. Box 141104, Gainesville, FL 32614-1104; calling 352-334-0911; or e-mailing info@nccer.org. More information may be found at our Web site, www.nccer.org.

HOW TO USE THIS ANNOTATED INSTRUCTOR'S GUIDE

Each page presents two sections of information. The larger section displays each page exactly as it appears in the Participant Module. The narrow column ties suggested participant and instructor actions to each page and provides icons (detailed below) to call your attention to material, safety, audiovisual, or testing requirements. The bottom of each page includes space for your notes.

The **Audiovisual** icon indicates an appropriate time to show a transparency or other audiovisual aid.

The **Classroom** icon prompts you to define a term, stress a point, ask participants to explain a concept, or give examples.

The **Demonstration** icon directs you to show participants how to perform tasks.

The **Examination** icon tells you to administer the written module examination.

The **Homework** icon is placed where you may wish to assign reading for the next class, to assign a project, or to advise participants to prepare for an examination.

The **Laboratory** icon is used when participants are to practice performing tasks.

The **Materials** icon is a reminder for you to gather materials needed for classes, labs, and testing.

The **Performance Testing** icon tells you to administer a performance test or a portion thereof.

The **Safety** icon is used to emphasize safety issues. It is often keyed to *Caution* and *Warning* statements in the Participant Module.

The **Teaching Tip** icon indicates additional guidance is available, such as how to conduct an exercise, get the most educational value from a field trip, or encourage class participation. Teaching Tips may expand on a feature (*Think About It*, *Did You Know?*) or provide Quick Quizzes or similar exercises. You will be referred to the Teaching Tips section at the back of the module if there is additional material.

The **Combination** icon indicates that the laboratory listed corresponds with a performance task. If desired, you can note the proficiency of the participants during the laboratory and use it to satisfy performance testing requirements.

PREPARATION

Before teaching this module, you should review the Objectives, Performance Tasks, Materials and Equipment List, and the Module Outline. Be sure to allow ample time to prepare your own training or lesson plan and gather all required materials and equipment.

MODULE OVERVIEW

In this module, participants will learn the components of a site-specific ES&H plan, and learn to identify and assess potential hazards.

PREREQUISITES

Please refer to the Course Map in the Participant Module. Prior to training with this module, it is recommended that the participant shall have successfully completed the following:

Field Safety; Safety Technology, Modules 75201-03 through 75205-03.

OBJECTIVES

Upon completion of this module, the participant will be able to:

1. Evaluate hazard risks based on probability and consequences of outcome.
2. Identify specific job-site hazards and requirements using existing pre-bid planning checklists.
3. Modify your existing company Environmental Safety and Health (ES&H) program or Safety and Loss Prevention Manual to meet specific job conditions.
4. Describe coordination needed to implement your company's ES&H plan with other entities.
5. Describe and explain administrative controls needed to make the plan effective.

PERFORMANCE TASKS

Under the supervision of the instructor, the participant should be able to:

1. Use the risk assessment matrix to determine the risk of a given situation.
2. Modify an existing site safety plan to meet specific job conditions.

MATERIALS AND EQUIPMENT LIST

Overhead projector and screen
Transparencies
Whiteboard/chalkboard
Markers/chalk
Blank acetate sheets
Transparency pens
Pencils and scratch paper
Appropriate personal protective equipment
Module Examinations*
Performance Profile Sheets*

* Located in the Test Booklet.

SAFETY CONSIDERATIONS

Ensure that participants are equipped with appropriate personal protective equipment.

ADDITIONAL RESOURCE

This module is intended to present thorough resources for task training. The following reference work is suggested for both instructors and motivated participants interested in further study. This is optional material for continued education rather than for task training.

Web site: www.osha.gov

TEACHING TIME FOR THIS MODULE

An outline for use in developing your lesson plan is presented below. Approximately 2½ hours are suggested to cover *Site-Specific ES&H Plans*. This includes 10 minutes at the beginning of the session for administrative tasks and one 10-minute break. You may need to adjust the time required for hands-on activities and testing based on your class size and resources. Because laboratories often correspond to Performance Tasks, the proficiency of the participants may be noted during these exercises for Performance Testing purposes.

Topic	**Planned Time**
A. Introduction	__________
B. Pre-Bid Safety Planning	__________
1. Hazard Identification	__________
2. Risk Assessment	__________
3. Hazard Control	__________
4. Laboratory	__________
Provide the participants with a scenario and ask them to use the risk assessment matrix to determine the risk of the situation. This laboratory corresponds with Performance Task 1.	
C. The Site Safety Plan	__________
1. Site Location and Layout	__________
2. Scope of Work	__________
3. Personnel Health and Safety	__________
4. Safety Coordination	__________
5. Administration	__________
D. Preparing a Site-Specific ES&H Plan	__________
1. Laboratory	__________
Provide the participants with an existing site safety plan and ask them to modify it to meet specific job conditions. This laboratory corresponds with Performance Task 2.	
E. Process Safety Management	__________
F. Review	__________
G. Module Examination	__________
1. Participants must score 70% or higher to receive recognition from the NCCER.	
2. Record the testing results on Craft Training Report Form 200 and submit the results to the Training Program Sponsor.	
H. Performance Testing	__________
1. Participants must perform each task to the satisfaction of the instructor to receive recognition from the NCCER. If applicable, proficiency noted during laboratory exercises can be used to satisfy the Performance Testing requirements.	
2. Record the testing results on Craft Training Report Form 200 and submit the results to the Training Program Sponsor.	

Module 75206-03

Site-Specific ES&H Plans

Instructor's Notes:

COURSE MAP

This course map shows all of the modules in Safety Technology. The suggested training order begins at the bottom and proceeds up. The local Training Program Sponsor may adjust the training order.

Assign reading of Module 75206-03.

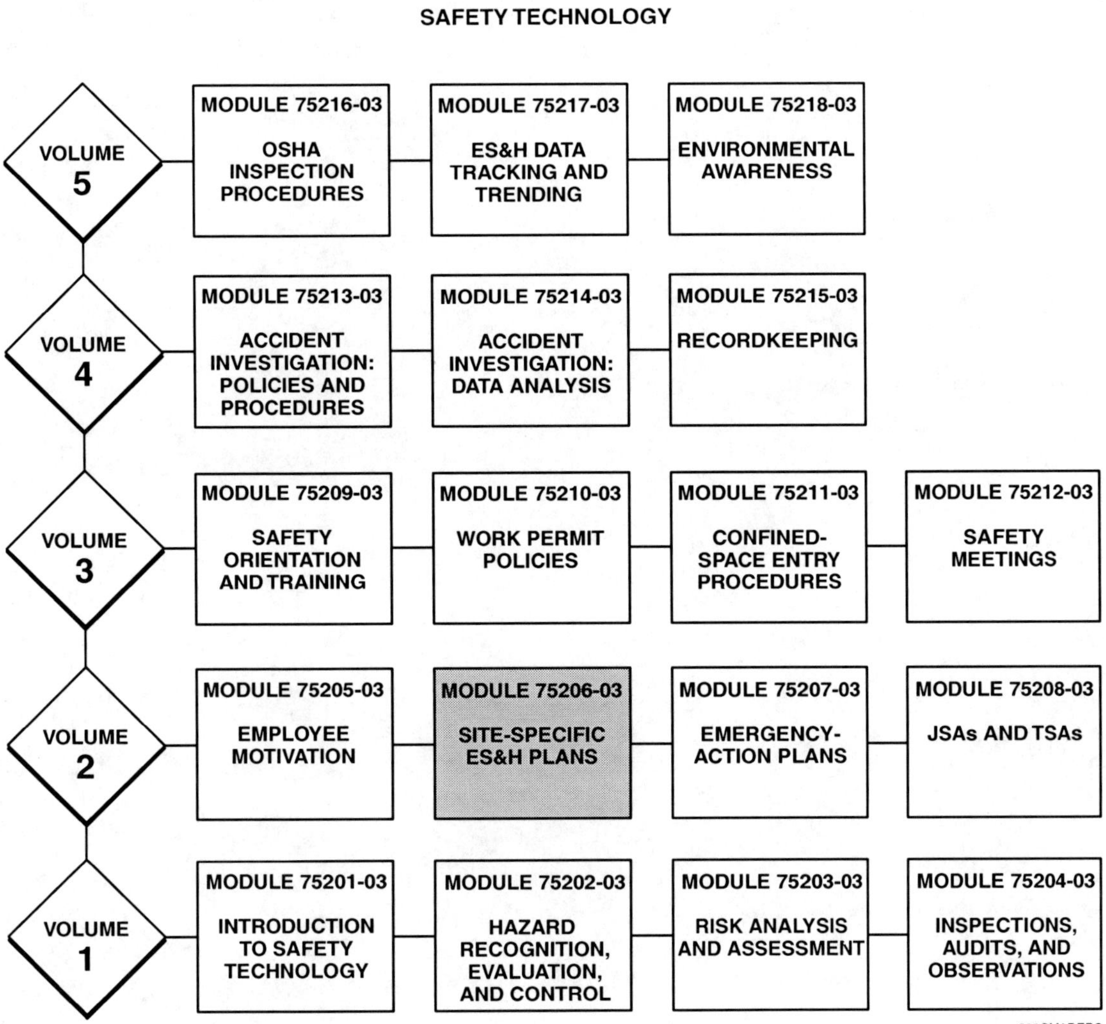

206CMAP.EPS

SITE-SPECIFIC ES&H PLANS 6.iii

Instructor's Notes:

MODULE 75206-03 CONTENTS

Figures

SITE-SPECIFIC ES&H PLANS 6.v

Instructor's Notes:

MODULE 75206-03

Site-Specific ES&H Plans

Objectives

When you have completed this module, you will be able to do the following:

1. Evaluate hazard risks based on probability and consequences of outcome.
2. Identify specific job-site hazards and requirements using existing pre-bid planning checklists.
3. Modify your existing company Environmental Safety and Health (ES&H) program or Safety and Loss Prevention Manual to meet specific job conditions.
4. Describe coordination needed to implement your company's ES&H plan with other entities.
5. Describe and explain administrative controls needed to make the plan effective.

Prerequisites

Before you begin this module, it is recommended that you successfully complete the following: Field Safety; Safety Technology, Modules 75201-03 through 75205-03.

Required Materials

1. Pencil and paper
2. Appropriate personal protective equipment
3. Copy of *29 CFR 1926 OSHA Construction Industry Regulations*

1.0.0 ◆ INTRODUCTION

Every project is different and each project will have different safety issues. The hazards present on each job site must be identified and evaluated. The risk associated with these hazards must be assessed and the consequences minimized.

General safety policies are not enough to cover unusual jobs or sites. Policies must be customized for each job. This is known as a site-specific environmental safety and health (ES&H) plan or a site safety plan. A site safety plan will vary in size and complexity depending on the scope and location of the project.

At a minimum, the site safety plan must include information on safety responsibilities and emergency procedures. It should have provisions for hazard communication, incident prevention, inspections, grounded electrical systems, recordkeeping, personal protective equipment, and housekeeping. Operation-specific safety procedures (for example, hoisting and rigging, or demolition) can be appended to a site safety plan.

The site safety plan must be coordinated with other groups. The client or general contractor will also have a safety plan. Roles, responsibilities, and procedures must be clear before the job starts. Some OSHA programs have strict rules for safety coordination. In addition, the site plan needs to be coordinated with any local emergency-response services that you may rely on.

In order for any plan to be effective, there must be adequate oversight and review. The site plan needs to include inspections, auditing, and reporting. The plan must be reviewed during the project to make sure safety goals are met.

2.0.0 ◆ PRE-BID SAFETY PLANNING

Planning safety into a project is just as important as planning for materials and equipment. Failure to do so may result in costly work slowdowns or stoppage due to serious accidents, injuries, or citation

Ensure that you have everything you need to teach the course. Check the Materials and Equipment list in the front of this module.

Explain that terms shown in bold are defined in the Glossary at the back of this module.

Review the goals of the course and explain what will be expected of the participant.

Show Transparency 1, Objectives.

Show Transparency 2, Performance Tasks.

Explain the need for and importance of a site-specific ES&H plan.

Explain the importance of planning safety into a project.

from regulatory agencies. It must be done before a bid is submitted, so all expenses and other provisions are included. Most **contractor pre-qualification** programs include safety requirements. A sample contractor pre-qualification form is included in *Appendix A*.

Pre-bid safety planning should be done for every job. Planning should analyze the work being performed and necessary safety precautions. You will need to develop accident prevention measures for each phase and component of the work. The plan should list hazard control measures needed. Some phases of the work may be subcontracted. The plan should include provisions for sub-contractors' safety.

Planning will identify all potential areas for loss including:

- People (employees, contractors, and the general public)
- Buildings or structures
- Equipment
- The environment

Pre-job safety planning should follow this general format:

Step 1 Identify the hazards.

Step 2 Evaluate the consequences.

Step 3 Rank the hazards by risk.

Step 4 Control or minimize prioritized hazards.

2.1.0 Hazard Identification

In order to plan safety into a job, certain information must be obtained and analyzed. The necessary information is extensive. Some of the components include:

- Site location and layout
- Scope of work
- Client and contract requirements
- Regulatory permits and requirements

It is essential that you do not miss any important aspects of operations. Several groups have developed **pre-bid checklists** to aid their membership. A sample pre-bid planning safety checklist is attached as *Appendix B*. You may need to modify the standard checklists to fit your operation. Specific checklists may be needed for different project phases. For example, you may want a checklist for initial site construction and a separate one for finish work.

2.2.0 Risk Assessment

Once the hazards are identified, you must assess the associated risks. The two aspects of **risk assessment** are consequences and probability. You need to determine what would happen in the case of an accident. These consequences can be categorized as follows:

I. *Catastrophic* – death or loss of facility

II. *Critical* – severe illness, injury, or property damage

III. *Marginal* – minor illness, injury, or property damage

IV. *Negligible* – low hazard but a violation of standards

Next you need to determine the potential for the event to happen. This is known as **probability of occurrence**. These can also be ranked as follows:

A – Likely to happen now or very soon

B – Probably will happen in time

C – Possibly will happen in time

D – Unlikely to happen

These two factors can be used to find an overall risk using the matrix shown in *Table 1*.

Dehydration is one example of a negligible hazard which may be very likely to occur during the summer. This would rate as a moderate risk. A fuel tank explosion could have catastrophic results. But it may be very unlikely to happen.

Table 1 Risk Assessment Matrix

	Consequence			
Probability	**I**	**II**	**III**	**IV**
A	Catastrophic	Critical	Serious	Moderate
B	Critical	Critical	Serious	Moderate
C	Serious	Moderate	Moderate	Minor
D	Moderate	Minor	Minor	Negligible

Instructor's Notes:

Explain the pre-bid safety planning process.

Explain the purpose of a contractor pre-qualification form. Refer the participants to Appendix A in the Participant Module.

Show Transparency 3, Potential Areas of Loss.

Show Transparency 4, Pre-Job Safety Planning.

Explain the components of a pre-bid plan that must be considered when planning for safety.

Describe the purpose and use of a pre-bid checklist. Refer the trainees to Appendix B in the Participant Module.

Explain that the two aspects of risk assessment are consequences and probability.

Show Transparency 5 (Table 1).

Provide the participants with a scenario and ask them to use the risk assessment matrix to determine the risk of the situation. Note the proficiency of each participant. This laboratory corresponds with Performance Task 1.

This would also rate as a moderate risk. The consequences and probability of different accidents must be determined given the situation.

> **NOTE**
> For more information on risk assessment, please refer to Volume 1, Module 75203-03, *Risk Analysis and Assessment.*

2.3.0 Hazard Control

You should establish proper safeguards to minimize the hazards. The hazards should be addressed in the order determined by the risk assessment discussed previously. Catastrophic hazards should be eliminated entirely. Major hazards considered critical or serious must be minimized. Finally, moderate, minor, and negligible hazards must be addressed.

These hazards can be minimized or eliminated in several ways. In order of preference, these controls include:

- Engineering controls or substitution to eliminate or reduce hazard
- Guarding or isolating the hazard
- Administrative controls (training, procedures, signs, alarms, instruction, or personnel rotation)
- Using personal protective equipment

Hazard controls should be balanced with your risk assessment. The best solution will eliminate the hazard. This is preferable for catastrophic or critical risks. The last choice, personal protective equipment, merely protects workers from the hazard. Minimal controls should only be used with minor or negligible risks.

3.0.0 ◆ THE SITE SAFETY PLAN

The site-specific safety plan includes safety procedures to deal with conditions at a particular job site. It is usually prepared by the safety manager and implemented with assistance from the safety technician. It can be a single document or a collection of documents and references.

It must be sufficiently detailed to cover the hazards posed on the job site. Each job site will include unique features that create safety challenges. For example, weather can be a factor in an outdoor environment, but it does not affect indoor jobs. Also, a job can be done in several phases, each with different safety hazards.

A site-specific safety plan helps create a safe work environment. It demonstrates a commitment to safety to employees, clients, and government inspectors. Safety briefings are more effective when they address specific conditions at a particular job site. A clear plan will list safety issues on a site. The safety technician can address these at the right time.

CASE HISTORY

Laborer Killed in Fall Through Roof

The employer was demolishing the roof of a warehouse. Work was done at night because the coal tar would release hazardous gases if disturbed during the day. The site had adequate lighting. None of the workers on the job were using fall protection.

After the roofing material was removed, 4' × 8' sheets of plywood were exposed. Workers were replacing damaged plywood. The helper's job was to follow the workers, pick up the damaged sheets, and dispose of them in a chute.

One worker removed a sheet of damaged plywood. However, he had run out of nails to attach the new plywood. He walked away to get more nails. He left the opening where the damaged plywood had been removed unguarded. The crew was not informed that the opening was temporarily unguarded. The opening was covered by silver-colored insulation inside the roof.

The helper came along, picked up the sheet of damaged plywood, and headed for the chute. He stepped into the opening, ripped through the insulation, and fell. He fell approximately 27 feet to the floor below and died.

The Bottom Line: Fall protection was not used because an event like this may have been considered unlikely to happen. The result was catastrophic. Use a risk assessment matrix to prioritize risk and minimize hazards.

Explain the criteria by which hazards should be addressed.

Identify the types of hazard controls that are commonly used.

Show Transparency 6, Hazard Controls.

Explain that hazard controls should be balanced with risk assessment to determine the best method for controlling hazards.

Explain the purpose and uses of a site safety plan.

Refer the participants to the Case History. Ask them how planning for safety could have prevented the accident.

Explain the federal, state, and local requirements for site safety plans. Emphasize the importance of OSHA compliance.

Explain the importance of site location and layout when planning for safety.

Describe the traffic issues that must be considered during site location and layout planning.

Explain how adjacent hazards can affect site safety planning. Emphasize that utilities are a significant safety issue.

Sometimes a site safety plan is mandatory. A general contractor may require site-specific safety plans from subcontractors. Your firm must have a site safety plan when working in plants covered by OSHA's **Process Safety Management (PSM)** rules. OSHA inspectors will thoroughly review your site safety plan. If it is not an adequately written plan, the inspectors will perform a more extensive inspection. This exposes the company to greater potential for citations and fines.

3.1.0 Site Location and Layout

The site safety plan must consider the site location and layout. The site location is an important factor in safety planning. You need to know if the area is rural, residential, commercial, or industrial. Each type of area poses different safety concerns. Pedestrian and vehicular traffic are greater concerns in urban or commercial areas. In rural areas, emergency services may be limited or distant.

Site location also affects site security. You need to know what safeguards are needed to protect the public, the client's employees, or other contractors from hazards on the job site. In the same way, you need to be aware of other contractors or client activities that will affect the safety of your personnel.

Material and tool storage is another feature of site security. Storage areas should be safe in addition to secure. Loading areas are high-hazard areas. You need to consider site access and material storage areas to minimize hazards from traffic and loading operations. Storage of chemicals and flammables need special consideration. This includes on-site vehicle fueling areas.

3.1.1 Traffic Patterns

In addition to loading areas, the general traffic patterns are an important part of site safety. You need to consider where employees will park and how vehicles and equipment will move through the site. Vehicle accidents are a major safety hazard. Traffic control is the most important feature of site layout. If possible, equipment and vehicles should move through the site in a one-way traffic pattern. This will reduce vehicle collisions.

Site access is also a factor. Job sites in congested city areas pose hazards not found on a rural site. You need to consider how vehicles and equipment will access the site. Public roads or sidewalks may need to be blocked. Permits are usually required. If you are on private property, the client's site may have restrictions which must be incorporated into your plan.

The placement of a field office and employee break areas affect safety. These areas tend to have a lot of pedestrian traffic. Walking routes may need to be designated, in addition to roadways.

3.1.2 Adjacent Hazards

The site location may include nearby hazards that will affect your site safety plan. These hazards can be natural or man-made. Significant slopes, water bodies, and other topological features must be considered. This is especially important during excavation and hazardous materials handling. A hill leading down to a river can turn a minor chemical spill into a major environmental problem.

Utilities can be a major safety issue. You must ensure that the locations of all utilities are marked (*Figure 1*). If possible, the site safety plan should include contact information for major utilities. These include:

- Water
- Electricity
- Gas
- Sewer
- Telephone and fiber-optic lines

GAS PIPELINE — FIBER-OPTIC CABLE — TELEPHONE CABLE

206F01.EPS

Figure 1 ◆ Standard underground markers.

Industrial activities on the client's site or adjacent areas can affect site safety. You need to be aware of area hazards. You need to consider interactions in both directions. How will your actions affect existing operations? How will existing operations affect your work? Construction activities can create noise, air, water, or ground pollution problems. These hazards must be minimized.

Instructor's Notes:

3.2.0 Scope of Work

The next significant consideration is the scope of work. You need to consider the safety precautions for different aspects of the project. The work may be done in phases with very different safety considerations. Safety for high-hazard work must be carefully planned. High-hazard work can include:

- Blasting
- Excavation
- Demolition
- Work at extreme heights
- Work in confined spaces
- Work over water
- Handling asbestos, chemicals, hazardous waste, or other regulated substances

Generally, high-hazard work is a smaller portion of a larger project. The timing of the various phases of the work plan must be included in the site safety plan. These operations may require special tools, equipment, and permits. Training, special precautions, and inspections should be scheduled in parallel with the work plan.

Some special tools or equipment may be needed at various times during the job. Fall-protection equipment, **intrinsically safe** electrical equipment, and air handling or testing equipment are a few examples. These should be specified in the site safety plan.

3.3.0 Personnel Health and Safety

The next phases of site safety planning combine both site and task considerations. A safety technician must list all safety equipment needed. This is similar to a foreman preparing a materials and tools breakout. You need to consider standard and specialized safety equipment. Standard safety equipment includes:

- Hard hats
- Safety glasses
- Gloves
- Hearing protection
- Boots

Specialized equipment may include:

- Fire-retardant clothing
- Chemical-resistant gloves or aprons
- Safety harnesses and fall-protection gear
- Debris and personnel nets
- Respiratory equipment

The site safety plan should include who will supply personal protective equipment. Some clients may provide equipment. Extra supplies may be needed. You will need to designate storage locations. The plan may include a system to issue gear and restock disposable supplies.

In addition to personal protective equipment, you must consider basic sanitary needs. If toilet facilities and drinking water are not available on-site they should be included in the site safety plan. Dehydration is a common hazard on job sites. Shower, changing, or decontamination areas may be needed. These may be needed for the duration or only for a specific phase of the project.

3.4.0 Safety Coordination

The site location and scope of work will provide a solid foundation for a site safety plan. In order to be effective, you must coordinate your plan with other parties. These include:

- The client
- Other contractors
- Emergency services

If the project is on an existing industrial site, the client will have a health and safety program. If several contractors are working on the site, or subcontractors are hired, you must coordinate your plans. Contractors must be aware of existing programs. The client's employees must be informed of new procedures that will affect their work.

Some of the site safety features must be coordinated with other parties. Coordination is essential for the following areas:

- Site security
- Vehicle and pedestrian traffic
- Communications
- Chemical handing
- Waste disposal
- Emergency response
- Work permit systems
- Lockout/tagout procedures
- Critical lifts
- Site-specific orientation and training

OSHA's policy on multi-employer worksites would apply. Your firm can be held responsible if other employees are harmed by your actions. Safety is a weakest link function. The site is only as safe as the least-trained worker.

Explain the significance of including safety when considering the scope of work.

Describe the types of high-hazard work that should be considered in the scope of work.

Transparency 7, High-Hazard Work.

Explain the personnel health and safety factors that must be considered in site safety planning. Discuss how this phase of planning involves both site and task consideration.

Show Transparency 8, Standard Safety Equipment.

Show Transparency 9, Specialized Safety Equipment.

Explain that the site safety plan should describe who is responsible for the care and distribution of safety equipment.

Explain that a site safety plan should be coordinated with other parties, including the client, other contractors, and emergency services.

Describe the areas for which safety coordination is essential. Refer the participants to the bulleted lists in the Participant Module.

Discuss the importance of emergency response coordination.

Explain that emergency response must be tailored to site conditions. Provide the participants with examples.

Describe the general procedures everyone involved in emergency response must know.

Show Transparency 10, General Emergency-Response Procedures.

Describe how site safety plans should be administered.

Explain that employee training is an important part of a site safety plan.

Explain the resources that are available to help create a site safety plan.

3.4.1 Emergency Procedures

Coordination is especially important in emergency response. A serious accident or fire can affect the entire site. All safety personnel must work together in an emergency. Issues that can affect the whole site include:

- First aid
- Emergency medical services
- Fire protection and prevention
- Hazardous materials incidents

Each site is different. Procedures will need to be tailored to site conditions. Site-specific emergency procedures may be needed for:

- Rescue from extreme heights, confined spaces, or trench collapse
- Tornadoes or extreme weather

The site safety plan must be coordinated with other parties. Often the most difficult aspect of an emergency is coordination and communication. All parties must understand each other's programs. You should consider drills for long-term projects. It is also a good idea to invite fire or emergency responders most likely to respond to an emergency on your site to tour the project on a regular basis. This has been shown to dramatically improve response time. All parties must understand general procedures for:

- Emergency lighting and loss of power
- Emergency reporting and notifications
- Emergency communication systems
- On-site medical facilities
- Off-site medical facilities
- Emergency assembly and evacuation routes
- Hazardous materials emergency response
- Fire procedures

It is imperative that all personnel understand any alarm systems. The site safety plan needs to include any emergency alarms, buzzers, or bells that will be used.

Emergency contact numbers should be on a posted list. A more extensive list that includes key personnel for all firms can be included in the site safety plan.

3.5.0 Administration

The site safety plan cannot be effective sitting on a desk. It must be put into action through training, notifications, and inspections. The plan must be communicated to the workers in several ways to be effective.

A safety bulletin board is a common tool for communicating safety information. The board must include mandatory OSHA posters and notices. It can also include emergency information and the location of the safety plan and MSDSs. Posting new information on a regular basis will increase effectiveness. Posting notices of high hazard activities or temporary safety precautions are a good way to encourage workers to check it regularly.

Safety meetings, audits, and inspections should be scheduled consistent with the work plan. The results of audits or inspections can also be posted to encourage good performance.

The site safety plan must also include any mandatory reporting requirements. These include safety incidents, accidents, and reports to regulatory agencies. The client or general contractor may also need to be notified. The contract should be reviewed for specific reporting requirements.

3.5.1 Personnel

The site safety plan must include a section on training. Training is required at several stages of employment. Training must include initial, periodic, and task-specific training. Records must be kept of all training. Some clients will need copies of training records.

At a minimum, the plan should include safety orientation for new employees. Current employees who are new to the job site should also receive a basic orientation.

Some clients or projects will require a medical monitoring program. Workers must be tested when they are hired, annually, and when they stop working for the firm. Any work-related medical evaluation should also be included For example, if the worker received first aid for an injury on the job, there must be a record. There are strict requirements for maintaining medical records. These requirements vary by state. You need to know your state's requirements.

4.0.0 ◆ PREPARING A SITE-SPECIFIC ES&H PLAN

The site safety plan should be prepared using the information gathered in the pre-bid checklist or other tools. The company's existing safety program or Safety and Loss Prevention Manual can be modified to address the specific safety issues on the site.

There are many professional safety firms who can assist your firm in preparing a site safety plan. State and federal OSHA offices also offer online guidance. The California DOSH offers a

Instructor's Notes:

model program for Incident and Illness Prevention. The guidelines can be downloaded from www.dir.ca.gov/dosh/puborder.asp. You must be familiar with the state and federal laws that apply to your project.

Another way to prepare a site-specific plan is to list site-specific information and use existing procedures as indicated in the work plan. Copies of relevant standard operating procedures must be appended. Typical sections would include:

- Site information (including site security and layout)
- Scope of work
- Key contacts
- Existing on-site and adjacent hazards
- Standard procedures used (HazCom, fall protection, welding, etc.)
- Additional procedures
- Emergency-response procedures

Appendix C is a generic site safety plan. This plan was developed by the U.S. Coast Guard for responding to hazardous chemical releases. It is used for sites covered under the OSHA *Hazardous Waste Site Operations Standard.* There are special site safety requirements for these sites. Typically, hazards on these sites are not known in advance. A construction site typically would have more information on existing hazards.

5.0.0 ◆ PROCESS SAFETY MANAGEMENT

Firms covered under OSHA Process Safety Management (PSM) regulations also have special requirements for site safety (*Figure 2*). *The Process Safety Management of Highly Hazardous Chemicals* standard can be found at *29 CFR 1910.119*. Generally, these regulations cover industrial facilities that handle large quantities of chemicals listed as highly hazardous chemicals. Chemical processing plants and refineries are two examples.

Construction contractors may work on a PSM site. Certain procedures must be followed for contractors who work in or around hazardous processes. General requirements include:

- The host company must pre-screen all contractors.
- Contractors must verify that their employees are trained for the tasks they will perform.

206F02.EPS

Figure 2 ◆ A plant covered under OSHA PSM rules.

- The contractors must inform the host company of hazards they may bring onto the site.
- The host employer must inform the contractors of any hazards near where they will be working.
- The host employer must control contractors' access to hazard areas.
- Contractors must be informed of site emergency-response procedures.
- If an incident involving highly hazardous chemicals occurs, a contractor representative must participate in the incident investigation.

Contractors who work on PSM sites may require additional training. The host employer and the contractor must work together to identify training needs. Records must be kept of training and plant safety orientation.

Describe the typical sections of a site safety plan.

Show Transparency 11, Typical Site Safety Plan Sections.

Refer the participants to Appendix C in the Participant Module for an example of a site safety plan. Explain each section.

Provide the participants with an existing site-specific plan and ask them to modify it to meet specific job conditions. Note the proficiency of each participant. This laboratory corresponds with Performance Task 2.

Explain OSHA's Process Safety Management regulations.

SITE-SPECIFIC ES&H PLANS 6.7

Review the material covered in the module.

Have the participants complete the Review Questions and go over the answers prior to administering the Module Examination.

Summary

It is important to coordinate your site safety plan with those of other contractors who will be working on the site. Certain aspects of site safety management, such as traffic control and emergency procedures, can be successful if done cooperatively.

The site safety plan should be compiled from pre-bid job checklists and existing safety programs. Special features and precautions should be noted in the plan. Site information must be recorded, especially emergency contact information.

The hazards must be analyzed for both probability and consequences. A risk assessment matrix is a useful tool for this analysis. You can balance high risk and low probability with low risk and high probability to prioritize hazards. Your site safety plan should minimize or eliminate these hazards through standard hazard control measures.

PSM sites involve highly hazardous chemicals and hazards. These are extreme examples of the need for site-specific environmental safety and health planning. These models can be used to create a comprehensive site safety plan for a typical construction site.

Administer the Module Examination. Record the results on Craft Training Report Form 200 and submit the results to the Training Program Sponsor.

Administer the Performance Test and fill out Performance Profile Sheets for each participant. If desired, participant proficiency noted during laboratory sessions may be used to complete the Performance Test. Record the results on Craft Training Report Form 200 and submit the results to the Training Program Sponsor.

Review Questions

1. Two aspects of risk assessment are _____.
   a. scope and size
   b. consequences and probability
   c. hazards and controls
   d. plans and procedures

2. All adjacent hazards are man-made.
   a. True
   b. False

3. All of the following are methods of minimizing hazards *except* _____.
   a. engineering controls
   b. personal protective equipment
   c. pre-bid checklists
   d. isolating hazards

4. The site safety plan can be a collection of documents.
   a. True
   b. False

5. The site safety plan must be coordinated with _____.
   a. the client, other contractors, and emergency services
   b. employees, safety managers, and foremen
   c. site security, doctors, and medical services
   d. the client, foremen, and doctors

6. Emergency drills should be incorporated into site safety plans for long-term projects.
   a. True
   b. False

7. A _____ is a common tool for communicating safety information.
   a. bullhorn
   b. radio
   c. site safety plan
   d. bulletin board

8. The site safety plan should be prepared using the information gathered in _____.
   a. audits
   b. the pre-bid checklist
   c. inspection checklists
   d. site visits

9. OSHA Process Safety Management rules apply to industrial firms that _____.
   a. employ contractors
   b. use highly hazardous chemicals
   c. have site safety plans
   d. have OSHA violations

10. On an OSHA PSM site, the host and the contractor must work together to identify training needs.
    a. True
    b. False

Instructor's Notes:

____

____

____

GLOSSARY

Trade Terms Introduced in This Module

Contractor pre-qualification: A process of screening contractors to allow them to bid on jobs for a specific company.

Intrinsically safe: An electric tool or device that is UL-rated to be explosion proof under normal uses. For example, a fuel pump must be intrinsically safe to prevent explosions sparked by static electricity.

Pre-bid checklist: A list of questions or issues that aids contractors in gathering the information need to prepare a bid package.

Probability of occurrence: The likelihood that a specific event will occur, usually expressed as the ratio of the number of actual occurrences to the number of possible occurrences.

Process Safety Management (PSM): *The Process Safety Management of Highly Hazardous Chemicals, 29 CFR 1910.119.* An OSHA standard that covers certain chemical plants and refineries.

Risk assessment: The process of qualifying or ranking hazards given their probability and consequences.

SITE-SPECIFIC ES&H PLANS 6.9

APPENDIX A

Contractor Prequalification Form

Instructor's Notes:

Contractor Prequalification Form (PQF)

GENERAL INFORMATION			
1. Company Name:		Telephone Number:	
		FAX Number:	
Street Address:		Mailing Address:	
Province:	Postal Code:	Province:	Postal Code:
2. Officers:			Years With Company
President:			
Vice-President:			
Treasurer:			
3. How many years has your organization been in business under your present firm name?			
4. Parent Company Name:			
City:		Province:	Postal Code:
Subsidiaries:			
5. Under Current Management Since (Date):			
6. Contact for Requesting Bids:			
Title:	Telephone:		Fax:
7. Contractor's PQF Completed By:			
Title:	Telephone:		Fax:

ORGANIZATION			
8. Form of Business	Sole Owner ☐	Partnership ☐	Corporation ☐
9. Percentage Owned:			

10. Describe Services Performed:

- ☐ Construction
- ☐ Construction Design
- ☐ Maintenance
- ☐ Manpower and Resources
- ☐ Original Equipment Manufacturer and Installer
- ☐ Original Equipment Manufacturer and Maintenance
- ☐ Project Maintenance
- ☐ Service Work (e.g., Janitorial, Clerical, etc.)
- ☐ Other

11. Describe the Additional Services Performed:

12. List other types of work within the services you normally perform that you subcontract to others, including brokers:

206A01.TIF

SITE-SPECIFIC ES&H PLANS 6.11

Contractor Prequalification Form (PQF)

13. Do you evaluate your subcontractor's health and safety program?			
14. Attach a list of the major equipment (e.g., cranes, forklifts, JLGs) your company has available for work at this facility, and the method of establishing the competencies to operate this equipment.			
15. Describe any affiliations with labour organizations.			
16. Annual Dollar Volume for the Past Three Years:	19 ____ $	19 ____ $	19 ____ $
17. Largest Job During the Last Three Years: $			
18. Your Firm's Desired Project Size:	Maximum:	Minimum:	
19. Financial Rating:	D and B:	Net Worth:	

COMPANY WORK HISTORY

20. Major jobs in Progress:

Customer/Location	Type of Work	Size $M	Customer Contact	Telephone	Fax

21. Major jobs Completed in the Past Three Years:

Customer/Location	Type of Work	Size $M	Customer Contact	Telephone	Fax

22. Are there any judgments, claims or suits pending or outstanding against your company? Yes ☐ No ☐
If yes, please attach details.

23. Are you now, or have you ever been, involved in any bankruptcy or reorganization proceedings? Yes ☐ No ☐
If yes to either of the above questions, please attach details.

HEALTH AND SAFETY PERFORMANCE

24. From the last three years (including subcontractors):	19 _____	19 _____	19 _____
• Number of fatalities?			
• Number of lost time incidents?			
• Number of medical aid injuries?			
• Do you have a modified work program?			

25. Please list your past three years' recordable injury incidence rate (including subcontractors):
____________, 19_____ ____________, 19_____ ____________, 19_____

$$\frac{\text{Number of Lost Time Accidents} \times 200{,}000}{\text{Total Employee Hours (Yearly)}}$$

206A02.TIF

Instructor's Notes:

Contractor Prequalification Form (PQF)

26. Man hours (including those of the subcontractors) worked in the last three years:	Year		19____	19____	19____
	Hours	Field			
		Total			

27. Please list your overall Worker's Compensation Rating for the past three years. Please attach your company's WCB summary.
____________, 19____ ____________, 19____ ____________, 19____

28. Have you received an Alberta Labour OH&S stop work order, or equivalent, from another province in the last three years? Yes ☐ No ☐
Describe ____
____

HEALTH AND SAFETY MANAGEMENT

29. Highest ranking safety professional in your organization:

Title:	Telephone:	Fax:

30. Do you have, or provide:
- A full-time health and safety representative? Yes ☐ No ☐
- A full-time site health and safety representative? Yes ☐ No ☐

HEALTH AND SAFETY PROGRAM AND PROCEDURES

31. Do you have a written Health and Safety Management Program? Yes ☐ No ☐

Does the program address the following key elements:
- Accountabilities and responsibilities for managers, supervisors, and employees? Yes ☐ No ☐
- Employee participation? Yes ☐ No ☐
- Hazard recognition and control? Yes ☐ No ☐
- Management commitment and expectations? Yes ☐ No ☐
- Periodic health and safety performance appraisals for all employees? Yes ☐ No ☐
- Resources for meeting health and safety requirements? Yes ☐ No ☐

32. Does the program include work practices and procedures such as:
- Accident/Incident Reporting? Yes ☐ No ☐
- Compressed Gas Cylinders? Yes ☐ No ☐
- Confined Space Entry? Yes ☐ No ☐
- Electrical Equipment Grounding Assurance? Yes ☐ No ☐
- Emergency Preparedness, including an Evacuation Plan? Yes ☐ No ☐
- Equipment Lockout and Tag Out (LOTO)? Yes ☐ No ☐
- Fall Protection? Yes ☐ No ☐
- Housekeeping? Yes ☐ No ☐
- Injury and Illness Recording? Yes ☐ No ☐
- Personal Protective Equipment (PPE)? Yes ☐ No ☐
- Portable Electrical/Power Tools? Yes ☐ No ☐
- Powered Industrial Vehicles (Cranes, Forklifts, JLGs, etc.)? Yes ☐ No ☐
- Unsafe Condition Reporting? Yes ☐ No ☐
- Vehicle Safety? Yes ☐ No ☐
- Waste Disposal? Yes ☐ No ☐

206A03.TIF

SITE-SPECIFIC ES&H PLANS 6.13

Contractor Prequalification Form (PQF)

33.	Do you have written programs for the following: • Hearing Conservation? • Respiratory Protection? Where applicable, have employees been: ☐ Fit Tested? ☐ Medically Approved? ☐ Trained? • WHMIS?	 Yes ☐ No ☐ Yes ☐ No ☐ Yes ☐ No ☐
34.	Do you have a Substance Abuse Program? If yes, does it include the following: • Preemployment? • Random Testing? • Testing for Cause?	Yes ☐ No ☐ Yes ☐ No ☐ Yes ☐ No ☐ Yes ☐ No ☐
35.	Medical: Do you conduct medical examinations for: • Preemployment? • Pulmonary? • Replacement Job Capability? • Respiratory? Describe how you will provide First Aid and other medical services for your employees while on site. Specify who will provide this service: ____________ Do you have personnel trained to perform First Aid and CPR?	 Yes ☐ No ☐ Yes ☐ No ☐ Yes ☐ No ☐ Yes ☐ No ☐ Yes ☐ No ☐
36.	Do you hold site safety meetings for: • Employees? Yes ☐ No ☐ Frequency ________ • Field Supervisors? Yes ☐ No ☐ Frequency ________ • New Hires? Yes ☐ No ☐ Frequency ________ • Subcontractors? Yes ☐ No ☐ Frequency ________ Are the health and safety meetings documented?	
37.	Personal Protection Equipment (PPE): Is applicable PPE provided for employees? Do you have a program to ensure PPE is inspected and maintained?	 Yes ☐ No ☐ Yes ☐ No ☐
38.	Do you have a corrective action process for addressing individual health and safety performance deficiencies?	Yes ☐ No ☐
39.	Equipment and Materials: • Do you conduct inspections on operating equipment (e.g., cranes, forklifts, JLGs, etc.) in compliance with the regulatory requirements? • Do you have a system for establishing the applicable health, safety, and environmental specifications for the acquisition of materials and equipment? • Do you maintain operating equipment in compliance with the regulatory requirements? • Do you maintain the applicable inspection and maintenance certification records for operating equipment?	 Yes ☐ No ☐ Yes ☐ No ☐ Yes ☐ No ☐ Yes ☐ No ☐

206A04.TIF

Instructor's Notes:

Contractor Prequalification Form (PQF)

40. Subcontractors:
Do you evaluate the ability of subcontractors to comply with applicable health and safety requirements as part of the selection process? Yes ☐ No ☐
Do you include your subcontractors in:
    - Audits? Yes ☐ No ☐
    - Health and Safety Meetings? Yes ☐ No ☐
    - Health and Safety Orientation? Yes ☐ No ☐
    - Inspections? Yes ☐ No ☐

    Do your subcontractors have a written Health and Safety Management Program? Yes ☐ No ☐
    Do you use health and safety performance criteria in the selection of subcontractors? Yes ☐ No ☐

41. Inspections and Audits:
    - Are corrections of the deficiencies documented? Yes ☐ No ☐
    - Do you conduct health and safety inspections? Yes ☐ No ☐
    - Do you conduct Health and Safety Management Program audits? Yes ☐ No ☐

42. Craft Training:
    - Are employees' job skills certified, where required, by regulatory or industry consensus standards? Yes ☐ No ☐
    - Have employees been trained in the appropriate job skills? Yes ☐ No ☐
    - List crafts which have been certified:

    ____

    ____

    ____

HEALTH AND SAFETY TRAINING

43. Safety Orientation Program:

	New Hires	Supervisors
• Do you have a Health and Safety Management Orientation Program for new hires and newly hired or promoted supervisors?	Yes ☐ No ☐	Yes ☐ No ☐
• Does this program provide instruction on the following:		
• Emergency Procedures?	Yes ☐ No ☐	Yes ☐ No ☐
• Fire Protection and Prevention?	Yes ☐ No ☐	Yes ☐ No ☐
• First Aid Procedures?	Yes ☐ No ☐	Yes ☐ No ☐
• Incident Investigation?	Yes ☐ No ☐	Yes ☐ No ☐
• New Worker Orientation?	Yes ☐ No ☐	Yes ☐ No ☐
• Safe Work Practices?	Yes ☐ No ☐	Yes ☐ No ☐
• Safety Intervention?	Yes ☐ No ☐	Yes ☐ No ☐
• Safety Supervisors?	Yes ☐ No ☐	Yes ☐ No ☐
• Toolbox Meetings?	Yes ☐ No ☐	Yes ☐ No ☐
• WHMIS Training?	Yes ☐ No ☐	Yes ☐ No ☐
• How long is the orientation program?	_______ Hours	_______ Hours

44. Health and Safety Training Program:
    - Do you have a specific Health and Safety Training Program
    - for supervisors? Yes ☐ No ☐
    - Do you know the regulatory health and safety training requirements for your employees? Yes ☐ No ☐
    - Have your employees received the required health and safety training and retraining? Yes ☐ No ☐

206A05.TIF

____

____

____

____

SITE-SPECIFIC ES&H PLANS 6.15

Contractor Prequalification Form (PQF)

45. Training Records:
    - Do you have health and safety, and crafts training records for your employees? Yes ☐ No ☐
    - Do the training records include the following:
        - Date of Training? Yes ☐ No ☐
        - Employee Identification? Yes ☐ No ☐
        - Method Used to Verify Understanding? Yes ☐ No ☐
        - Name of Trainer? Yes ☐ No ☐
    - How do you verify understanding of the training? (Check all that apply)
      ☐ Job Monitoring ☐ Written Test
      ☐ Oral Test ☐ Other (List) ______________________
      ☐ Performance Test ______________________

INFORMATION SUBMITTAL

Please provide copies of checked (√) items with the completed contractor's PQF:

_____ Accident/Incident Investigation Procedure.
_____ Example of Employee Health and Safety Training Records.
_____ Health and Safety Audit Procedure or Form.
_____ Health and Safety Incentive Program.
_____ Health and Safety Inspection Form.
_____ Health and Safety Orientation Outline.
_____ Health and Safety Program.
_____ Health and Safety Training for Supervisors (Outline).
_____ Health and Safety Training Program (Outline).
_____ Health and Safety Training Schedule (Sample).
_____ Housekeeping Policy.
_____ Respiratory Protection Program.
_____ Substance Abuse Program.
_____ Unsafe Conditions Reporting Procedure.
_____ WHMIS Program.

Note: Owner checks items to be provided with the contractor's PQF.

Individual to contact for clarification or additional information:

Name: ______________________ Telephone: ______________________

FAX: ______________________

OWNER'S USE ONLY

DO NOT FILL OUT - OWNER'S USE ONLY

Contractor is: ☐ Acceptable for Approved Contractors' List.

☐ Conditionally acceptable for Approved Contractors List.

Conditions: ____

____

____

Reviewed By: ______________________ Date: ______________________

206A06.TIF

Instructor's Notes:

____

____

____

APPENDIX B

Pre-Job Planning Safety Checklist

SITE-SPECIFIC ES&H PLANS 6.17

`Pre-Job Planning Safety Checklist

1. **General Information:**

   Job Number: ______________ Client: ____________________________

   Location: ____

   Client Contact: _________________Phone: __________Fax: ___________

   Start Date: ________________ Completion Date: _________________

2. **Scope of Work:** Briefly describe the project and your scope of work. (Type and size of project, materials of construction and construction methods)

   Peak Employment: Company _______________Subcontractors ___________

   Will the job involve any unusual or high risk work? If so, specify the nature of the potential problem(s) and the proposed solutions.

   ____
   ____
   ____
   ____

   Will the job involve any of the following? If so, describe.

   _____ Blasting
   _____ Pile driving
   _____ Tunnelling or major excavations
   _____ De-watering
   _____ Demolition
   _____ Work at extreme heights
   _____ Work over water
   _____ Underpinning
   _____ Handling or exposure to asbestos or hazardous wastes, lead or OSHA regulated substances
   _____ Work in, on or adjacent to equipment handling flammable, toxic or otherwise hazardous chemicals

   ____
   ____

206A07.TIF

Instructor's Notes:

____
____
____

3. **Hazardous Processes, Materials or Equipment:**

Identify processes, materials or equipment that may expose employees to hazardous conditions either in routine work or emergency situations. Obtain Material Safety Data Sheets or Hazardous Waste Sheets on all hazardous materials to which employees may reasonably be exposed. Find out how the client is complying with the OSHA Hazard Communications Standard and how this information will be conveyed to contractor employees. Find out if the proposed work falls under the OSHA Process Safety Management Standard 1910.119.

4. **Client Safety Rules & Procedures:**

Obtain copies of all client safety rules, procedures or manuals that contractors are expected to follow. Do not overlook emergency plans or procedures. List any unusual or special safety requirements.

5. **Medical Surveillance / Industrial Hygiene Monitoring Requirements:**

List any medical surveillance or industrial hygiene monitoring requirements for contractor personnel. Determine who will be expected to provide these services. List any special requirements such as no beard policies, drug screening tests, showering facilities and decontamination facilities.

6. **Regulatory Permit Requirements and Inspection Schedules:**

List all required permits and inspection schedules and who is responsible for obtaining these permits and coordinating the necessary inspections?

206A08.TIF

SITE-SPECIFIC ES&H PLANS 6.19

7. **Site Location & Layout:**

A. Briefly describe site conditions and layout. (Rural, residential, commercial, industrial, congested, etc.)

B. Describe adjacent exposures. How might they be affected by the project? Is there any need to take photos of the adjacent structures or have a physical inspection made by an independent third party?

C. Utilities: List all utilities available at or near the site. If possible, obtain the names and phone numbers of area utility representatives.

1. Water:

2. Electricity:

3. Gas:

4. Sewer:

5. Other:

6. Have all existing utilities been located and marked appropriately?

7. Will an assured grounding program or ground fault circuit interrupters be used to protect temporary electrical circuits? Who is responsible for coordinating this activity?

D. Field office, storage and laydown areas:

1. Any designated areas?

206A09.TIF

Instructor's Notes:

2. Accessible by trucks, forklifts and cranes? Will vehicles be able to drive through as opposed backing out?

3. Any restricted or congested roads / areas? If so, describe.

4. Parking facilities for employees, contractors and visitors?

5. Any special requirements for temporary buildings with respect to location, size, materials of construction, and etc.?

8. Personnel Health and Safety:

A. Personal protective equipment: What equipment is required and who will provide it?

_____ Hard Hats
_____ Safety Shoes
_____ Chemical Resistant Boots
_____ Fire Retardant Clothing
_____ Plano Safety Glasses, with or without side shields
_____ Prescription Safety Glasses, with or without side shields
_____ Gloves, specify _______________
_____ Safety Harnesses and Lanyards, specify type: ________________
_____ Chemical Resistant Clothing, specify: _______________
_____ Life Lines
_____ Personnel Nets
_____ Debris Nets
_____ Face Shields / Goggles
_____ Respiratory Protection (Specify what type? _____________________)
_____ Other; specify: ____
____

B. What temporary toilet facilities will be needed? Will any showering facilities be required?

C. Is a drinking water supply available on site?

D. What method(s) will be used to protect wall and floor openings?

206A10.TIF

E. Any special scaffolding requirements? If so, specify:

F. Any materials or personnel hoists or elevators? If so, who will operate, inspect and maintain them?

G. Overhead protection required at building entrances?

9. Emergency Reporting & Response:

A. List all emergency communication systems and onsite phone numbers.

B. How will onsite medical and First Aid be handled? Who will provide what type and level of care? What First Aid supplies and rescue equipment will be required / provided?

C. Location and telephone number of the nearest off-site medical facilities and ambulance service companies.

D. Name, address and phone number of company physician(s):

E. Obtain information on the client's emergency assembly areas and evacuation routes.

206A11.TIF

Instructor's Notes:

F. What onsite emergency notification and response procedures will be used? Obtain information on the client's emergency alarm system and signals. If possible, obtain lists of emergency alarm codes. Try to get a tape recording of the alarm signals.

G. What emergency escape equipment (if any) is needed? Who will provide it and where will it be located?

H. What offsite emergency reporting and response procedures will be used?

I. Name, address and phone number of fire department.

J. Will contractor personnel be expected to serve on a Hazardous Materials Emergency Response Team? If so, specify the level and extent of contractor involvement.

10. Fire Protection and Prevention:

A. Is there an adequate number of active fire hydrants on site? What provisions will be made to keep them unobstructed and accessible?

B. List the number, type and size of portable fire extinguishers to be provided and by whom:

C. Will standpipes and / or sprinkler systems be required to follow the building up floor by floor? ___Yes ___No ___N / A. Briefly describe type of systems, maintenance and inspection requirements and responsibilities during the construction phase.

206A12.TIF

SITE-SPECIFIC ES&H PLANS 6.23

D. Is there a site Fire Brigade? Will contractors be expected to serve on this brigade?

E. Will temporary heating be used: ___Yes ___No ___N / A. If so, what type.

F. Requirements for the storage and handling of flammable liquids. Any secondary containment required? Describe:

G. How, where and at what frequency will trash be removed and disposed?

H. How will hazardous waste be handled?

11. Administration:

A. Who will be responsible for collecting and disseminating Material Safety Data Sheets?

B. Who will be responsible for hiring site personnel?

C. What screening and placement procedures will be used?

_____ Written applications
_____ Reference checks
_____ DMV Records Check
_____ Medical exams
_____ Drug Screening
_____ Other, specify:

D. Who will be responsible for New Employee / Contractor Orientation and what materials and / or handouts will be used?

E. What types of safety inspections are required, at what frequency and who will conduct them?

206A13.TIF

Instructor's Notes:

___

___

___

F. What types of safety meetings will be held, at what frequency and with whom?

12. Special Tools or Equipment Requirements:

Will any special tools or equipment be required? If so, specify types and brand of:

A. Explosion proof lighting

B. Non-sparking tools

C. Air operated tools or equipment

D. Ground fault circuit interrupters

E. Extension cords

F. Air movers or other gas freeing equipment

G. Gas testing equipment

H. Other (specify)

13. Work Permit Requirements:

Obtain as much information as possible concerning the client's work permitting procedures. Are permits required for the following types of work?

A. Hot / Hazardous Work

B. Cold / Safe Work

C. Vessel Entry (Confined Space)

D. Gas Testing

E. Equipment Isolation Including Lockout / Tag Out

F. Excavations

206A14.TIF

SITE-SPECIFIC ES&H PLANS 6.25

G. Vehicle Entry

H. Critical Lifts

I. Scaffolding

J. Other (specify)

14. Training Requirements:*

List any special safety training requirements such as:

A. Respirator

B. Hazard Communication

C. Handling of certain hazardous materials (i.e., asbestos, lead). Specify the material(s).

D. Hazardous Waste Site

E. Response to spills or releases of hazardous materials. Specify what level of response will be required by contractor personnel.

F. Certification for operators of Fork Trucks, Cranes, Manlifts, etc.

G. Other (specify)

* **Important Note:** Find out who will perform and document this training, the client or the contractor. What training aids such as video tapes, films, slide / tape presentations are available from the client for contractor use.

15. Reporting Requirements:

List any special reporting requirements the client may have, such as:

A. Accident, injuries, illnesses or near misses

206A15.TIF

6.26 SAFETY TECHNOLOGY — PARTICIPANT MODULE 75206-03

Instructor's Notes:

B. Safety meetings

C. Safety inspections

D. Regulatory agency visits

E. Other (specify)

16. Site Security:

A. Describe the security measures required for the job during normal working hours and after hours.

B. If controlled access is required, how will it be handled and by whom?

C. Will any burglar alarm system, guard dog or watchman services be used? ___Yes___No___N / A. If so,describe.

D. Is / will site lighting be adequate?

17. Public Liability:

A. Will any sidewalks or streets have to be blocked? ___Yes ___No ___N / A. If so, describe:

B. Any overhead protection and / or lighting required: ___Yes ___No ___N / A. If so, describe:

C. Briefly describe any barricading, lighting, signs, traffic control devices or flagmen that may be required:

206A16.TIF

18. Environmental Hazards & Controls:

A. EPA, State or local Right to Know Requirements: Describe:

B. If asbestos or lead is suspected or anticipated, what special precautions, monitoring and training will be required? Who will coordinate and pay for these activities?

C. Will any wastes generated or leaving the site be classified as hazardous and require disposal as a hazardous waste. ___Yes ___No ___N / A. If so, will it be necessary to obtain any types of generator numbers and or permits? Who will obtain the necessary permits?

D. Will there be any open burning or on site landfills? ___Yes ___No___N / A. If so, describe:

E. Will any site activities create noise, air, water or ground pollution problems? ___Yes ___No ___ N / A. If so, describe potential problems and proposed solutions.

F. Is secondary containment required for flammable / hazardous material storage?

19. Cranes / Hoists and Lifting Devices:

A. Will there be any cranes, hoists or lifting devices on site? ___Yes ___No. If so describe:

B. Who will operate, maintain and inspect these devices? How will they be certified?

206A17.TIF

Instructor's Notes:

C. Any heavy, unusual or critical lifts to be made? ___Yes ___No. If so, describe:

D. Any requirements for inspections, certifications or load tests? ___Yes ___No. If so, describe:

E. Any requirements for critical lift plans? ___Yes ___No. If so, describe:

F. Any planned use or prohibition on use of crane suspended personnel baskets? ___Yes ___No. Explain:

20. Signs, Posters and Bulletin Boards:

A. List all signs, posters, notices required for the project:

_____ OSHA Employee Rights poster
_____ Worker's Compensation notice
_____ Company / jobsite rules
_____ Emergency phone numbers
_____ Location of MSDS's
_____ Employee access to medical & exposure records
_____ Others; specify:

B. Where will these notices and signs be posted?

21. Contract Specifications:

A. Bonding requirements: (Specify)

206A18.TIF

SITE-SPECIFIC ES&H PLANS 6.29

B. Insurance requirements:

1. Types of coverages & policy limits:

2. Who will provide what coverages?

3. Any special requirements needing insurance carrier approval?

4. Have certificates of insurance been requested and received from?

   a. Client / owner
   b. Prime or general contractor
   c. Sub contractors
   d. Others

5. Names, addresses and phone numbers of key insurance company representatives:

   a. Agent / broker:

   b. Claims representative:

   c. Loss control or safety representative:

6. Any hold harmless or indemnity agreements? If so, obtain copies.

7. Any unusual safety and loss prevention requirements, such as:

   a. Site specific safety and health plan required?
   b. Site safety representative required?
   c. Drug, alcohol and substance abuse policy required?
   d. Any pre-employment and or follow up medical exams required?
   e. Other, specify:

206A19.TIF

Instructor's Notes:

22. **Additional Information or Comments:**

Source of Information:

Name: ______________________ Title ______________ Date ________

Name: ______________________ Title ______________ Date ________

Report Prepared by:

Name: ______________________ Title ______________ Date ________

Report Reviewed by:

Name: ______________________ Title ______________ Date ________

206A20.TIF

SITE-SPECIFIC ES&H PLANS **6.31**

APPENDIX C

Generic Site Safety Plan

Instructor's Notes:

This appendix provides a generic plan based on a plan developed by the U.S. Coast Guard for responding to hazardous chemical releases.[1] This generic plan can be adapted for designing a Site Safety Plan for hazardous waste site cleanup operations. It is not all inclusive and should only be used as a guide, not a standard.

A. SITE DESCRIPTION

Date__________ Location__________

Hazards__________

Area affected__________

Surrounding population__________

Topography__________

Weather conditions__________

Additional information__________

B. ENTRY OBJECTIVES - The objective of the initial entry to the contaminated area is to (describes actions, tasks to be accomplished; i.e., identify contaminated soil; monitor conditions, etc.)

C. ONSITE ORGANIZATION AND COORDINATION - The following personnel are designated to carry out the stated job functions on site. (Note: One person may carry out more than one job function.)

PROJECT TEAM LEADER__________

SCIENTIFIC ADVISOR__________

SITE SAFETY OFFICER__________

PUBLIC INFORMATION OFFICER__________

SECURITY OFFICER__________

RECORDKEEPER__________

FINANCIAL OFFICER__________

FIELD TEAM LEADER__________

FIELD TEAM MEMBERS__________

[1] U.S. Coast Guard. Policy Guidance for Response to Hazardous Chemical Releases. USCG Pollution Response COMDTINST-M16465.30.

206A21.TIF

FEDERAL AGENCY REPS (i.e., EPA, NIOSH) ____________________

STATE AGENCY REPS ____________________

LOCAL AGENCY REPS ____________________

CONTRACTOR(S) ____________________

All personnel arriving or departing the site should log in and out with the Recordkeeper. All activities on site must be cleared through the Project Team Leader.

D. ONSITE CONTROL

(Name of individual or agency ______ has been designated to coordinate access control and security on site. A safe perimeter has been established at (distance or description of controlled area) ____________________

No unauthorized person should be within this area.

The onsite Command Post and staging area have been established at ____________________

The prevailing wind conditions are ______________. This location is upwind from the Exclusion Zone.

Control boundaries have been established, and the Exclusion Zone (the contaminated area), hotline, Contamination Reduction Zone, and Support Zone (clean area) have been identified and designated as follows: (describe boundaries and/or attach map of controlled area) ____________________

These boundaries are identified by: (marking of zones, i.e., red boundary tape - hotline; traffic cones - Support Zone; etc.) ____________________

206A22.TIF

Instructor's Notes:

____

____

____

E. HAZARD EVALUATION

The following substance(s) are known or suspected to be on site. The primary hazards of each are identified.

Substances Involved	Concentrations (If Known)	Primary Hazards
(chemical name)		(e.g., toxic on inhalation)

The following additional hazards are expected on site: (i.e., slippery ground, uneven terrain, etc.)

Hazardous substance information form(s) for the involved substance(s) have been completed and are attached.

F. PERSONAL PROTECTIVE EQUIPMENT

Based on evaluation of potential hazards, the following levels of personal protection have been designated for the applicable work areas or tasks:

Location	Job Function	Level of Protection
Exclusion Zone	______	A B C D Other
	______	A B C D Other
	______	A B C D Other
	______	A B C D Other
Contamination Reduction Zone	______	A B C D Other
	______	A B C D Other
	______	A B C D Other
	______	A B C D Other

Specific protective equipment for each level of protection is as follows:

Level A Fully-encapsulating suit
SCBA
(disposable coveralls)

Level C Splash gear (type)
Full-face canister resp.

Level B Splash gear (type)
SCBA

Level D

Other

206A23.TIF

The following protective clothing materials are required for the involved substances:

Substance	Material
(chemical name)	(material name, e.g., Viton)
______	______
______	______
______	______
______	______

If air-purifying respirators are authorized, __(filtering medium)__ is the appropriate canister for use with the involved substances and concentrations. A competent individual has determined that all criteria for using this type of respiratory protection have been met.

NO CHANGES TO THE SPECIFIED LEVELS OF PROTECTION SHALL BE MADE WITHOUT THE APPROVAL OF THE SITE SAFETY OFFICER AND THE PROJECT TEAM LEADER.

G. ONSITE WORK PLANS

Work party(s) consisting of ____ persons will perform the following tasks:

Project Team Leader	(name)	(function)
		______
		______
Work Party #1	______	______
		______
		______
		______
Work Party #2	______	______
		______
		______
		______
Rescue Team (required for entries to IDLH environments)	______	______
		______
		______
		______
		______
Decontamination Team	______	______
		______
		______
		______
		______

The work party(s) were briefed on the contents of this plan at ______________.

206A24.TIF

Instructor's Notes:

______

______

______

H. COMMUNICATION PROCEDURES

Channel ____ has been designated as the radio frequency for personnel in the Exclusion Zone. All other onsite communications will use channel ____.

Personnel in the Exclusion Zone should remain in constant radio communication or within sight of the Project Team Leader. Any failure of radio communication requires an evaluation of whether personnel should leave the Exclusion Zone.

__(Horn blast, siren, etc.)__ is the emergency signal to indicate that all personnel should leave the Exclusion Zone. In addition, a loud hailer is available if required.

The following standard hand signals will be used in case of failure of radio communications:

Hand gripping throat --------------- Out of air, can't breathe
Grip partner's wrist or ------------ Leave area immediately
both hands around waist
Hands on top of head --------------- Need assistance
Thumbs up -------------------------- OK, I am all right, I understand
Thumbs down ------------------------ No, negative

Telephone communication to the Command Post should be established as soon as practicable. The phone number is ______________________.

I. DECONTAMINATION PROCEDURES

Personnel and equipment leaving the Exclusion Zone shall be thoroughly decontaminated. The standard level _____ decontamination protocol shall be used with the following decontamination stations: (1) ______________
(2) ______________ (3) ______________ (4) ______________ (5) ______________
(6) ______________ (7) ______________ (8) ______________ (9) ______________
(10) ______________ Other ____

Emergency decontamination will include the following stations: ______________
____
____
____

The following decontamination equipment is required: ______________________
____
____
____

__(Normally detergent and water)__ will be used as the decontamination solution.

J. SITE SAFETY AND HEALTH PLAN

1. ______(name)______ is the designated Site Safety Officer and is directly responsible to the Project Team Leader for safety recommendations on site.

206A25.TIF

____
____
____
____

2. Emergency Medical Care

(names of qualified personnel) ______ are the qualified EMTs on site. (medical facility names) __________, at (address) __________, phone ____________________ is located _____ minutes from this location. (name of person) ____________ was contacted at (time) ____ and briefed on the situation, the potential hazards, and the substances involved. A map of alternative routes to this facility is available at (normally Command Post) ______.

Local ambulance service is available from ________________________ at phone _________________. Their response time is _____ minutes. Whenever possible, arrangements should be made for onsite standby.

First-aid equipment is available on site at the following locations:

First-aid kit ____________________
Emergency eye wash ____________________
Emergency shower ____________________
(other) ____________ ____________________

Emergency medical information for substances present:

Substance	Exposure Symptoms	First-Aid Instructions

List of emergency phone numbers:

Agency/Facility	Phone #	Contact
Police		
Fire		
Hospital		
Airport		
Public Health Advisor		

3. Environmental Monitoring

The following environmental monitoring instruments shall be used on site (cross out if not applicable) at the specified intervals.

Combustible Gas Indicator - continuous/hourly/daily/other ________
O_2 Monitor - continuous/hourly/daily/other ________
Colorimetric Tubes - continuous/hourly/daily/other ________
(type) ____________ ____________________
____________ ____________________
____________ ____________________
HNU/OVA - continuous/hourly/daily/other ________
Other ____________ - continuous/hourly/daily/other ________
____________ - continuous/hourly/daily/other ________

206A26.TIF

Instructor's Notes:

____
____
____

4. Emergency Procedures (should be modified as required for incident)

The following standard emergency procedures will be used by onsite personnel. The Site Safety Officer shall be notified of any onsite emergencies and be responsible for ensuring that the appropriate procedures are followed.

Personnel Injury in the Exclusion Zone: Upon notification of an injury in the Exclusion Zone, the designated emergency signal ____________________ shall be sounded. All site personnel shall assemble at the decontamination line. The rescue team will enter the Exclusion Zone (if required) to remove the injured person to the hotline. The Site Safety Officer and Project Team Leader should evaluate the nature of the injury, and the affected person should be decontaminated to the extent possible prior to movement to the Support Zone. The onsite EMT shall initiate the appropriate first aid, and contact should be made for an ambulance and with the designated medical facility (if required). No persons shall reenter the Exclusion Zone until the cause of the injury or symptoms is determined.

Personnel Injury in the Support Zone: Upon notification of an injury in the Support Zone, the Project Team Leader and Site Safety Officer will assess the nature of the injury. If the cause of the injury or loss of the injured person does not affect the performance of site personnel, operations may continue, with the onsite EMT initiating the appropriate first aid and necessary follow-up as stated above. If the injury increases the risk to others, the designated emergency signal ____________________ shall be sounded and all site personnel shall move to the decontamination line for further instructions. Activities on site will stop until the added risk is removed or minimized.

Fire/Explosion: Upon notification of a fire or explosion on site, the designated emergency signal ____________________ shall be sounded and all site personnel assembled at the decontamination line. The fire department shall be alerted and all personnel moved to a safe distance from the involved area.

Personal Protective Equipment Failure: If any site worker experiences a failure or alteration of protective equipment that affects the protection factor, that person and his/her buddy shall immediately leave the Exclusion Zone. Reentry shall not be permitted until the equipment has been repaired or replaced.

Other Equipment Failure: If any other equipment on site fails to operate properly, the Project Team Leader and Site Safety Officer shall be notified and then determine the effect of this failure on continuing operations on site. If the failure affects the safety of personnel or prevents completion of the Work Plan tasks, all personnel shall leave the Exclusion Zone until the situation is evaluated and appropriate actions taken.

206A27.TIF

SITE-SPECIFIC ES&H PLANS 6.39

The following emergency escape routes are designated for use in those situations where egress from the Exclusion Zone cannot occur through the decontamination line: (describe alternate routes to leave area in emergencies)

In all situations, when an onsite emergency results in evacuation of the Exclusion Zone, personnel shall not reenter until:

1. The conditions resulting in the emergency have been corrected.
2. The hazards have been reassessed.
3. The Site Safety Plan has been reviewed.
4. Site personnel have been briefed on any changes in the Site Safety Plan.

5. Personal Monitoring

The following personal monitoring will be in effect on site:

Personal exposure sampling: (describe any personal sampling programs being carried out on site personnel. This would include use of sampling pumps, air monitors, etc.)

Medical monitoring: The expected air temperature will be (°F) . If it is determined that heat stress monitoring is required (mandatory if over 70°F) the following procedures shall be followed:

(describe procedures in effect, i.e., monitoring body temperature, body weight, pulse rate)

All site personnel have read the above plan and are familiar with its provisions.

Site Safety Oficer (name) (signature)
Project Team Leader
Other Site Personnel

206A28.TIF

Instructor's Notes:

Additional Resources

This module is intended to present thorough resources for task training. The following reference works are suggested for further study. These are optional materials for continued education rather than for task training.

www.osha.gov

www.asse.org

Figure Credits

Brigid R. McKenna	206F01
Photodisc	206F02
Construction Owners Associate of Alberta	Appendix A

SITE-SPECIFIC ES&H PLANS 6.41

MODULE 75206-03 — ANSWERS TO REVIEW QUESTIONS

Answer		Section
1.	b	2.2.0
2.	b	3.1.2
3.	c	2.3.0
4.	a	3.0.0
5.	a	3.4.0
6.	a	3.4.1
7.	d	3.5.0
8.	b	4.0.0
9.	b	5.0.0
10.	a	5.0.0

CONTREN™ LEARNING SERIES — USER UPDATES

The NCCER makes every effort to keep these textbooks up-to-date and free of technical errors. We appreciate your help in this process. If you have an idea for improving this textbook, or if you find an error, a typographical mistake, or an inaccuracy in NCCER's Contren™ textbooks, please write us, using this form or a photocopy. Be sure to include the exact module number, page number, a detailed description, and the correction, if applicable. Your input will be brought to the attention of the Technical Review Committee. Thank you for your assistance.

Instructors – If you found that additional materials were necessary in order to teach this module effectively, please let us know so that we may include them in the Equipment/Materials list in the Instructor's Guide.

Write: Curriculum Revision and Development Department
National Center for Construction Education and Research
P.O. Box 141104, Gainesville, FL 32614-1104

Fax: 352-334-0932

E-mail: curriculum@nccer.org

Craft ______________ Module Name ______________

Copyright Date ______________ Module Number ______________ Page Number(s) ______________

Description ______________

____

____

____

(Optional) Correction ______________

____

____

(Optional) Your Name and Address ______________

____

____

Module 75207-03

Emergency-Action Plans

NCCER STANDARDIZED CRAFT TRAINING PROGRAM

The National Center for Construction Education and Research (NCCER) provides a standardized national program of accredited craft training. Key features of the program include instructor certification, competency-based training, and performance testing. The program provides trainees, instructors, and companies with a standard form of recognition through a National Craft Training Registry. The program is described in full in the *Guidelines for Accreditation*, published by the NCCER. For more information on standardized craft training, contact the NCCER by writing us at P.O. Box 141104, Gainesville, FL 32614-1104; calling 352-334-0911; or e-mailing info@nccer.org. More information may be found at our Web site, www.nccer.org.

HOW TO USE THIS ANNOTATED INSTRUCTOR'S GUIDE

Each page presents two sections of information. The larger section displays each page exactly as it appears in the Participant Module. The narrow column ties suggested participant and instructor actions to each page and provides icons (detailed below) to call your attention to material, safety, audiovisual, or testing requirements. The bottom of each page includes space for your notes.

The **Audiovisual** icon indicates an appropriate time to show a transparency or other audiovisual aid.

The **Classroom** icon prompts you to define a term, stress a point, ask participants to explain a concept, or give examples.

The **Demonstration** icon directs you to show participants how to perform tasks.

The **Examination** icon tells you to administer the written module examination.

The **Homework** icon is placed where you may wish to assign reading for the next class, to assign a project, or to advise participants to prepare for an examination.

The **Laboratory** icon is used when participants are to practice performing tasks.

The **Materials** icon is a reminder for you to gather materials needed for classes, labs, and testing.

The **Performance Testing** icon tells you to administer a performance test or a portion thereof.

The **Safety** icon is used to emphasize safety issues. It is often keyed to *Caution* and *Warning* statements in the Participant Module.

The **Teaching Tip** icon indicates additional guidance is available, such as how to conduct an exercise, get the most educational value from a field trip, or encourage class participation. Teaching Tips may expand on a feature (*Think About It*, *Did You Know?*) or provide Quick Quizzes or similar exercises. You will be referred to the Teaching Tips section at the back of the module if there is additional material.

The **Combination** icon indicates that the laboratory listed corresponds with a performance task. If desired, you can note the proficiency of the participants during the laboratory and use it to satisfy performance testing requirements.

PREPARATION

Before teaching this module, you should review the Objectives, Performance Tasks, Materials and Equipment List, and the Module Outline. Be sure to allow ample time to prepare your own training or lesson plan and gather all required materials and equipment.

MODULE OVERVIEW

In this module, participants will learn the elements of an emergency-action plan as well as how to deal with the media after an emergency situation.

PREREQUISITES

Please refer to the Course Map in the Participant Module. Prior to training with this module, it is recommended that the participant shall have successfully completed the following:

Field Safety; Safety Technology, Modules 75201-03 through 75206-03

OBJECTIVES

Upon completion of this module, the participant will be able to:

1. Describe the types of emergencies that can occur on construction sites and at industrial facilities.
2. Describe the fundamental elements of an emergency-action plan.
3. Identify the correct procedures for dealing with the media.

PERFORMANCE TASKS

There are no performance tasks for this module.

MATERIALS AND EQUIPMENT LIST

Overhead projector and screen
Transparencies
Whiteboard/chalkboard
Markers/chalk
Blank acetate sheets
Transparency pens
Pencils and scratch paper
Portable radio unit (walkie talkie)
Cell phone
Megaphone
Safety glasses
Gloves
Ear plugs
Hard hat
Safety shoes
Fall protection equipment
Chemical-resistant clothing
TV (optional)
VCR (optional)
The MARCOM Group Ltd. video *Emergency Planning* (optional)
Module Examinations*

* Located in the Test Booklet.

SAFETY CONSIDERATIONS

Ensure that participants are equipped with appropriate personal protective equipment.

ADDITIONAL RESOURCE

This module is intended to present thorough resources for task training. The following reference work is suggested for both instructors and motivated participants interested in further study. This is optional material for continued education rather than for task training.

Web site: www.osha.gov/SLTC/etools/evacuation/index.html

TEACHING TIME FOR THIS MODULE

An outline for use in developing your lesson plan is presented below. Approximately 2½ hours are suggested to cover *Emergency-Action Plans*. This includes 10 minutes at the beginning of the session for administrative tasks and one 10-minute break. You may need to adjust the time required for hands-on activities and testing based on your class size and resources.

Topic	Planned Time
A. Introduction	________
B. Emergency-Action Plans	________
1. Chain of Command	________
2. Communications	________
3. Accounting for Personnel	________
4. Emergency-Response Teams	________
5. Training	________
6. Personal Protection	________
7. Medical Assistance	________
8. Security	________
C. Pre-Planning For Specific Types of Emergencies	________
1. Trapped Workers	________
2. Severe Weather	________
3. Bomb Threats	________
4. Fire Emergencies	________
D. Dealing with the Media	________
E. Review	________
F. Module Examination	________
1. Participants must score 70% or higher to receive recognition from the NCCER.	
2. Record the testing results on Craft Training Report Form 200 and submit the results to the Training Program Sponsor.	

Module 75207-03

Emergency-Action Plans

Instructor's Notes:

COURSE MAP

This course map shows all of the modules in Safety Technology. The suggested training order begins at the bottom and proceeds up. The local Training Program Sponsor may adjust the training order.

Assign reading of Module 75207-03.

SAFETY TECHNOLOGY

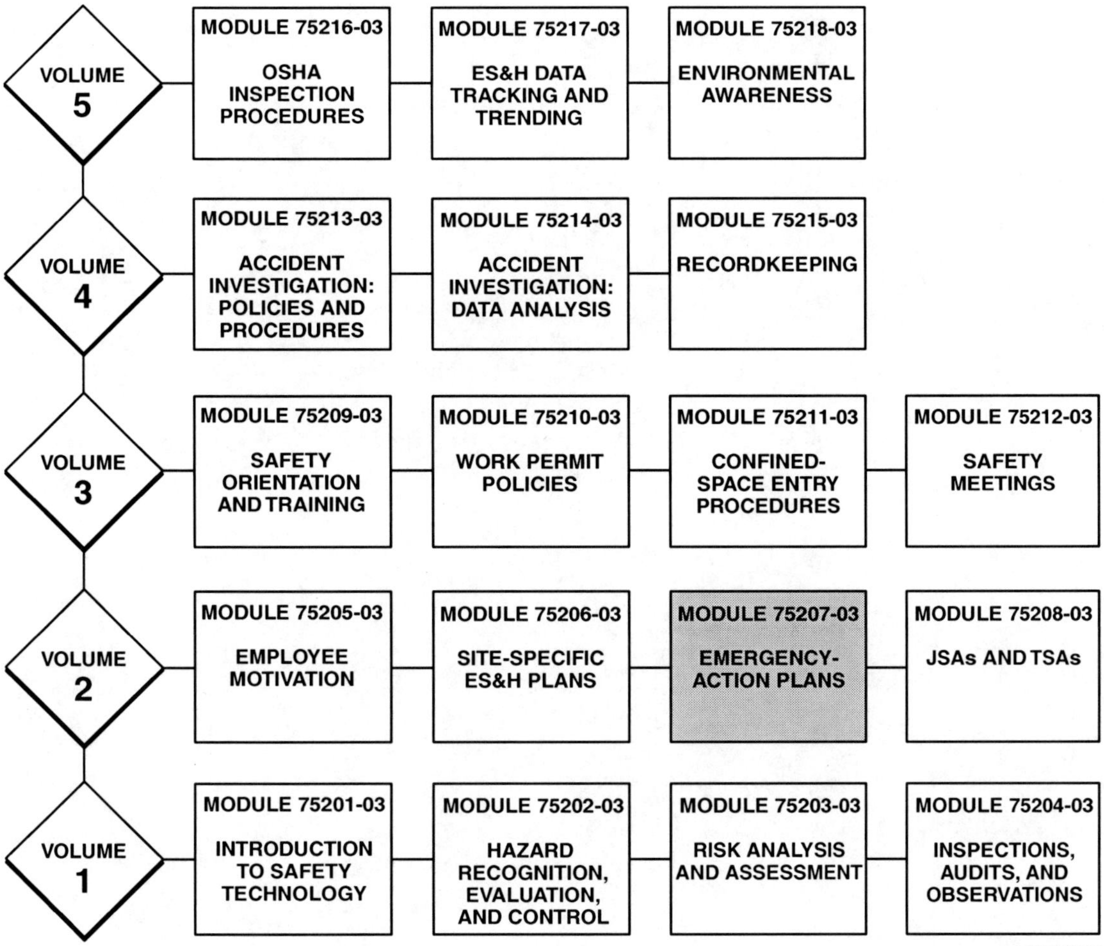

207CMAP.EPS

EMERGENCY-ACTION PLANS 7.iii

Instructor's Notes:

MODULE 75207-03 CONTENTS

Figures

Table

EMERGENCY-ACTION PLANS 7.v

Instructor's Notes:

MODULE 75207-03

Emergency-Action Plans

Objectives

When you have completed this module, you will be able to do the following:

1. Describe the types of emergencies that can occur on construction sites and at industrial facilities.
2. Describe the fundamental elements of an emergency-action plan.
3. Identify the correct procedures for dealing with the media.

Prerequisites

Before you begin this module, it is recommended that you successfully complete the following: Field Safety; Safety Technology, Modules 75201-03 through 75206-03.

Required Materials

1. Pencil and paper
2. Appropriate personal protective equipment
3. Copy of *29 CFR 1926, OSHA Construction Industry Regulations.*

1.0.0 ◆ INTRODUCTION

Proper planning for emergencies is necessary to ensure an injured employee receives proper medical treatment as soon as possible after an accident. Emergencies can occur on both construction and industrial sites. The types of emergencies that commonly occur on both types of sites include:

- Personal injury
- Releases of toxic gases
- Chemical spills
- Fires
- Explosions
- Trench cave-ins
- Confined-spaces accidents
- Accidents at extreme heights

Emergency-action plans can cut down response time to job-site emergencies. In many cases, advance planning can even minimize or reduce the severity of the accident. For example, if the names, addresses, and phone numbers of the nearest medical, fire, police, and emergency-response agencies are posted at each job-site phone, it is easier to get the help you need. Also, as part of the emergency-action plan, everyone should be familiar with the site emergency reporting and response procedures. Another part of the emergency-action plan is the availability of prompt access to first aid and follow-up medical care. Emergency-action plans should also require that at least two people on each job site be trained in basic first aid and cardiopulmonary resuscitation (CPR). Lives are saved when all of these elements are incorporated into emergency-action plans.

2.0.0 ◆ EMERGENCY-ACTION PLANS

The effectiveness of response during emergencies depends on the amount of planning and training that have been done. Management must show support for safety programs and the importance of emergency planning. If management is not interested in employee protection and minimizing property loss, little can be done to promote a safe workplace. It is therefore management's responsibility to develop and implement an emergency-action plan than can be adapted to meet job-site conditions. The input and support of all

Ensure that you have everything you need to teach the course. Check the Materials and Equipment list in the front of this module.

Explain that terms shown in bold are defined in the Glossary at the back of this module.

Review the goals of the course and explain what will be expected of the participant.

Show Transparency 1, Objectives.

Describe the different types of emergencies that can occur on a site.

Show Transparency 2, Types of Emergencies.

Explain the benefits of emergency-action plans.

Discuss the roles and responsibilities of both management and the safety technician in relation to emergency-action plans.

Identify the basic elements of an emergency-action plan. Include information about how the plan will be distributed or displayed for all personnel on site.

Show Transparency 3, Basic Elements of an Emergency-Action Plan.

Identify the elements of a comprehensive emergency-action plan.

Show Transparency 4, Elements of a Comprehensive Emergency-Action Plan.

Explain that a chain of command is established for decision-making during an emergency.

Describe the duties of an emergency-response coordinator.

Show Transparency 5, Duties of an Emergency-Response Coordinator.

employees must be obtained to ensure an effective program. Emergency-action plans should be developed for each job and should be comprehensive enough to deal with all types of emergencies. When emergency-action plans are required by a particular OSHA standard, the plan must be in writing. If the company has 10 or fewer employees, the plan can be spoken, rather than written.

Emergency-action plans must have, as a minimum, the following elements:

- Emergency escape procedures and emergency escape route assignments
- Procedures to be followed by employees who remain to perform (or shut down) critical operations before they evacuate
- Procedures to account for all employees after emergency evacuation has been completed
- Rescue and medical duties for those employees who are to perform them
- The preferred means for reporting fires and other emergencies
- Names or regular job titles of persons or departments to be contacted for further information or explanation of duties under the plan

The emergency-action plan should address all potential emergencies that can be expected in the workplace. Therefore, it will be necessary to perform a pre-job hazard assessment to determine what types of emergencies could reasonably occur. This should be done during the pre-job planning phase. For work done at existing facilities, the contractor and the host employer should coordinate their emergency reporting and response procedures before the job starts.

For information on chemicals, contact the manufacturer or supplier to obtain material safety data sheets (MSDSs). These forms describe the hazards that a chemical may present, list precautions to take when handling, storing, or using the substance, and outline emergency and first-aid procedures.

The plan should list in detail the procedures to be taken by those employees who must remain behind to care for essential operations or until their evacuation becomes absolutely necessary. This can include crane and mobile equipment operators, riggers, and other personnel who may be needed during the emergency.

For emergency evacuation, the plan should include the use of floor plans or workplace maps that clearly show the emergency escape routes and safe or refuge areas. All employees must be told what actions they are to take in the emergency situations that may occur in the workplace.

The job-site emergency-action plan should be reviewed with employees initially when the plan is developed, whenever the employees' responsibilities under the plan change, and whenever the plan is altered.

In addition to the elements previously discussed, effective emergency-action plans must provide detailed information about:

- Chain of command
- Communications
- Accounting for personnel
- Emergency-response teams
- Training procedures
- Personal protection
- Medical assistance
- Security

2.1.0 Chain of Command

A chain of command must be established so that employees will have no doubt about who has authority for making decisions. At most existing industrial facilities, an emergency-response team coordinator is selected to coordinate the work of the emergency-response team. That person is the first person in the chain of command. In larger organizations, there may be a plant coordinator in charge of plant-wide operations, public relations, and ensuring that outside aid is called in. Under these circumstances, when there is great importance in all of these functions, additional backup must be arranged so that trained personnel are always available. The duties of the emergency-response team coordinator should include the following:

- Assessing the situation and determining whether an emergency exists that requires activating the emergency procedures
- Directing all efforts in the area, including evacuating personnel and minimizing property loss
- Ensuring that outside emergency services such as medical aid and local fire departments are called in when necessary
- Directing the shutdown of plant operations when necessary

On grass roots construction projects, the contractor will have to implement a chain of command as part of the emergency-action plan. This chain of command must be flexible and work in conjunction with local law enforcement and emergency response personnel.

Instructor's Notes:

2.2.0 Communications

During a major emergency involving a fire or explosion, it may be necessary to evacuate all portions of the site, including offices. Also, normal services, such as electricity, water, and telephones, may be unavailable. Under these conditions, it may be necessary to have an alternate area to which employees can report or that can act as a focal point for incoming and outgoing calls. Because time is an essential element for adequate response, the person designated as being in charge should make this the alternate headquarters so that he or she can be easily reached.

Emergency communications equipment *(Figure 1)* such as portable radio units (walkie talkies), cell phones, or public address systems should be present for notifying employees of the emergency and for contacting local authorities, such as law enforcement officials, the fire department, ambulance services, and emergency-response contractors.

207F01.EPS

Figure 1 ◆ Emergency communications equipment.

A method of communication also is needed to alert employees to the evacuation or to take other action as required in the plan. Alarms should be audible or visible by all people on site, and should have an auxiliary power supply in the event electricity is affected. The alarm should be distinctive and recognizable as a signal to evacuate the work area or perform actions designated under the emergency-action plan. The employer should explain to each employee the means for reporting emergencies. Emergency phone numbers should be posted on or near telephones, on employees' notice boards, or in other obvious locations. The warning plan should be in writing and management must be sure each employee knows what it means and what action is to be taken.

An updated written list should be kept of key personnel, listed in order of priority. Also, site personnel should be provided with phone numbers of key personnel in the main office, including daytime and after-hours numbers.

2.3.0 Accounting for Personnel

Site management will need to know when all personnel have been accounted for. This can be difficult on a construction project. Someone on site should be appointed to account for personnel and to inform police or emergency-response personnel of those persons believed missing.

2.4.0 Emergency-Response Teams

Emergency-response teams are the first line of defense in emergencies. Before assigning personnel to these teams, the contractor must assure that employees are physically capable of performing the duties that may be assigned to them. Depending on the size of the facility, there may be one or several teams trained in the following areas:

- Use of various types of fire extinguishers
- First aid, including CPR
- Confined-space rescues
- Evacuation procedures
- Chemical spill control procedures
- Use of self-contained breathing apparatus (SCBA)
- Emergency rescue procedures for trenches and elevated locations

The type and extent of the emergency will depend on the site conditions. The response will vary according to the type of emergency, the material handled, the number of employees, and the availability of outside resources. Emergency-response teams should be trained in the types of possible emergencies and the emergency actions to be performed. They should be informed about special hazards, such as storage and use of flammable materials, toxic chemicals, radioactive sources, and water-reactive substances, to which they may be exposed during fire and other emergencies.

Explain how a chain of command is used on a smaller site versus how it is used on a large site.

Explain why a communications area should be designated in an emergency-action plan.

Describe the communication equipment used during an emergency, including alarms. Pass around an example of portable communication devices, such as a portable radio unit (walkie talkie), cell phone, or megaphone.

Discuss the methods for reporting emergencies, gathering contact information, and posting emergency phone numbers.

Emphasize the importance of designating someone on site to account for all personnel.

Describe the role and responsibility of an emergency-response team during an emergency. Explain that team members must be physically capable of performing emergency response.

Explain the areas in which emergency-response team members are typically trained.

EMERGENCY-ACTION PLANS 7.3

Discuss the importance of knowing when an emergency-response team should intervene and when professional emergency-response personnel should be called.

Show Transparency 6 (Table 1).

Discuss the training requirements for all site personnel.

Explain when training should be provided.

> **NOTE**
> On a typical construction site, contractors will not have organized emergency-response teams. Instead, the contractors will usually reply on off-site emergency response personnel. One exception would be on some large projects, where contractors may be asked to serve on emergency-response teams.

It is important to determine when not to intervene (*Table 1*). For example, team members must be able to determine if the fire is too large for them to handle or whether emergency rescue procedures should be performed. If there is a possibility of the emergency-response team members receiving fatal or incapacitating injuries, they should wait for professional firefighters or emergency-response groups.

2.5.0 Training

Training is important for the effectiveness of an emergency-action plan. Before implementing an emergency-action plan, a sufficient number of persons must be trained to assist in the safe and orderly evacuation of employees. Training for each type of disaster response is necessary so that employees know what actions are required.

In addition to the specialized training for emergency-response team members, all employees should be trained in the following:

- Evacuation plans
- Alarm systems
- Reporting procedures for personnel
- Shutdown procedures
- Types of potential emergencies

These training programs should be provided as follows:

- Initially when the plan is developed
- For all new employees
- When new equipment, materials, or processes are introduced
- When procedures have been updated or revised
- When exercises show that employee performance must be improved
- At least annually

Table 1 Evacuate or Fight?

SHOULD EMPLOYEES EVACUATE OR BE PREPARED TO FIGHT A SMALL FIRE?		
CHOOSING TO EVACUATE THE WORKPLACE RATHER THAN PROVIDE FIRE EXTINGUISHERS FOR EMPLOYEE USE IN FIGHTING FIRES WILL MOST EFFECTIVELY MINIMIZE THE POTENTIAL FOR FIRE-RELATED INJURIES TO EMPLOYEES. IN ADDITION, TRAINING EMPLOYEES TO USE FIRE EXTINGUISHERS AND MAINTAINING THEM REQUIRES CONSIDERABLE RESOURCES. HOWEVER, OTHER FACTORS, SUCH AS THE AVAILABILITY OF A PUBLIC FIRE DEPARTMENT OR THE VULNERABILITY OF EGRESS ROUTES, WILL ENTER INTO THIS DECISION.		
OPTION 1	**OPTION 2**	**OPTION 3**
TOTAL EVACUATION OF EMPLOYEES FROM THE WORKPLACE IMMEDIATELY WHEN ALARM SOUNDS. NO ONE IS AUTHORIZED TO USE AVAILABLE PORTABLE FIRE EXTINGUISHERS.	DESIGNATED EMPLOYEES ARE AUTHORIZED TO USE PORTABLE FIRE EXTINGUISHERS TO FIGHT FIRES. ALL OTHER EMPLOYEES MUST EVACUATE WORKPLACE IMMEDIATELY WHEN ALARM SOUNDS.	DESIGNATED EMPLOYEES ARE AUTHORIZED TO USE PORTABLE FIRE EXTINGUISHERS TO FIGHT FIRES.
REQUIREMENT	**REQUIREMENT**	**REQUIREMENT**
ESTABLISH AN EMERGENCY-ACTION AND FIRE-PREVENTION PLAN AND TRAIN EMPLOYEES ACCORDINGLY. IF FIRE EXTINGUISHERS ARE LEFT IN THE WORKPLACE, THEY MUST BE REGULARLY INSPECTED, TESTED, AND MAINTAINED.	ESTABLISH AN EMERGENCY-ACTION AND FIRE-PREVENTION PLAN AND TRAIN EMPLOYEES ACCORDINGLY. MEET ALL GENERAL FIRE EXTINGUISHER REQUIREMENTS PLUS ANNUALLY TRAIN DESIGNATED EMPLOYEES TO USE FIRE EXTINGUISHERS. FIRE EXTINGUISHERS IN THE WORKPLACE MUST BE REGULARLY INSPECTED, TESTED, AND MAINTAINED.	IF ANY EMPLOYEES WILL BE EVACUATING, ESTABLISH AN EMERGENCY-ACTION AND FIRE-PREVENTION PLAN AND TRAIN EMPLOYEES ACCORDINGLY. MEET ALL GENERAL FIRE EXTINGUISHER REQUIREMENTS PLUS ANNUALLY TRAIN ALL EMPLOYEES TO USE FIRE EXTINGUISHERS. FIRE EXTINGUISHERS IN THE WORKPLACE MUST BE REGULARLY INSPECTED, TESTED, AND MAINTAINED.

207T01.EPS

Instructor's Notes:

The emergency-control procedures should be written in concise terms and made available to all personnel. A drill should be held for all personnel, at random intervals at least annually, and an evaluation of performance made immediately by management and employees. When possible, drills should include groups supplying outside services such as fire and police departments. On jobs with multiple contractors, the emergency plans should be coordinated with other contractors and employees on site. Finally, the emergency plan should be reviewed periodically and updated to maintain adequate response personnel and program efficiency.

> **NOTE**
> On small jobs or short-term jobs, drills may not be practical. However, it is a good idea to have local law enforcement and fire department personnel visit your site early in the project for pre-planning purposes.

2.6.0 Personal Protection

Effective personal protection is essential for any person who may be exposed to potentially hazardous substances. In emergency situations, employees may be exposed to a wide variety of hazards, including:

- Chemical splashes or contact with toxic materials
- Falling objects and flying particles
- Falls from elevations
- Unknown atmospheres that may contain toxic gases, vapors or mists, or inadequate oxygen to sustain life
- Fires and electrical hazards

It is extremely important that employees be adequately protected in these situations.

Some of the safety equipment *(Figure 2)* that may be used includes:

- Safety glasses, goggles, or face shields for eye protection
- Hard hats and safety shoes for head and foot protection
- Respiratory protection for breathing protection
- Chemical-resistant suits, gloves, hoods, and boots for body protection from chemicals
- Body protection for abnormal environmental conditions such as extreme temperatures
- Personal fall-arrest systems

The equipment selected should be approved jointly by the Mine Safety and Health Administration (MSHA) and the National Institute for

Explain how and why drills are an important part of emergency control procedures.

Explain the hazards workers can be exposed to as the result of an emergency situation.

Show Transparency 7, Hazardous Conditions.

Describe the types of personal protection that is required at different sites. Show the participants examples of commonly used personal protective equipment.

Figure 2 ◆ Personal protective equipment.

EMERGENCY-ACTION PLANS 7.5

Explain that personal protective equipment must be approved by the Mine Safety and Health Administration (MSHA) and the National Institute for Occupational Safety and Health (NIOSH) or should meet the standards set by the American National Standards Institute (ANSI).

Describe the four basic categories of respirators. Refer the participants to Figure 3 in the Participant Module.

Explain the conditions that must be met before using or assigning respiratory equipment.

Show Transparency 8 (Figure 4).

Explain why a SCBA offers the best protection when controlling eme-rgency situations.

Describe the conditions that would require using SCBA equipment. Refer the participants to the bulleted list in the Participant Module.

Explain the dangers of entering a confined space.

Occupational Safety and Health (NIOSH), or should meet the standards set by the American National Standards Institute (ANSI). The choice of proper equipment is not a simple matter; health and safety professionals should be consulted before making any purchases. Manufacturers and distributors of health and safety products may be able to answer questions if they have enough information about the potential hazards involved.

> **NOTE**
> For more information about personal protective equipment, please refer the *Personal Protective Equipment* module in Field Safety.

2.6.1 Respirators

Professional consultation will most likely be needed in providing adequate respiratory protection. Respiratory protection is necessary for toxic atmospheres of dusts, mists, gases, or vapors, and for oxygen-deficient atmospheres. There are four basic categories of respirators *(Figure 3)*:

- Air-purifying devices such as filters, gas masks, and chemical cartridges remove contaminants from the air but cannot be used in oxygen-deficient atmospheres.
- Air-supplied respirators should not be used in atmospheres that are **immediately dangerous to life or health (IDLH)** unless operated in the positive pressure mode and equipped with an escape bottle.
- SCBA are required for unknown atmospheres, oxygen-deficient atmospheres, or atmospheres immediately dangerous to life or health.
- Escape masks are temporary devices used to make an emergency exit from an unsafe atmosphere.

Before assigning or using respiratory equipment, the following conditions must be met:

- A medical evaluation must be made to determine if the employees are physically able to use the respirator.
- Written procedures *(Figure 4)* must be prepared covering safe use and proper care of the equipment, and employees must be trained in these procedures and the use and maintenance of respirators.
- A fit test must be made to determine a proper match between the face piece of the respirator and the face of the wearer. This testing must be repeated periodically. Training must provide the employee an opportunity to handle the respirator, have it fitted properly, test its face piece-to-face seal, wear it in normal air for a familiarity period, and wear it in a test atmosphere.
- A regular maintenance program must be instituted, including cleaning, inspecting, and testing of all respiratory equipment. Respirators used for emergency response must be inspected after each use and at least monthly to assure that they are in satisfactory working condition. A written record of inspection must be maintained.
- Distribution areas for equipment used in emergencies must be readily accessible to employees.

A SCBA offers the best protection to employees involved in controlling emergency situations. It should have a minimum service life rating of 30 minutes. Conditions that require an SCBA include the following:

- Leaking cylinders or containers, smoke from chemical fires, or chemical spills that indicate high potential for exposure to toxic substances
- Atmospheres with unknown contaminants or unknown contaminant concentrations, confined spaces that may contain toxic substances, or oxygen-deficient atmospheres

2.6.2 Confined Spaces

Emergency situations may involve entering confined spaces to rescue employees who are overcome by hazardous atmospheres. These confined spaces include tanks, vaults, pits, sewers, pipelines, and vessels. Entry into confined spaces can expose employees to a variety of hazards, including toxic gases, flammable atmospheres, oxygen deficiency, electrical hazards, and hazards created by mixers and impellers that have not been deactivated and locked out. Personnel should never enter a confined space under normal circumstances unless the atmosphere has been tested for adequate oxygen, combustibility, and toxic substances. Conditions in a confined space must be considered immediately dangerous to life and health unless shown otherwise. If a confined space must be entered in an emergency, these precautions must be followed:

- All lines containing inert, toxic, flammable, or corrosive materials must be disconnected or valved off before entry.
- All impellers, agitators, or other moving equipment inside the vessel must be locked out.

Instructor's Notes:

SELF-CONTAINED BREATHING APPARATUS

SUPPLIED AIR MASK

HALF MASK

FULL FACE PIECE MASK

207F03.EPS

Figure 3 ◆ Four types of respirators.

- Employees must wear appropriate personal protective equipment when entering the vessel. Mandatory use of safety harnesses should be stressed.
- Rescue procedures must be specifically designed for each entry. When there is an atmosphere immediately dangerous to life or health, or a situation that has the potential for causing injury or illness to an unprotected worker, a trained confined-space attendant should be present. This person should be assigned a fully charged, positive-pressure, SCBA with a full face piece. The confined-space attendant must maintain unobstructed lifelines and communications to all workers within the confined space and be prepared to summon rescue personnel if

Describe the precautions that must be taken if entry into a confined space is required.

Transparency 9, Emergency Entry Precautions.

The MARCOM Group Ltd. video, *Emergency Planning*, can be used to enhance the topics covered in this module.

EMERGENCY-ACTION PLANS 7.7

Describe the medical and first aid services that contractors must provide. Refer the participants to the bulleted list in the Participant Module.

Explain the actions that contractors should consider when trying to meet medical and first-aid services requirements. Refer the participants to the bulleted list in the Participant Module.

BEFORE USING A RESPIRATOR YOU MUST DETERMINE THE FOLLOWING:

1. THE TYPE OF CONTAMINANT(S) FOR WHICH THE RESPIRATOR IS BEING SELECTED
2. THE CONCENTRATION LEVEL OF THE CONTAMINANT(S)
3. WHETHER THE RESPIRATOR CAN BE PROPERLY FITTED ON THE WEARER'S FACE

ALL RESPIRATOR INSTRUCTIONS, WARNINGS, AND USE LIMITATIONS CONTAINED ON EACH PACKAGE MUST ALSO BE READ AND UNDERSTOOD BY THE WEARER BEFORE USE.

207F04.EPS

Figure 4 ◆ Procedures to follow before using a respirator.

necessary. The confined-space attendant should not enter the confined space. Instead, he or she should assist workers leaving the space, including using hoisting equipment to retrieve the entrants from outside the space. This is called an external rescue.

A more complete description of procedures to follow while working in confined spaces may be found in NIOSH, *Publication Number 80-106, Criteria for a Recommended Standard: Working in Confined Spaces.*

NOTE

For more information on confined spaces, please refer to Volume 3, Module 75211-03, *Confined-Space Entry Procedures.*

2.7.0 Medical Assistance

In a major emergency, time is a critical factor in minimizing injuries. Most contractors do not have a formal medical program, but they are required to have the following medical and first-aid services:

- In the absence of an infirmary, clinic, or hospital in close proximity to the workplace that can be used for the treatment of all injured employees, the employer must ensure that personnel are adequately trained to render first aid.
- Where the eyes or body of any employee may be exposed to harmful corrosive materials, eye washes or suitable equipment for quick drenching or flushing must be provided in the work area for immediate emergency use. Employees must be trained to use the equipment.
- The employer must ensure the ready availability of medical personnel for advice and consultation on matters of employee health. This does not mean that health care must be provided, but rather that, if health problems develop in the workplace, medical help will be available to resolve them.

To fulfill the requirements just listed, the following actions should be considered:

- Survey the medical facilities near the job site and make arrangements to handle routine and emergency cases. A written emergency medical procedure should then be prepared for handling accidents with minimum confusion.
- If the job is located far from medical facilities, at least one, and preferably more employees on each shift must be adequately trained to render first aid. The American Red Cross, some insurance carriers, local safety councils, fire departments, and others may be contacted for this training.
- Inform all personnel about the hazards of bloodborne pathogens and the need to follow **universal precautions** when exposed to blood or other bodily fluids that could contain blood.
- First-aid supplies should be provided for emergency use. This equipment should be ordered through consultation with a physician.
- Emergency phone numbers should be posted in conspicuous places near or on telephones. On the emergency phone list, post the physical address for the job site with cross streets and roads.
- Sufficient ambulance service should be available to handle any emergency. This requires advance contact with ambulance services to ensure they become familiar with plant location, access routes, and hospital locations.
- Evaluate the need for automated external **defibrillators** (AEDs).

NOTE

Many contractors have established relationships with local occupational or industrial medical clinics to provide services for handling off-site medical emergencies that are not life threatening. These firms also provide other services such as performing medical evaluations, drug and alcohol testing, and respirator fit testing.

Instructor's Notes:

DID YOU KNOW?

Automated External Defibrillators Can Save Lives During Cardiac Emergencies

Automated External Defibrillators (AEDs) improve survival after an out-of-hospital cardiac arrest. Their presence reduces the critical time for treatment. Less time to defibrillation improves victims' chances of survival. Having the devices appropriately located in a business or workplace improves the survival rate of people experiencing a cardiac crisis.

Why should employers make AEDs available to employees?
There are 300,000–400,000 deaths per year in the United States from cardiac arrest. Most cardiac arrest deaths occur outside the hospital. Current out-of-hospital survival rates are 1 to 5%. In 1999 and 2000, 815 of 6,339 workplace fatalities reported to OSHA were caused by cardiac arrest.

Jobs with shift work, high stress, and exposure to certain chemicals and electrical hazards increase the risks of heart disease and cardiac arrest.

What causes cardiac arrest, and how does an AED improve survival rates?
Abnormal heart rhythms, with ventricular fibrillation (VF) being the most common, cause cardiac arrest. Treatment of VF with immediate electronic defibrillation can increase survival to more than 90%. With each minute of delay in defibrillation, 10% fewer victims survive.

Is AED equipment expensive?
The average initial cost for an AED ranges from $3,000 to $4,500.

Are AEDs difficult to use?
AEDs are easy to use. In mock cardiac arrest, untrained sixth-grade children were able to use AEDs without difficulty.

The Bottom Line:
AEDs are effective, easy to use, and relatively inexpensive.

Source: Occupational Safety and Health Administration (OSHA)

2.8.0 Security

During an emergency, it is often necessary to secure the area to prevent unauthorized access and to protect vital records and equipment. An off-limits area must be established by cordoning off the area with ropes and signs. It may be necessary to notify local law enforcement personnel or to employ private security personnel to secure the area and prevent the entry of unauthorized personnel.

Certain records also may need to be protected, such as essential accounting files, legal documents, and lists of employees' relatives to be notified in case of emergency. These records should be stored in duplicate outside the plant or in protected secure locations within the plant.

3.0.0 ◆ PRE-PLANNING FOR SPECIFIC TYPES OF EMERGENCIES

Pre-planning is an essential part of an emergency-action plan. It helps to ensure that all personnel know what to do during specific types of emergencies. Some of the specific types of emergencies that can occur on a site include:

- Trapped workers
- Severe weather
- Bomb threats
- Fire emergencies

3.1.0 Trapped Workers

Rescuing trapped workers requires specially trained and equipped personnel. Some examples of the types of conditions and work locations in which workers can get trapped include:

- Extreme heights (scaffolds, structural steel)
- Confined spaces (sewers, manholes, process vessels, tanks, etc.)
- Trench cave-ins
- Cranes in contact with energized power lines

Many contractors erroneously assume that the local fire department can handle all of these emergencies. In many instances, they cannot. If your job

Refer to the Did You Know? and discuss automatied external defibrillators. Ask the participants if they have ever worked at a site that had an AED. Have them share stories from their own experience.

Explain the required steps for securing the scene of an emergency and protecting vital records.

Identify the types of emergencies for which emergency pre-planning is essential.

Describe the work locations in which workers are at risk of being trapped. Emphasize the importance of finding out which local emergency personnel are equipped to deal with this type of emergency.

EMERGENCY-ACTION PLANS 7.9

Describe the problems of working in an area where severe weather occurs. Emphasize the importance of establishing and following a severe weather plan.

Explain that bomb threats should be considered during emergency response planning.

Discuss the OSHA requirements for fire protection at construction sites. Refer the participants to the bulleted list in the Participant Module.

could reasonably involve emergencies like these, it is critical that plans be worked out in advance. Find out what the local fire department can and cannot do and make your plans accordingly.

3.2.0 Severe Weather

Severe weather can be another problem. If your job is in an area where major thunderstorms or tornadoes often occur, you need a plan that notifies all site personnel and provides information for appropriate sheltering locations. On very remote sites, a weather alert radio may be needed.

3.3.0 Bomb Threats

Don't overlook the potential for bomb threats and civil disturbances. On jobs with labor unrest, bomb threats often occur. Make sure your emergency-action plan addresses these issues. Failure to do so can result in disorder on the site and ultimately injuries or deaths.

3.4.0 Fire Emergencies

Fires are one of the more common types of emergencies on construction projects. Failure to prepare for fires can result in catastrophic consequences. The OSHA Construction Standards call for a fire protection and prevention plan throughout all phases of the construction, repair, alteration, or demolition work. Fire protection and suppression equipment is required. An adequate number of portable fire extinguishers are mandated. Fire extinguishers can be supplemented with fire hoses but training is necessary if personnel are expected to use the equipment.

Listed here are some key OSHA requirements for fire protection at construction sites:

- Access to all available firefighting equipment shall be maintained at all times.
- All firefighting equipment shall be periodically inspected and maintained in operating condition. Defective equipment shall be immediately replaced.
- As warranted by the project, the employer shall provide a trained and equipped firefighting organization (Fire Brigade) to ensure adequate protection to life.
- A temporary or permanent water supply, of sufficient volume, duration, and pressure, required to properly operate the firefighting equipment shall be made available as soon as combustible materials accumulate.
- Where underground water mains are to be provided, they shall be installed, completed, and made available for use as soon as practicable.
- A fire extinguisher, rated not less than 2A, shall be provided for each 3,000 square feet of the protected building area, or major fraction thereof. Travel distance from any point of the protected area to the nearest fire extinguisher shall not exceed 100 feet.
- One or more fire extinguishers, rated not less than 2A, shall be provided on each floor. In multi-story buildings, at least one fire extinguisher shall be located adjacent to stairway.
- A fire extinguisher, rated not less than 10B, shall be provided within 50 feet of wherever more than five gallons of flammable or combustible liquids or five pounds of flammable gas are being used on the job site.
- Portable fire extinguishers shall be inspected periodically and maintained in accordance with *Maintenance and Use of Portable Fire Extinguishers, NFPA No. 10A-1970.*
- During demolition involving combustible materials, charged hose lines, supplied by hydrants, water tank trucks with pumps, or equivalent, shall be made available.
- If the facility being constructed includes the installation of automatic sprinkler protection, the installation shall closely follow the construction and be placed in service as soon as applicable laws permit following completion of each story.
- During demolition or alterations, existing automatic sprinkler installations shall be retained in service as long as reasonable.
- In all structures in which standpipes are required, or where standpipes exist in structures being altered, they shall be brought up as soon as applicable laws permit, and shall be maintained as construction progresses in such a manner that they are always ready for fire protection use. The standpipes shall be provided with Siamese fire department connections on the outside of the structure, at the street level, which shall be conspicuously marked. There shall be at least one standard hose outlet at each floor.
- An alarm system shall be established by the employer whereby employees on the site and the local fire department can be alerted for an emergency.
- The alarm code and reporting instructions shall be conspicuously posted at phones and at employee entrances.

Instructor's Notes:

4.0.0 ◆ DEALING WITH THE MEDIA

In the event of an emergency or accident, it is very possible that the news media will try to get the story for newspapers or television reports. They will want to know:

- What happened?
- Was anyone hurt?
- Why did it happen?
- How will the community be affected?
- How is the construction company going to fix it?

Keep in mind that you may not have the answers to all of these questions right away. In this instance, it is acceptable to tell the media an investigation is currently underway and you, or the appropriate company representative, will report the findings as soon as they are known.

Additional guidelines for dealing with the media include:

- Know your company policy. Only provide information that you are allowed to give. If there is a company spokesperson, be polite, and advise the press to speak to that person.
- Express concern for the safety and well being of any injured personnel and their families.
- Be prepared if you are to provide information and be certain to:
- Provide only facts. Do not lie or offer opinions.
- Say you don't know when you don't know.
- Be polite and helpful.
- Maintain control; don't answer leading questions.
- Be professional; don't make jokes.
- Never make "off-the-record" statements.
- Never release the names of workers involved.
- Avoid discussion of legal questions.
- Do not accept responsibility or admit liability for the incident.
- Stress your company's commitment to safety.
- Keep reporters out of hazardous areas.
- Keep track of who you talked to.
- End the interview when you have covered the subject.

Summary

Emergency-action plans save lives. They must be detailed and comprehensive in order to accomplish this goal. Emergency-action plans should contain specific elements such as an established chain of command, good communications, a method to account for all personnel, emergency-response personnel, thorough training for all employees, the use of personal protective equipment, and information on how to get medical assistance.

Another important element of an emergency-action plan is knowing how to deal with the media. Make sure all employees know who is responsible for dealing with press and that they also know the company's policy regarding the press.

Explain that emergencies often get the attention of the media.

Explain the six things reporters will want to know: who, what, when, where, how, and why.

Discuss the guidelines for dealing with media, including knowing your company's policy regarding who is allowed to talk to the media.

Review the material covered in the module.

Have the participants complete the Review Questions and go over the answers prior to administering the Module Examination.

Administer the Module Examination. Record the results on Craft Training Report Form 200 and submit the results to the Training Program Sponsor.

EMERGENCY-ACTION PLANS 7.11

Review Questions

1. Training on the site emergency-action plan should be provided to employees upon initial assignment and _____.
   a. weekly
   b. monthly
   c. quarterly
   d. when the plan is changed

2. Workers can avoid injury from exposure to harmful or corrosive chemicals by using _____.
   a. hard hats
   b. safety shoes
   c. chemical-resistant personal protective equipment
   d. personal fall-arrest systems

3. Before personnel are allowed to wear tight-fitting respiratory protection, they must have a medical evaluation and be _____.
   a. drug tested
   b. fit tested
   c. alcohol tested
   d. competency tested

4. Each of the following is a duty of a confined-space attendant *except* _____.
   a. keeping watch over personnel in the space
   b. assisting with external rescue
   c. summoning help in an emergency
   d. making an internal rescue alone

5. When exposed to blood or other bodily fluids, personnel should follow _____.
   a. good surgical practices
   b. good environmental practices
   c. universal precautions
   d. sterilization procedures

6. OSHA requires prompt access to first aid and follow-up medical care for job-related injuries and illnesses.
   a. True
   b. False

7. Most fire departments are properly trained and equipped to handle rescues from trench cave-ins and extreme heights.
   a. True
   b. False

8. According to OSHA Fire Protection requirements, the maximum travel distance to a Class A fire extinguisher is _____.
   a. 25 feet
   b. 50 feet
   c. 75 feet
   d. 100 feet

9. When flammable or combustible liquids in storage exceed five gallons, the maximum travel distance to a Class B fire extinguisher shall *not* exceed _____.
   a. 25 feet
   b. 50 feet
   c. 75 feet
   d. 100 feet

10. It is acceptable to make "off-the-record" statements to trustworthy members of the press.
    a. True
    b. False

Instructor's Notes:

GLOSSARY

Trade Terms Introduced in This Module

Defibrillator: An electronic device that administers an electric shock of preset voltage to the heart through the chest wall in an attempt to restore the normal rhythm of the heart during ventricular fibrillation.

Immediately dangerous to life and health (IDLH): A situation that poses a threat of exposure to airborne contaminants when that exposure is likely to cause death or immediate or delayed permanent adverse health effects or prevent escape from such an environment.

Universal precautions: A set of precautions designed to prevent transmission of human immunodeficiency virus (HIV), hepatitis B virus (HBV), and other bloodborne pathogens when providing first aid or health care. Under universal precautions, blood and certain body fluids of all patients are considered potentially infectious for HIV, HBV, and other bloodborne pathogens.

EMERGENCY-ACTION PLANS 7.13

Instructor's Notes:

Additional References

This module is intended to present thorough resources for task training. The following reference works are suggested for further study. These are optional materials for continued education rather than for task training.

www.osha.gov

www.asse.org

Figure Credits

Anchor Audio	207F01
Bacou-Dalloz	207F02 (Glasses)
Bullard Classic Head Protection	207F02 (Hard hat)
Bon Tool Company	207F02 (Gloves)
Anna Meade	207F02 (Safety Shoe)
North Safety Products	207F02 (Ear plugs), 207F03

EMERGENCY-ACTION PLANS 7.15

MODULE 75207-03 — ANSWERS TO REVIEW QUESTIONS

Answer	Section
1. d	2.5.0
2. c	2.6.0
3. b	2.6.1
4. d	2.6.2
5. c	2.7.0
6. a	2.7.0
7. b	3.1.0
8. d	3.4.0
9. b	3.4.0
10. b	4.0.0

CONTREN™ LEARNING SERIES — USER UPDATES

The NCCER makes every effort to keep these textbooks up-to-date and free of technical errors. We appreciate your help in this process. If you have an idea for improving this textbook, or if you find an error, a typographical mistake, or an inaccuracy in NCCER's Contren™ textbooks, please write us, using this form or a photocopy. Be sure to include the exact module number, page number, a detailed description, and the correction, if applicable. Your input will be brought to the attention of the Technical Review Committee. Thank you for your assistance.

Instructors – If you found that additional materials were necessary in order to teach this module effectively, please let us know so that we may include them in the Equipment/Materials list in the Instructor's Guide.

Write: Curriculum Revision and Development Department
National Center for Construction Education and Research
P.O. Box 141104, Gainesville, FL 32614-1104

Fax: 352-334-0932

E-mail: curriculum@nccer.org

Craft ______ Module Name ______

Copyright Date ______ Module Number ______ Page Number(s) ______

Description ______

______

______

______

(Optional) Correction ______

______

______

(Optional) Your Name and Address ______

______

______

Module 75208-03

JSAs and TSAs

NCCER STANDARDIZED CRAFT TRAINING PROGRAM

The National Center for Construction Education and Research (NCCER) provides a standardized national program of accredited craft training. Key features of the program include instructor certification, competency-based training, and performance testing. The program provides trainees, instructors, and companies with a standard form of recognition through a National Craft Training Registry. The program is described in full in the *Guidelines for Accreditation*, published by the NCCER. For more information on standardized craft training, contact the NCCER by writing us at P.O. Box 141104, Gainesville, FL 32614-1104; calling 352-334-0911; or e-mailing info@nccer.org. More information may be found at our Web site, www.nccer.org.

HOW TO USE THIS ANNOTATED INSTRUCTOR'S GUIDE

Each page presents two sections of information. The larger section displays each page exactly as it appears in the Participant Module. The narrow column ties suggested participant and instructor actions to each page and provides icons (detailed below) to call your attention to material, safety, audiovisual, or testing requirements. The bottom of each page includes space for your notes.

The **Audiovisual** icon indicates an appropriate time to show a transparency or other audiovisual aid.

The **Classroom** icon prompts you to define a term, stress a point, ask participants to explain a concept, or give examples.

The **Demonstration** icon directs you to show participants how to perform tasks.

The **Examination** icon tells you to administer the written module examination.

The **Homework** icon is placed where you may wish to assign reading for the next class, to assign a project, or to advise participants to prepare for an examination.

The **Laboratory** icon is used when participants are to practice performing tasks.

The **Materials** icon is a reminder for you to gather materials needed for classes, labs, and testing.

The **Performance Testing** icon tells you to administer a performance test or a portion thereof.

The **Safety** icon is used to emphasize safety issues. It is often keyed to *Caution* and *Warning* statements in the Participant Module.

The **Teaching Tip** icon indicates additional guidance is available, such as how to conduct an exercise, get the most educational value from a field trip, or encourage class participation. Teaching Tips may expand on a feature (*Think About It*, *Did You Know?*) or provide Quick Quizzes or similar exercises. You will be referred to the Teaching Tips section at the back of the module if there is additional material.

The **Combination** icon indicates that the laboratory listed corresponds with a performance task. If desired, you can note the proficiency of the participants during the laboratory and use it to satisfy performance testing requirements.

PREPARATION

Before teaching this module, you should review the Objectives, Performance Tasks, Materials and Equipment List, and the Module Outline. Be sure to allow ample time to prepare your own training or lesson plan and gather all required materials and equipment.

MODULE OVERVIEW

In this module, participants will learn how to conduct a job safety analysis (JSA) and a task safety analysis (TSA).

PREREQUISITES

Please refer to the Course Map in the Participant Module. Prior to training with this module, it is recommended that the participant shall have successfully completed the following:

Field Safety; Safety Technology, Modules 75201-03 through 75207-03.

OBJECTIVES

Upon completion of this module, the participant will be able to:

1. Define job safety analysis.
2. Describe how to conduct a job safety analysis.
3. Describe the purpose of a task safety analysis.
4. Explain the difference between a job safety analysis and a task safety analysis.

PERFORMANCE TASKS

Under the supervision of the instructor, the participant should be able to:

1. Demonstrate how to properly complete a JSA.
2. Demonstrate how to properly complete a TSA.

MATERIALS AND EQUIPMENT LIST

Overhead projector and screen
Transparencies
Whiteboard / chalkboard
Markers / chalk
Blank acetate sheets
Transparency pens
Pencils and scratch paper

Appropriate personal protective equipment
Job safety analysis (JSA) form
Task safety analysis (TSA) form
Hazard analysis chart
Module Examinations*
Performance Profile Sheets*

* Located in the Test Booklet.

SAFETY CONSIDERATIONS

Ensure that participants are equipped with appropriate personal protective equipment.

ADDITIONAL RESOURCE

This module is intended to present thorough resources for task training. The following reference work is suggested for both instructors and motivated participants interested in further study. This is optional material for continued education rather than for task training.

Web site: www.osha.gov

TEACHING TIME FOR THIS MODULE

An outline for use in developing your lesson plan is presented below. Approximately 2½ hours are suggested to cover *JSAs and TSAs*. This includes 10 minutes at the beginning of the session for administrative tasks and one 10-minute break. You may need to adjust the time required for hands-on activities and testing based on your class size and resources. Because laboratories often correspond to Performance Tasks, the proficiency of the participants may be noted during these exercises for Performance Testing purposes.

Topic	**Planned Time**
A. Introduction	________
B. Reasons for Conducting a Job Safety Analysis	________
1. Recognizing and Reducing the Risk of Potential Hazards	________
2. Improve Hazard Awareness	________
3. Standardized Work Practices	________
4. Facilitate Job Training	________
C. Conducting a Job Safety Analysis	________
1. Selecting Jobs or Tasks to be Analyzed	________
2. Preparing for a Job Safety Analysis	________
3. Collecting Data	________
4. Identify Hazards and Risk Factors	________
5. Develop Solutions	________
6. Common Errors	________
7. Laboratory Provide the participants with a job scenario and a copy of a JSA form. Have the participants complete the JSA form for the suggested job. Note the proficiency of each participant. This laboratory corresponds with Performance Task 1.	________
D. Task Safety Analysis	________
1. Laboratory Provide the participants with a task scenario and a copy of a TSA form. Have the participants complete the TSA form for the suggested task. Note the proficiency of each participant. This laboratory corresponds with Performance Task 2.	
E. Task Safety Analysis vs. Job Safety Analysis	________
F. Review	________
G. Module Examination	________
1. Participants must score 70% or higher to receive recognition from the NCCER.	
2. Record the testing results on Craft Training Report Form 200 and submit the results to the Training Program Sponsor.	
H. Performance Testing	________
1. Participants must perform each task to the satisfaction of the instructor to receive recognition from the NCCER. If applicable, proficiency noted during laboratory exercises can be used to satisfy the Performance Testing requirements.	
2. Record the testing results on Craft Training Report Form 200 and submit the results to the Training Program Sponsor.	

Module 75208-03

JSAs and TSAs

Instructor's Notes:

COURSE MAP

This course map shows all of the modules in Safety Technology. The suggested training order begins at the bottom and proceeds up. The local Training Program Sponsor may adjust the training order.

Assign reading of Module 75208-03.

SAFETY TECHNOLOGY

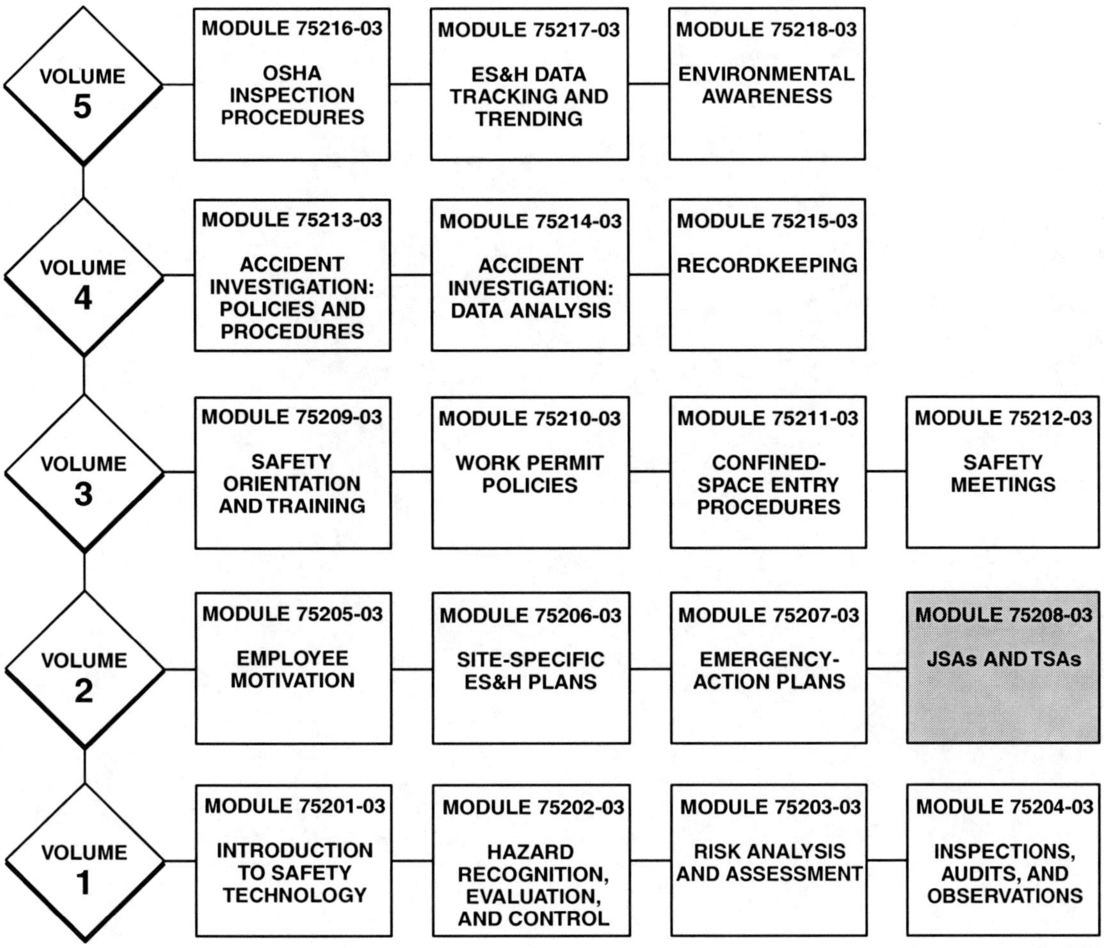

208CMAP.EPS

JSAs AND TSAs 8.iii

Instructor's Notes:

MODULE 75208-03 CONTENTS

Figures

JSAs AND TSAs 8.v

Instructor's Notes:

MODULE 75208-03

JSAs and TSAs

Ensure that you have everything you need to teach the course. Check the Materials and Equipment list in the front of this module.

Explain that terms shown in bold are defined in the Glossary at the back of this module.

Review the goals of the course and explain what will be expected of the participant.

Show Transparency 1, Objectives.

Show Transparency 2, Performance Tasks.

Objectives

When you have completed this module, you will be able to do the following:

1. Define job safety analysis.
2. Describe how to conduct a job safety analysis.
3. Describe the purpose of a task safety analysis.
4. Explain the difference between a job safety analysis and a task safety analysis.

Prerequisites

Before you begin this module, it is recommended that you successfully complete the following: Field Safety; Safety Technology, Modules 75201-03 through 75207-03.

Required Material

1. Pencil and paper
2. Appropriate personal protective equipment

1.0.0 ◆ INTRODUCTION

Define job safety analysis (JSA) and task safety analysis (TSA). Describe the uses of each.

Developing methods and controls to prevent accidents from occurring can dramatically increase the safety of a **job** or **task**. *Figure 1* shows the relationship between a job and a task. Conducting job safety analyses (JSAs) and task safety analyses (TSAs) will allow site personnel to reduce the risk of accidents. A job safety analysis is a careful study of a job to find all of the associated hazards. A task safety analysis is a careful study of a task to find all of the associated hazards. Both analyses will also identify procedures and practices for reducing the risks associated with the jobs and tasks. It's important to note that JSAs are a routine part of any operation, whereas TSAs are less common.

Time spent doing JSAs and TSAs should be seen as an investment. Each helps to reduce the number of accidents, improve productivity, and aid in developing job training.

> **NOTE**
> OSHA has extensive information on both of these topics on their Web site (www.osha.gov).

Show Transparency 3 (Figure 1)

JOB: CHANGE BURNED OUT LIGHT BULB
TASK 1: LOCATE BURNED OUT BULB.
TASK 2: SET UP EQUIPMENT TO REACH BULB.
TASK 3: DETERMINE BULB TYPE.
TASK 4: REMOVE BURNED OUT BULB.
TASK 5: INSTALL WORKING BULB.

208F01.EPS

Figure 1 ◆ The job/task relationship.

2.0.0 ◆ REASONS FOR CONDUCTING A JOB SAFETY ANALYSIS

Explain the overall benefits of conducting JSAs.

There are several reasons to conduct JSAs. Most of the benefits are directly related to safety. Other benefits are in the areas of productivity and training. JSAs benefit the employees, the company, and the workers.

> **NOTE**
> The terms *job safety analysis* and *job hazard analysis* are often used interchangeably. Each involves a methodical review of job steps identifying safety concerns. The JSA is more inclusive and covers all hazards including safety, health, and environmental issues.

Describe the benefits of conducting JSAs: risk reduction, improved hazard awareness, standardization of work practices, and the facilitation of job training.

Explain the five steps for conducting a JSA.

Show Transparency 4, Procedure for Conducting a JSA.

Describe the criteria used to select jobs for analysis.

Show Transparency 5, Criteria for Selecting a Job for Analysis.

2.1.0 Recognizing and Reducing the Risk of Potential Hazards

The main benefit to conducting a JSA is that it provides information on potential hazards associated with a job or task. Besides identifying hazards, a JSA will also help you determine ways to avoid or minimize unsafe or unhealthy conditions. Identifying hazards and determining the appropriate steps, equipment, and controls needed to reduce those hazards will minimize accidents and injuries.

2.2.0 Improve Hazard Awareness

Hidden or underestimated hazards may be uncovered through the careful study and analysis involved in a JSA. As JSAs are conducted on individual jobs, you will begin to develop a larger picture of the overall hazard level at the site. This information can be used to prevent accidents or illnesses and to reduce the level of risk throughout.

2.3.0 Standardized Work Practices

JSAs also provide a means for standardizing work practices. When conducting a JSA, a job is studied in great detail. Several workers may be observed, or even videotaped, while performing a job. The workers may also be interviewed to determine all of the steps involved in performing the job. This study produces a large volume of information on what practices work efficiently and safely and which do not. Documentation of this information can be used to provide standardized work practices.

2.4.0 Facilitate Job Training

Observing and analyzing a job and the tasks that go into performing that job will yield documentation and information that can be useful when developing training for the job. The training developed could be job-specific safety training, process training, or **structured on-the-job training (SOJT)**. Because the research that goes into developing technical and safety training often involves a similar type of job analysis, the data provided by a JSA may be easily adapted for this purpose.

3.0.0 ◆ CONDUCTING A JOB SAFETY ANALYSIS

Depending on the job to be analyzed, conducting a JSA can be a rigorous process. A thorough JSA involves five basic steps:

Step 1 Selecting the job to be analyzed

Step 2 Preparing for the JSA

Step 3 Collecting data

Step 4 Identifying hazards and risk factors

Step 5 Developing solutions

3.1.0 Selecting Jobs or Tasks To Be Analyzed

Every job has a potential to cause injury or illness, but not every job can be analyzed because JSAs can consume a lot of time and resources. A priority rating system must be used to select jobs for analysis. It is suggested that jobs be rated based on:

- History of accidents or injuries
- Potential for serious injuries
- Newness of the job or the equipment used
- A change in procedure or routine
- High turnover rates

Using these criteria, along with your experience and professional judgment, will help you to select jobs for analysis that will provide the greatest impact on worksite safety.

History of accidents or injuries – A job that is known to frequently result in accidents or injuries should be strongly considered for a JSA. For example, if workers who receive and open shipments of materials are frequently cutting themselves with box cutters or razors, a JSA may be in order to determine the causes and develop solutions.

Potential for serious injuries – A job that is more likely to result in serious injury or a fatality should be rated a higher priority for a JSA than one in which the injuries are less severe. If, for example, a job may result in an accident involving high-voltage electrocution or a fall from a great height, that job should receive a higher priority for a JSA than a job where the most likely accidents typically result in minor injuries.

Newness of the job or the equipment used – When a new job or process is started, it should be thoroughly analyzed to determine what risk factors it introduces to the worksite and the workers. If a new piece of equipment is introduced, all of the jobs for which that equipment will be used should be subjected to a JSA to make sure that the new equipment does not create new or different hazards.

A change in procedure or routine – Changing a procedure or routine may introduce new risks or hazards because workers are not used to new procedures or routines. Also, the new procedure or routine may create or aggravate hazards not present or as likely with the former procedure or routine.

High turnover rates – Jobs with high turnover rates warrant a JSA for two reasons. First of all, the unsafe condition or perceived unsafe condition of

Instructor's Notes:

the job may be a cause for a high turnover rate. Secondly, a high flow of new people through a job means that the workers have little experience and are more likely to have an accident or cause one. Clearly documented work processes and safety procedures help to reduce such accidents.

3.2.0 Preparing for a Job Safety Analysis

In order to work most efficiently, some preparation needs to go into the JSA process before data collection begins. First, take a look at the general conditions under which the job is being performed and develop a list of questions to be answered, or a checklist of items to evaluate, while conducting the JSA. Some things to look for include:

- Adequate lighting
- Trip hazards
- Live electrical hazards
- Fire hazards
- Tools or equipment in need of repair
- Availability of personal protective equipment

It is useful to create a form on which to record observations before conducting the JSA. Use of a form will simplify data collection, documentation, and recordkeeping. *Figure 2* shows a sample of a job safety analysis form.

3.3.0 Collecting Data

There are several methods of collecting data, including:

- Documentation review
- Direct observation
- Videotape and review
- Group discussion

When conducting a JSA, each method of data collection does not have to be used, but the more techniques used, the better the overall picture.

3.3.1 Documentation Review

The first resource for collecting data is the documentation associated with the job being analyzed. The documentation includes:

- Reports on accidents or incidents occurring on the job
- Job descriptions
- Equipment and tool maintenance logs
- Procedure manuals
- Material safety data sheets (MSDSs) for materials used on the job

Study and analysis of this data will provide important background information on how the job should be done and the necessary equipment and safeguards. This research may also reveal **trends**, **causal links**, or **patterns** associated with accidents. Reviewing the associated documentation will not typically provide all of the necessary detail, nor is it likely to provide solutions. It will, however, make the other aspects of data collection far more efficient and should not be overlooked as a key method of data collection.

3.3.2 Direct Observation

Direct observation, simply put, is going to the job site, watching workers do the job, and asking the workers about the job and its hazards. Direct observation is the preferred method because it gives you the best perspective on the job, the conditions under which it is performed, the tasks that need to be performed, and the risks associated with the job and its tasks. The drawbacks to direct observation are that it can be distracting to the workers and may cause the workers to change the way they work because they know they are being observed. The best way to work around these issues is to keep as low a profile as possible.

3.3.3 Videotape and Review

Videotape and review is a way to collect data in which a camera is placed at the job site to record how the job is done. The videotape is later reviewed to analyze the job. Videotape and review provides a few advantages as a method of data collection. Videotaping someone performing a job is typically less distracting than direct observation. Also, recording a job on videotape provides for a considerable amount of review after the fact, because the videotape provides a lasting record of the job. Finally, a videotape can be reviewed by different experts, either together or separately, to observe specific areas of interest such as hazmat handling procedures, equipment operation techniques, or job **ergonomics**. These many different perspectives give the JSA a broader view of the job and are likely to result in better solutions.

Videotaping does have some problems, however. A mounted video camera will not move or pan on its own to capture all of the activity. Therefore, consideration must be given to setting up the camera so that the captured image is large enough to be useful, but the angle is wide enough to cover all of the required information. Also, because the analyst is not present while the job is being performed, videotape and review does not provide the clearest view of the job and its risks.

Describe the steps for preparing for a JSA.

Explain that identifying areas that may need special attention during a JSA helps increase the accuracy and effectiveness of JSAs. Refer the participants to the bulleted list in Section 3.2.0.

Explain that writing a checklist or using a JSA form ensures that all areas are carefully analyzed.

Show Transparency 6 (Figure 2).

Identify the different methods for collecting data. Describe each in detail.

Provide the participants with a job scenario and a copy of a JSA form. Have the participants complete a JSA form for the suggested job. Note the proficiency of each participant. This laboratory corresponds with Performance Task 1.

JOB SAFETY ANALYSIS FORM

Job Title:	Date of Analysis:
Required PPE:	Conducted By:
	Duration:
Required Tools & Equipment:	
Materials Used:	

Task	Hazard	Quality Concern	Recommendation

208F02.EPS

Figure 2 ◆ An example of a job safety analysis form.

Instructor's Notes:

While direct observation is preferred over videotape and review alone, when the two methods are combined they can provide a vast amount of information about the job, the people doing the job, and the context in which it is performed.

3.3.4 Group Discussion

A group discussion is a structured interview with people related to the job being analyzed. This may include workers who perform the job, their supervisors, and engineers who design the processes the workers use. A group discussion may be used in advance to prepare for observation or videotaping, or it may be used to follow up on observing or videotaping. The group discussion should be conducted out of the immediate worksite, preferably in a conference room or classroom. The session should be videotaped and detailed notes should be taken so that the data provided can be revisited at a later time. Structure the questions to be asked in advance. These questions should be generated by the review of documentation, direct observation, or review of the videotape. If possible, use two people to run the group discussion. One person should ask the questions, encourage discussion, and document the feedback generated on flipcharts. The other person should quietly observe the discussion, take careful notes, and operate the video camera. This will help ensure that the meeting runs quickly and smoothly and that none of the important data gets lost or goes uncaptured.

3.4.0 Identify Hazards and Risk Factors

When identifying hazards and risk factors, carefully consider and analyze each task within the job, each tool or piece of equipment used, and all of the materials used so that you may determine what risks they pose to the workers. This is done to determine all of the safety hazards presented by a job and to develop solutions, preventive measures, and new procedures to eliminate or reduce those hazards.

Hazard and risk identification should be done in two steps. First, while making the initial observation, take notes about possible risks or hazards. To do this, ask these questions about each task:

- Is there a danger of striking against, being struck by, or otherwise making harmful contact with an object?
- Can employees be caught in, on, by, or between objects?
- Is there a potential for a slip, trip, or fall? If so, will it be on the same elevation or to a different elevation?
- Can workers strain themselves by pushing, pulling, lifting, bending, or twisting?
- Is the environment hazardous to anyone's safety or health?

These initial observations should be researched further and explored more thoroughly in a videotape review by various experts, or in a group discussion. After all of the potential hazards and risks have been identified, each should be thoroughly analyzed and documented. This may be done using a chart similar to the one in *Figure 3* or by writing a report covering the job, its tasks, and the associated risks.

A severity and likelihood score should be assigned to each hazard identified. The severity score indicates the expected severity of an injury resulting from that hazard. For example, if the hazard is a high-voltage electrical shock that could result in severe burns or a fatality, the severity score would be high. The likelihood score indicates how probable it is that the hazard would result in an accident. These scores will assist in assigning priorities to the hazards so that the analysis can be used to best advantage.

NOTE

For more information on risk assessment, please refer to the risk assessment matrix in the *Site-Specific ES&H Plans* module in this volume.

Job:	Hazard	Severity (0–3)	Likelihood (0–3)
TASK 1			
TASK 2			
TASK 3			
TASK 4			
TASK 5			

208F03.EPS

Figure 3 ◆ Hazard analysis chart.

Explain the two steps that should be used during hazard and risk identification.

Discuss the questions that should be asked when identifying hazard and risk factors.

Show Transparency 7, Identifying Questions.

Explain how the hazard analysis chart is used during hazard and risk identification.

Show Transparency 8 (Figure 3).

Complete a hazard analysis chart using an example from your own experiences.

Discuss the principal solutions for correcting hazards.

3.5.0 Develop Solutions

The final step in conducting a JSA is to develop practical methods of making the job safer. The goal is to prevent the occurrence of potential accidents and minimize the risk and severity of injury should an accident occur. The principal solutions are:

Show Transparency 9, Principal Solutions.

- Find a safer way to do the job.
- Change the physical conditions that create the hazard.
- Change work methods or procedures to eliminate hazards.
- Try to reduce the necessity or frequency of doing the job or some of its tasks.

Explain that the higher the severity and likelihood scores of a hazard are, the more that should be done to eliminate the risk.

The higher the severity and likelihood scores of a hazard are, the more that should be done to eliminate the risk. For example, if cleaning out a chemical tank is likely to result in death due to asphyxiation, then every effort should be made to minimize the risks associated with that job. These may include acquiring better personal protective equipment, providing better job safety training, or purchasing automated equipment to clean the tank rather than risking workers' safety.

Developing solutions should be done with a team that includes safety experts, process engineers, and other experts such as ergonomists, fire experts, or medical doctors. Most changes will also require a commitment on the part of supervisory and management personnel. For that reason, it is a good idea to involve them at an appropriate point in the process.

Discuss the common errors made during JSAs.

3.6.0 Common Errors

When doing JSAs, there are some common errors that should be avoided. Some of the most common errors are:

Show Transparency 10, Common JSA Errors.

- Making the breakdown so detailed that an unnecessarily large number of tasks are listed
- Making the job so general that basic tasks are not covered
- Failing to identify the education and experience levels of the target audience
- Conducting JSAs for all jobs instead of identifying jobs that require JSAs

The best way to avoid those mistakes is to take the time up front to make a clear plan on how the JSA will be conducted, who will be a part of the JSA process, and what the likely outcomes and uses will be for the results of the JSA.

Describe how TSAs are done.

4.0.0 ◆ TASK SAFETY ANALYSIS

Performing a TSA is a process for identifying and evaluating potential hazards associated with a given task or work assignment. Once the hazards have been identified, methods of eliminating or controlling the hazards should be developed and incorporated into the job plan.

A TSA is typically performed just prior to starting the task. The evaluation may be performed individually or collectively. The ideal situation is to have the entire work group involved in the process. Some companies mandate that the supervisor coordinate the process while others encourage individual crew members to lead the process. When employees are working alone, they are encouraged to conduct a personal TSA.

Show Transparency 11 (Figure 4).

Most companies provide a pocket-size form with a list of items to consider. Two sample forms are shown in *Figure 4*. Potential hazards are checked off or noted on the form. Personal protective equipment requirements are usually indicated. In most cases, there are spaces for the workers to sign indicating they have reviewed the form and that they understand and agree with the precautions listed.

Some pre-task analysis forms incorporate the JSA process, which requires listing each step, the associated hazards, and the appropriate safeguards.

Completed pre-task safety analysis forms are typically reviewed by the supervisor at the start of the workday or during the day. The forms are also used to audit safe work practices during the day.

Upon completion of the task, completed forms are usually collected and retained for a predetermined length of time.

Pre-task safety analyses are conducted each time a new task is assigned. It is possible for a craftsman to complete or review two or more pre-task safety analysis forms on any given day.

Some firms use the TSA form for pre- and post-job safety briefings. In some cases, workers initial the completed form at the end of the day indicating the job was completed safely and without incident.

5.0.0 ◆ TASK-SAFETY ANALYSIS VS. JOB SAFETY ANALYSIS

TSA and JSA are both tools for identifying and controlling potential hazards.

Instructor's Notes:

PRE-JOB SAFETY BRIEFING / TASK HAZARD ANALYSIS

Task/work to be performed ____________________
____

Tools/equipment/materials involved
____
____

Physical Hazards

θ Falls on same elevation
θ Falls from elevation
θ Pinch points
θ Rotating/moving equipment
θ Electrical hazards
θ Hot/cold substances/surfaces
θ Strains/sprains/repetitive motion
θ Struck by falling/flying objects
θ Sharp objects

Hazardous Chemicals/Substances

θ Flammable materials
θ Reactive materials
θ Corrosive chemicals
θ Toxic chemicals
θ Oxidizers
θ Hazardous wastes
θ Biohazards
θ Radiation hazards
θ Other

Energy Sources

θ Where is the energy?
θ What is the magnitude of the energy?
θ What could happen or go wrong to release the energy?
θ How can it be eliminated or controlled?
θ What am I going to do to avoid contact?

PRE-JOB SAFETY BRIEFING / TASK HAZARD ANALYSIS

Permits Required

θ Welding & Burning
θ Lockout/Tagout
θ Excavation
θ Line Entry
θ Electrical Hot Work
θ Confined Space Entry
θ Critical Lift
θ Vehicle Entry
θ Other

PPE Requirements

θ Safety Glasses
θ Goggles
θ Face Shield
θ SCBA
θ Respirator
θ Gloves
θ Chemical Suit
θ Head Protection
θ Safety Shoes/Boots
θ Full Body Harness
θ Lanyard
θ Lifeline
θ Other
θ Special Precautions

Employee Signatures

____
____
____
____

208F04.TIF

Figure 4 ◆ Sample task safety analysis forms.

Provide the participants with a task scenario and a copy of a TSA. Have the participants complete the TSA based on the information in the scenario. Note the proficiency of each participant. This laboratory corresponds with Performance Task 2.

Explain the differences between a JSA and a TSA.

The three major differences are timing, level of detail, and use. TSAs are typically done prior to starting a task and do not include a high level of detail. JSAs are typically done in advance and contain significantly more detail. JSAs are more formal and include key job steps, while TSAs are more general and usually do not include each job step.

JSAs are typically done for recurring and/or repetitive tasks and are usually retained and referenced. The completed JSA often becomes the basis for standard operating or maintenance procedures. JSAs are often used as training tools while TSAs are primarily used to raise awareness and maintain focus. However, when performing a TSAs, one should ask, "Do we have any JSAs or

JSAs AND TSAs 8.7

Go to OSHA's Web site and download the *OSHA Hazard Awareness Advisor* (http://www.osha.gov/dts/osta/oshasoft/hazexp.html). Create several scenarios for the participants to use in entering data into the advisor. Discuss the outcome of each hazard analysis.

Review the material covered in the module.

Have the participants complete the Review Questions and go over the answers prior to administering the Module Examination.

Administer the Module Examination. Record the results on Craft Training Report Form 200 and submit the results to the Training Program Sponsor.

Administer the Performance Test and fill out Performance Profile Sheets for each participant. If desired, participant proficiency noted during laboratory sessions may be used to complete the Performance Test. Record the results on Craft Training Report Form 200 and submit the results to the Training Program Sponsor.

standard operating procedures that would cover the job at hand?" If so, the applicable JSA or procedure may need to be reviewed as part of the TSA.

Both processes can be helpful in accident prevention, but the intent and use of each must be addressed in company procedures and clearly understood by all. Both processes require training on hazard recognition, evaluation, and control, as well as how to perform and document the specific analysis.

Summary

JSAs can be very beneficial if they are performed correctly. They not only result in a safer worksite, they also increase productivity and reduce waste. Besides the obvious safety benefits, the data provided in a JSA can help streamline processes, develop and document work practices, and design training. To make sure that JSAs maintain impact, the highest priority jobs should be analyzed first. The results will have an immediate and powerful impact and provide the basis to use this important tool in creating and maintaining a safe work environment.

Review Questions

1. A job is composed of several tasks.
   a. True
   b. False

2. Each of the following is a benefit of doing a JSA *except* _____.
   a. increased risk awareness
   b. increased turnover
   c. reduced risk
   d. standardized work practices

3. JSAs help reduce risk by _____.
   a. reporting hazardous situations to OSHA
   b. revealing hazardous situations to the media
   c. identifying hazards and determining appropriate steps to reduce them
   d. identifying ways to conceal unsafe conditions

4. The research that goes into conducting a JSA is similar to the research necessary for developing training.
   a. True
   b. False

5. When selecting jobs for analysis, the highest priority should be given to jobs with the _____.
   a. largest number of affected workers
   b. highest potential for serious injuries
   c. greatest impact on productivity
   d. longest history of performance

6. Jobs with high turnover rates should be analyzed _____.
   a. to improve hiring techniques
   b. because unsafe conditions may be the cause of the high turnover rate
   c. because management attention will make the workers want to stay
   d. to determine if personality conflicts are the cause of high turnover

7. When preparing to do a JSA you should first _____.
   a. look at the general conditions under which the job is performed
   b. ask management what they would like you to find
   c. break the job down into as many tasks as possible
   d. determine which risks you'd like to find

8. Reviewing documentation _____.
   a. is the preferred method of data collection
   b. will typically provide all of the necessary job detail
   c. provides important background information
   d. should include reviewing employee medical records

9. One drawback to direct observation is _____.
   a. it does not provide the information about the conditions under which a job is performed
   b. the workers may find it distracting
   c. it takes two people to do it correctly
   d. it raises more questions than it answers

10. A common error when conducting a JSA is _____.
   a. spending time reviewing documentation
   b. placing too much emphasis on risky jobs
   c. calling in other experts to help develop solutions
   d. breaking the job into too many tasks

Instructor's Notes:

GLOSSARY

Trade Terms Introduced in This Module

Causal links: A relationship between two events or occurrences in which one is the cause of the other.

Ergonomics: The applied science of equipment design, as for the workplace, intended to maximize productivity by reducing operator fatigue and discomfort.

Job: A regular activity performed to achieve some end.

Pattern: An indication of how predictable the reoccurrence of an event is.

Structured on-the-job training (SOJT): Training that takes place during the performance of the *job* with the guidance of documentation or a facilitator.

Task: A discrete step or portion of a *job*.

Trends: The tendency to take a particular direction.

JSAs AND TSAs 8.9

Additional References

This module is intended to present thorough resources for task training. The following reference works are suggested for further study. These are optional materials for continued education rather than for task training.

www.osha.gov

www.asse.org

Job Hazard Analysis: A Guide to Identifying Risks in the Workplace, 2001. George Swartz. Rockville, MD: Government Institutes.

Figure Credit

Professional Safety Associates, Inc.	208F01

Instructor's Notes:

MODULE 75208-03 — ANSWERS TO REVIEW QUESTIONS

Answer		Section
1.	a	1.0.0
2.	b	2.1.0, 2.2.0, 2.3.0, 2.4.0
3.	c	2.1.0
4.	a	2.4.0
5.	b	3.1.0
6.	b	3.1.0
7.	a	3.2.0
8.	c	3.3.1
9.	b	3.3.2
10.	d	3.6.0

CONTREN™ LEARNING SERIES — USER UPDATES

The NCCER makes every effort to keep these textbooks up-to-date and free of technical errors. We appreciate your help in this process. If you have an idea for improving this textbook, or if you find an error, a typographical mistake, or an inaccuracy in NCCER's Contren™ textbooks, please write us, using this form or a photocopy. Be sure to include the exact module number, page number, a detailed description, and the correction, if applicable. Your input will be brought to the attention of the Technical Review Committee. Thank you for your assistance.

Instructors – If you found that additional materials were necessary in order to teach this module effectively, please let us know so that we may include them in the Equipment/Materials list in the Instructor's Guide.

Write: Curriculum Revision and Development Department
National Center for Construction Education and Research
P.O. Box 141104, Gainesville, FL 32614-1104

Fax: 352-334-0932

E-mail: curriculum@nccer.org

Craft ______ Module Name ______

Copyright Date ______ Module Number ______ Page Number(s) ______

Description ______

______

______

______

(Optional) Correction ______

______

______

(Optional) Your Name and Address ______

______

______

Module 75209-03

Safety Orientation and Training

NCCER STANDARDIZED CRAFT TRAINING PROGRAM

The National Center for Construction Education and Research (NCCER) provides a standardized national program of accredited craft training. Key features of the program include instructor certification, competency-based training, and performance testing. The program provides trainees, instructors, and companies with a standard form of recognition through a National Craft Training Registry. The program is described in full in the *Guidelines for Accreditation*, published by the NCCER. For more information on standardized craft training, contact the NCCER by writing us at P.O. Box 141104, Gainesville, FL 32614-1104; calling 352-334-0911; or e-mailing info@nccer.org. More information may be found at our Web site, www.nccer.org.

HOW TO USE THIS ANNOTATED INSTRUCTOR'S GUIDE

Each page presents two sections of information. The larger section displays each page exactly as it appears in the Participant Module. The narrow column ties suggested participant and instructor actions to each page and provides icons (detailed below) to call your attention to material, safety, audiovisual, or testing requirements. The bottom of each page includes space for your notes.

The **Audiovisual** icon indicates an appropriate time to show a transparency or other audiovisual aid.

The **Classroom** icon prompts you to define a term, stress a point, ask participants to explain a concept, or give examples.

The **Demonstration** icon directs you to show participants how to perform tasks.

The **Examination** icon tells you to administer the written module examination.

The **Homework** icon is placed where you may wish to assign reading for the next class, to assign a project, or to advise participants to prepare for an examination.

The **Laboratory** icon is used when participants are to practice performing tasks.

The **Materials** icon is a reminder for you to gather materials needed for classes, labs, and testing.

The **Performance Testing** icon tells you to administer a performance test or a portion thereof.

The **Safety** icon is used to emphasize safety issues. It is often keyed to *Caution* and *Warning* statements in the Participant Module.

The **Teaching Tip** icon indicates additional guidance is available, such as how to conduct an exercise, get the most educational value from a field trip, or encourage class participation. Teaching Tips may expand on a feature (*Think About It*, *Did You Know?*) or provide Quick Quizzes or similar exercises. You will be referred to the Teaching Tips section at the back of the module if there is additional material.

The **Combination** icon indicates that the laboratory listed corresponds with a performance task. If desired, you can note the proficiency of the participants during the laboratory and use it to satisfy performance testing requirements.

PREPARATION

Before teaching this module, you should review the Objectives, Performance Tasks, Materials and Equipment List, and the Module Outline. Be sure to allow ample time to prepare your own training or lesson plan and gather all required materials and equipment.

MODULE OVERVIEW

In this module, participants will learn how to coordinate and conduct effective training programs.

PREREQUISITES

Please refer to the Course Map in the Participant Module. Prior to training with this module, it is recommended that the participant shall have successfully completed the following:

Field Safety; Safety Technology, Modules 75201-03 through 75208-03

OBJECTIVES

Upon completion of this module, the participant will be able to:

1. Effectively train all employees on a job site about safety policies and procedures.
2. Coordinate safety training programs.

PERFORMANCE TASKS

There are no performance tasks for this module.

MATERIALS AND EQUIPMENT LIST

Overhead projector and screen
Transparencies
Whiteboard/chalkboard
Markers/chalk
Blank acetate sheets
Transparency pens
Pencils and scratch paper
Appropriate personal protective equipment
Copy of *29 CFR 1926, OSHA Construction Industry Regulations*
TV (optional)
VCR (optional)
The Video Arts video *You'll Soon Get The Hang Of It* (optional)
Module Examinations*

* Located in the Test Booklet.

SAFETY CONSIDERATIONS

Ensure that participants are equipped with appropriate personal protective equipment.

ADDITIONAL RESOURCES

This module is intended to present thorough resources for task training. The following reference works are suggested for both instructors and motivated participants interested in further study. These are optional materials for continued education rather than for task training.

Web site: www.osha.gov

Web site: www.asse.org

The Psychology of Safety Handbook, 2001. E. Scott Geller, Ph.D. Boca Raton, FL: CRC/Lewis Publishers.

The Participation Factor—How to Increase Involvement in Occupational Safety, 2002. E. Scott Geller, Ph.D. Des Plaines, IL: The American Society of Safety Engineers (ASSE).

TEACHING TIME FOR THIS MODULE

An outline for use in developing your lesson plan is presented below. Approximately 2½ hours are suggested to cover *Safety Orientation and Training*. This includes 10 minutes at the beginning of the session for administrative tasks and one 10-minute break. You may need to adjust the time required for hands-on activities and testing based on your class size and resources.

Topic	Planned Time
A. Introduction	________
B. Conducting Training	________
1. New Employee Orientation Training	________
2. Job-Specific Safety Training	________
3. Supervisory Training	________
4. Job Instructional Training (JIT)	________
C. Coordinating the Training Program	________
1. Identify and Invite the Correct People	________
2. Obtain and Secure a Classroom	________
3. Coordinate Room and Participant Availability	________
D. Methods of Training	________
1. Sequence the Training Logically	________
2. Know-Show-Do	________
3. What-Why-How	________
4. Hands-On Practice and Demonstration	________
5. Using Audiovisual Training Aids	________
E. Preparing to Train	________
1. Classroom Preparation	________
F. Delivering the Training	________
1. Course and Instructor Introduction	________
2. Class Administration	________
3. Ice-Breaking Activities	________
4. Review the Class Objectives	________
5. Classroom Management	________
6. Course Closure	________
G. Review	________
H. Module Examination	________

1. Participants must score 70% or higher to receive recognition from the NCCER.
2. Record the testing results on Craft Training Report Form 200 and submit the results to the Training Program Sponsor.

Module 75209-03

Safety Orientation and Training

Instructor's Notes:

COURSE MAP

This course map shows all of the modules in Safety Technology. The suggested training order begins at the bottom and proceeds up. The local Training Program Sponsor may adjust the training order.

Assign reading of Module 75209-03.

SAFETY TECHNOLOGY

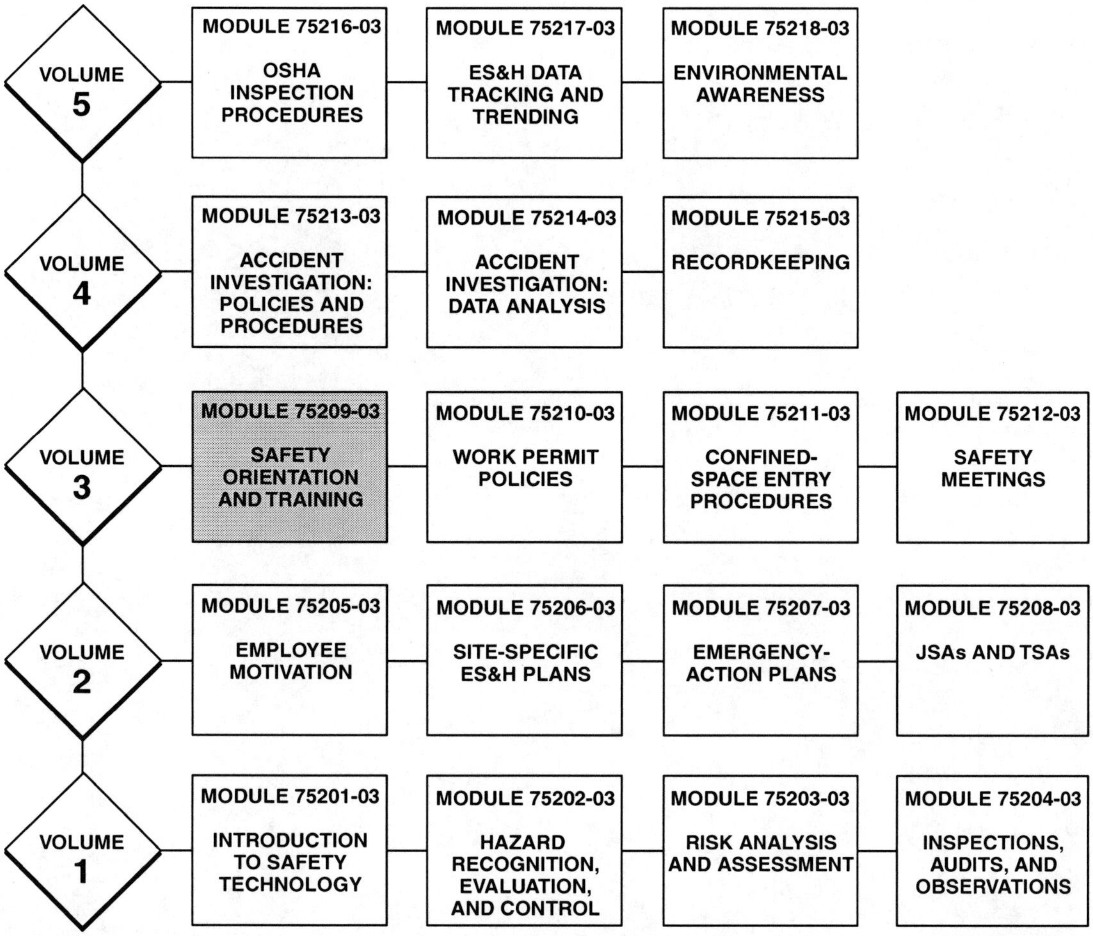

209CMAP.EPS

SAFETY ORIENTATION AND TRAINING 9.iii

Instructor's Notes:

MODULE 75209-03 CONTENTS

SAFETY ORIENTATION AND TRAINING 9.v

Figures

Instructor's Notes:

MODULE 75209-03

Safety Orientation and Training

Ensure that you have everything you need to teach the course. Check the Materials and Equipment list in the front of this module.

Explain that terms shown in bold are defined in the Glossary at the back of this module.

Review the goals of the course and explain what will be expected of the participant.

Objectives

When you have completed this module, you will be able to do the following:

1. Effectively train all employees on a job site about safety policies and procedures.
2. Coordinate safety training programs.

Prerequisites

Before you begin this module, it is recommended that you successfully complete the following: Field Safety; Safety Technology, Modules 75201-03 through 75208-03.

Required Materials

1. Pencil and paper
2. Appropriate personal protective equipment
3. Copy of *29 CFR 1926, OSHA Construction Industry Regulations*

Show Transparency 1, Objectives.

1.0.0 ◆ INTRODUCTION

In order to have a safe work site, you need to have a safe working environment, management that supports safety programs, and workers who are motivated to work safely and who have the knowledge and skills necessary to do so. Safety orientation and training help you ensure that the workforce is knowledgeable about safe work practices, company policies, and safety rules and regulations. Another way is through craft training. If workers know the proper policies and procedures for doing their job, they are already ahead of those who have not had any craft training, in terms of safety. In your role as safety technician, the responsibility for conducting and coordinating safety orientation and training will fall to you.

Explain the impact safety orientation and training have on an organization.

Describe the types of training that are typically done in the workplace.

2.0.0 ◆ CONDUCTING TRAINING

Safety training is one of the basic elements of a safety program. The intent of a safety training program is to affect workers' behavior by increasing their knowledge and/or improving their skills.

There are two kinds of training that occur in the workplace: informal and formal. Informal training is not structured or planned, but occurs when an opportunity for training avails itself. These opportunities may arise in safety meetings, in one-on-one conversations with workers, or as you observe the workers performing various job duties and tasks. Formal training is a planned and organized activity, prearranged with a group of workers to teach them a specific group of skills or give them specific knowledge. While informal training is an important tool, formal training programs will be the focus of this module.

Formal safety training can be broken down into three major categories:

- New employee orientation
- Job-specific safety training
- Supervisory training

> **NOTE**
> If you find that some workers have limited reading or English speaking skills, assign a helper or translator who speaks and reads English well to the worker.

SAFETY ORIENTATION AND TRAINING 9.1

Discuss when new employee orientation should take place.

Explain the topics that are covered in new employee orientation.

2.1.0 New Employee Orientation Training

When new employees are hired they need to receive safety orientation. This training comes in two forms: new employee orientation and site-specific orientation.

As new employees, workers receive a lot of new information all at the same time, so you should plan on breaking their training into two sections. The first section, new employee orientation, should be a general overview of company policies and safety rules. The second section, site-specific orientation, covers the high-risk areas or duties that exist on the job site, preventive measures to minimize the risk of accidents, and contingency plans in the case of an accident. In addition, you may want to cover subjects such as:

Show Transparency 2, Topics for New Employee Orientation.

- The company chain of command
- Where to go for help in case of an emergency
- A walk-through tour of the job site showing safety equipment and escape routes
- The use of standard personal protective equipment
- A description of any hazardous materials used on the site
- Emergency reporting and response procedures
- Where to find MSDSs

After the general safety orientation has been completed, employees will be ready for more specific safety training on the job for which they were hired.

> **NOTE**
> Follow-up training with new employees should be done within 30 to 60 days to ensure that workers have retained the information received during training and are doing the job safely.

Explain who should receive job-specific safety training.

Describe the topics that should be covered in job-specific safety training.

2.2.0 Job-Specific Safety Training

Job-specific safety training should be offered to any new employees, anyone acquiring a new job or job duties, or anyone whose safety practices need improvement. This training must cover:

Show Transparency 3, Topics for Job-Specific Safety Training.

- Specific and correct work procedures
- Care, use, maintenance, and limitations of all pieces of required personal protective equipment
- The risks associated with any harmful or hazardous materials used, including warning properties, symptoms of overexposure, first aid, and cleanup procedures for spills

Job-specific safety training should include as much demonstration and hands-on practice as possible.

> **NOTE**
> Emergency response personnel should be present when demonstrating the use of any hazardous material or action.

Explain OSHA's requirement regarding job-specific safety training.

OSHA 1926.21 requires that, "The employer shall instruct each employee in the recognition and avoidance of unsafe conditions and the regulations applicable to his work environment to control or eliminate any hazards or other exposures to illness or injury." This regulation is commonly known as the right-to-know rule. It requires employers to inform employees of how to recognize, avoid, and work safely around any and all hazards they will encounter on the job site. In addition to this general statement, OSHA specifies job-specific safety training for a wide variety of areas.

Describe the best practices for conducting supervisory training. Include the objectives that should be met during supervisory training.

2.3.0 Supervisory Training

Supervisors play a large role in ensuring a safe work place. To maximize their effectiveness, safety training should be provided for supervisors. Supervisory safety training should accomplish the following objectives:

- Reinforce the company's commitment to a safe work environment.
- Review in detail company safety policies and procedures.
- Explain how to analyze a job to look for potential health and safety risks.
- Explain how to conduct a safety meeting.
- Describe what to do if an unsafe condition exists.
- Describe the policy and procedure to follow after an accident.
- Explain how to coach, counsel, and discipline employees.
- Explain how to investigate and document an accident.

Supervisors have the ability to set the tone for their workers. A supervisor who looks out for workers' safety will help keep accidents and health risks to a minimum and reduce the impact of an accident by following procedures and plans. To do these things effectively, every supervisor should go through specific supervisor safety training.

Instructor's Notes:

2.4.0 Job Instructional Training (JIT)

A large portion of construction safety training is done in the field by a crew leader or supervisor. One technique for on-the-job training is Job Instructional Training (JIT). JIT is particularly helpful for training workers how to do a particular task. The following steps are used to prepare for and present JIT:

Step 1 Prepare the participants by using these techniques.

- Put them at ease.
- Explain the job in detail.
- Define the quality standards.
- Explain the importance of doing the job properly.

Step 2 Present the operation.

- Demonstrate each step of the job.
- Explain why the sequence is important.
- Encourage questions.
- Stress key points.

Step 3 Try out performance.

- Watch the participant do the job.
- Have the participant explain each step.
- Be patient.
- Try to anticipate questions.

Step 4 Follow up.

- Make observations after training.
- Give praise for proper activities.
- Give suggestions when necessary.
- Encourage questions.

3.0.0 ◆ COORDINATING THE TRAINING PROGRAM

Coordinating the training means inviting the right people to be participants, getting a classroom that can accommodate all of the people, and making sure that the room, any necessary equipment, and all of the participants are available at the same time.

3.1.0 Identify and Invite the Correct People

Talk to the crew supervisor or foreman and work out together who should be invited to the class. If there are a large number of participants, you may want to divide them up into more than one class.

It is important to establish entrance requirements, because you want to be sure that everyone attending a class has received the necessary **prerequisite training**. If you are teaching a confined-space course, for example, you may want to require that all persons taking the course have received prior training in selecting and using respirators. If you have to stop the class to instruct one or two people on respirators, the rest of the class may not receive all of the training they need because the class will take longer than planned.

The required prerequisite training should be included in your course announcement, and you should verify that all entrants have received that training. If some haven't, you should consider providing them with the required training before the full class starts. It is better to do that than to deny an applicant who may need the course.

3.2.0 Obtain and Secure a Classroom

Even if the majority of your class is going to be a hands-on course with equipment, you will still need a classroom area to conduct your class, especially if you plan to use transparencies, flipcharts, or a whiteboard. Work with the administrative staff, facilities crew, or supervisor to obtain and secure classroom space as close as possible to the equipment on which you will be training. Some facilities have classrooms or conference rooms available; occasionally training must be conducted in a lunchroom, break room, or office. It is important to make sure that the area you have for a classroom is going to be large enough to accommodate all of the participants, that it is as comfortable as possible, and that it is as free of distractions as possible.

3.3.0 Coordinate Room and Participant Availability

Perhaps the trickiest part of setting up a training class is trying to coordinate your schedule with the schedules of each of the participants, and then finding an available room for just that time. You may need to offer the class more than once to accommodate different schedules or shifts. While this increases the number of times you have to teach the class, it does make the classroom coordination much easier.

4.0.0 ◆ METHODS OF TRAINING

There is more to effective training than simply telling a room full of people what you know about the subject matter or talking about one transparency after another. Participants generally retain:

- 10% of what they read
- 20% of what they hear
- 30% of what they see
- 50% of what they see and hear
- 70% of what they see, hear, and respond to
- 90% of what they see, hear, and do

Explain the purpose and use of Job Instructional Training (JIT).

Discuss the four steps used to prepare for and present JITs.

Show Transparency 4, Steps for Preparing and Presenting JITs.

Describe how a successful training program is coordinated.

Explain how to identify and select an appropriate audience, including the development of prerequisites.

Identify the resources that should be used to obtain and secure a classroom.

Discuss circumstances in which a location other than a classroom would be more appropriate for conducting training.

Explain how to schedule training so that participants are available to attend and classroom space is available.

Identify the most effective methods of conducting training.

Show Transparency 5, Rates of Retention.

SAFETY ORIENTATION AND TRAINING 9.3

Describe how to sequence the training logically. Give an example of how to logically organize material using the topic and subtopic approach.

Explain that the know-show-do method is most effective for procedural training. Give an example of how to use this method.

Explain that the what-why-how method is most effective for training that requires the participant to have background knowledge and information about a task or job. Give an example of how to use this method.

Explain that the hands-on practice and demonstration method is the best way to teach workers how to operate equipment, perform tasks, or follow procedures. Describe how this method is typically used.

Therefore, your challenge as an instructor is to engage the participants in as many ways as possible. The following paragraphs describe strategies for doing that.

> **DID YOU KNOW?**
>
> *Construction Accidents*
>
> The most recent Construction User's Round Table A-3 Report indicates that 25% of all construction accidents involve personnel who have been on the job less than 30 days.

4.1.0 Sequence the Training Logically

Training sessions often present a lot of information over a relatively short period of time. The best way to prevent the information from becoming overwhelming is to teach the material logically, starting with what the participants already know and using that to lead them into the new material. For example, to begin a conversation about the Fire Triangle, start off with a question asking what kinds of flammable materials the participants see in the room. After the participants have listed some flammable items, you can begin talking about how those items might start burning and what is needed to start, maintain, and put out a fire. Starting with a piece of information that is common knowledge allows the workers to put the new information into a useful context.

4.2.0 Know-Show-Do

Know-show-do is a very effective method for teaching procedures or tasks. To employ the know-show-do strategy, you first provide the participants with the information that they need to know in order to perform the task. For example, if you are teaching them how to select and use a fire extinguisher, you will want the participants to know where fire extinguishers are located, the symbols and codes used to indicate the types of fires on which an extinguisher will work, and the parts of an extinguisher. After the participants have the necessary background information, show them how to operate an extinguisher. The demonstration may be most effective if you start a small, controlled fire outside in a safe area and then immediately put it out with a fire extinguisher. Finally, have the participants use a fire extinguisher several times until they have mastered the procedure. Again, if it can be done safely, you may have them practice extinguishing small controlled fires.

> **WARNING!**
>
> Only use demonstrations involving fire or other hazards if you have all the appropriate personal protective equipment and other necessary gear available. Ideally you should have emergency-response team members on hand.

4.3.0 What-Why-How

What-why-how is a great way to break down factual information for a class. Use these three key questions consistently as you describe safety rules, equipment operation, and use of personal protective equipment. For example, you might use the what-why-how method to teach the use of hearing protection by first showing and discussing the different types of hearing protection devices. After thoroughly reviewing what equipment is used to protect one's hearing, discuss why it is important to use hearing protection equipment. Finally, teach the workers how to use hearing protective equipment. The "how to" portion of the lesson would probably lead to a more detailed session using the know-show-do method to teach the participants when and how to select and use hearing protection.

4.4.0 Hands-On Practice and Demonstration

Hands-on practice and demonstrations are the best way to teach people how to operate equipment, perform tasks, or follow procedures. The most effective training course will dedicate a large portion of its time to allowing the participants to use the actual equipment or practice the desired procedure. At this time, you should assess the ability of each participant to accomplish the task.

For one reason or another, there may be a situation where demonstrating or allowing hands-on practice on the actual equipment is not practical. For example, it may be too risky or costly to have new forklift drivers begin by operating a forklift. In this situation, the training should include other training strategies such as **simulators**, videos, or models. In some instances, a videotape of the procedure can be used to train employees on how to do something, although this is much less effective than hands-on practice.

4.5.0 Using Audiovisual Training Aids

Proper use of audiovisual equipment can greatly improve the effectiveness of the class. When

Instructor's Notes:

preparing to teach, you should plan on using some audiovisuals to help the participants learn. These may include transparencies projected by an overhead projector, a computer presentation displayed by a projector, flipcharts, posters, whiteboards or chalkboards, videos, models, and samples. Each of these can be used to great effect. However, overuse or misuse of audiovisuals can detract from the training by being distracting or overwhelming to the participants. *Figure 1* shows a laptop and projector. *Figure 2* shows an example of a portable screen.

Table 1 lists some common audiovisual aids, their uses, and their advantages and disadvantages. This information is given to help you decide when and how to use some types of audiovisual equipment.

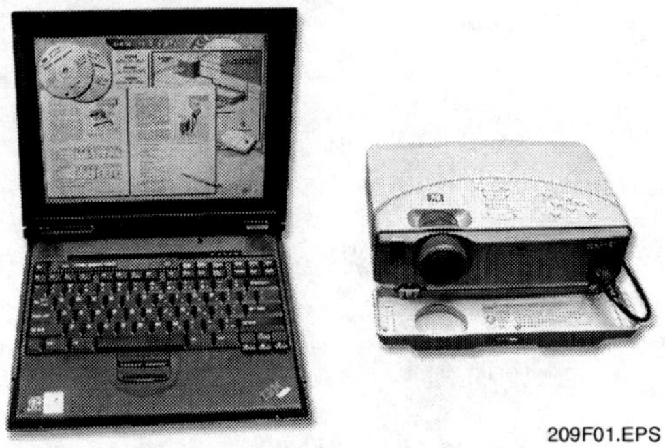

209F01.EPS

Figure 1 ◆ Laptop and projector.

209F02.EPS

Figure 2 ◆ Portable screen.

Audiovisuals are tools used to enhance your instruction, but should not be relied upon to carry the training for you. As a rule of thumb, you should be able to conduct an effective course even if you lose or can't use the audiovisuals you've prepared.

Explain the importance of using audiovisual training aids. Identify commonly used audiovisual training aids. Refer the participants to Figure 2 in the Participant Module.

5.0.0 ◆ PREPARING TO TRAIN

As the instructor, the class participants will see you as an expert. While this does not necessarily mean that you must know everything about the topic, you should be sufficiently prepared to answer most questions that arise and know where to look up the answers to questions for which you don't know the answer. Therefore you must prepare yourself in advance of conducting the course, rather than relying on prepared materials and your general knowledge to carry you through.

Show Transparency 6 (Table 1).

When preparing to conduct a training course, make sure that you are well versed and familiar with the equipment, procedures, or tasks on which you will be training. To do this, you should study any available documentation and regulations, practice the procedures and tasks yourself, and talk to the workers and their supervisors about their training needs and expectations.

Explain the techniques instructors commonly use to prepare for training.

To begin preparing yourself for the training class, obtain and study copies of all relevant documents. This may include as-built drawings of equipment, maintenance manuals, operation manuals, blueprints, company policies, OSHA regulations, or ANSI standards. This documentation will often form the foundation for the training, as well as provide valuable content.

You should practice any procedures and tasks to be taught to make sure that you know how to safely and correctly perform them. For example, if you need to teach a group of workers how to set up and use a new band saw, you should practice with the saw beforehand to make sure that you can accurately demonstrate the correct procedure.

Also while practicing, prepare your class notes on what you will say about the procedure and each step. Plan how you will teach key points such as setup, operation, troubleshooting, and repair/replace procedures. Make notes so that you will remember what you want to say when you are teaching the class.

Make sure that you talk to the workers and their supervisors about the training. You will want to know what kind of personnel will be in the class, what their roles are, and how much experience they have. This information will help you to tailor the information to meet the needs of the workers and the company.

SAFETY ORIENTATION AND TRAINING 9.5

Discuss the importance of arriving well enough in advance to prepare the classroom for training. Explain that this includes setting up and testing audiovisual equipment, arranging classroom seating, and setting up participant materials.

Table 1 Common Uses of Visual Aids

Audiovisual Aid	Use	Advantages	Disadvantages	Requirements
Overhead transparencies	Used to show non-moving (static) text and graphics to a large group.	You can annotate information on the transparency being shown to clarify a point in real time.	Many instructors rely on transparencies to guide the entire flow of instruction. Overuse can become tiresome and distracting to the trainees.	A set of transparencies or blank sheets of acetate. Electricity to run the projector. A screen or blank wall in an area big enough to project the image.
Computer and projector	Used to show both static and moving (dynamic) text and graphics to a large group.	Presentation can provide a mixture of animations, graphics, text, and sound. This often makes training more effective, since it will appeal to the learners on many different levels.	Can be overused at the expense of other important teaching strategies, like hands-on training.	Programmed material created in advance. Electricity to run the projector and computer. A screen or blank wall in an area big enough to project the image.
Video/DVD	Used to show presentations incorporating both static and dynamic images to a group.	Excellent for showing procedural tasks being peformed. Real reception.	Can be overused at the expense of other important teaching strategies. Unless specifically filmed locally in support of your instruction, videos may not accurately address the specific equipment being covered in the class.	Programmed video cassette or DVD. VCR or DVD player and television set. Electricity to power the player and television.
Flipcharts	Used to present or capture information so that it can be presented to a large group and preserved for later reference.	Flipcharts provide a useful means for capturing ideas from brainstorming sessions and group discussions. Flipchart pages can be posted in the classroom and saved for later reference.	Can be hard to read if writing is illegible or too small. It takes some practice to summarize and annotate information in real time.	Pads of flipchart paper, flipchart stands, large markers, and masking tape.
Whiteboard/ chalkboard	Used to present or capture information, illustrate a principle, or work out a problem with a large group.	Good for capturing ideas from brainstorming sessions or group discussions. Transparencies of schematics, drawings, or prints can be projected onto a whiteboard so that circuit paths can be traced or key points can be highlighted.	Whiteboards/ chalkboards can be hard to read if writing is illegible or too small. It takes some practice to summarize and annotate information in real time. Its use is not practical if it is desired to keep the recorded information for future reference.	Whiteboard/chalkboard, dry erase markers/chalk, and erasers.

209T01.EPS

5.1.0 Classroom Preparation

You should arrive at the training site well enough in advance to set up your classroom. You will need to set up an instructor area, arrange the classroom tables and chairs, set up your audiovisual equipment, and set out the participant materials.

Set up your training materials so that you can run the class smoothly and without a lot of pauses. You must set up the area from which you will do most of your speaking. You will need to keep your notes handy, along with any frequently used materials like manuals, markers, or a

Instructor's Notes:

pointer. If space permits, you should set up an area to stage your supplies and materials. This may include any models or samples, extra participant materials, pens, pencils, pads, markers, and reference materials. These should be kept off to the side where you can get to them easily, but not where they will be in the way or distracting.

Arrange the participant tables and chairs the way you want them. Typically, if there is enough space, the tables are pulled together into a U-shape with the participants sitting around the outer perimeter. This arrangement allows for better interaction between the participants and the instructor. *Figure 3* shows a typical U-shaped seating strategy.

If you are using any audiovisual equipment, you should set that up in advance as well. Any projectors should be set up, plugged in, and tested. If you are going to use a laptop to show a computer-driven slide presentation, make sure that you boot up the computer and get the presentation running. It can be very embarrassing and time wasting to arrive in a classroom to find that a projector bulb is burned out or that your laptop isn't able to drive the multimedia projector.

If you are planning to use audio, make sure the speakers provide enough volume for the classroom. Laptop speakers are generally audible only to people within a couple of feet of the laptop. If you plan on using flipcharts, you should make sure that you have enough flipchart stands and markers.

Finally, set up the participant materials. At a minimum, the participants should receive a notepad, pencil, and pen. Typically, participants receive copies of handouts and manuals. You may also want to distribute tent cards or name tags.

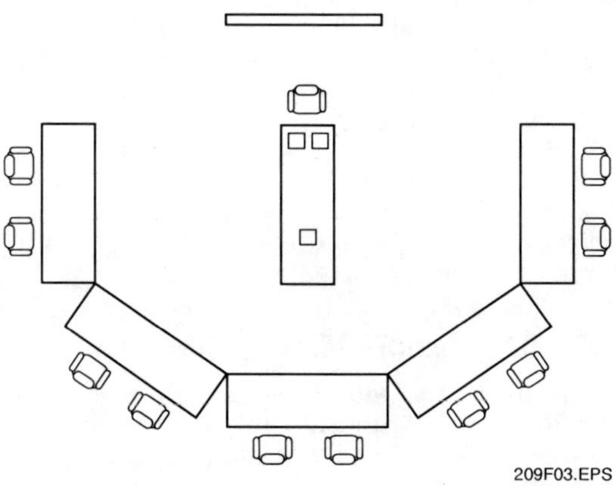

Figure 3 ◆ U-shaped seating.

These will be useful to you as you conduct the training so that you can address each participant by his or her name. You may want to set each participant's area up with all of the class materials or you may want to give them just the notebook, tent card, pencil, and pen initially, and then distribute the other class materials as you get to them during the training.

Arriving at the training site early to prepare the classroom and set up the materials and audiovisual equipment is essential to having a well-organized and successful training session. By getting all of the materials organized and the audiovisual equipment set up ahead of time you won't have to think about those details as you conduct the class. This will allow you to focus on conducting the training.

6.0.0 ◆ DELIVERING THE TRAINING

After adequate preparation and planning, you are now ready to conduct the training session. Even though conducting a training session may seem like a daunting or frightening situation, with proper preparation the class will run smoothly. As you get more experience conducting training courses, you will be more comfortable in front of a class.

6.1.0 Course and Instructor Introduction

It is important to set the proper tone in class right from the opening. Practice handling the opening a few times before the class so that it runs as smoothly and efficiently as possible. This will help build your credibility from the outset. Your opening should include the course and instructor introductions, class administration, a class icebreaker, and a review of the class objectives.

As the class starting time approaches, stand at the doorway to greet the participants as they arrive. If they are unfamiliar with the training site show them where the restrooms are, where to keep their jackets, and where to get coffee, and then invite them to find their seats.

When you begin the class, introduce yourself. Tell the class your name, your company name, your job title, and your background. Even if you know some or all of the class participants, it is important to reinforce your credibility and let them know why you are up in front of the classroom.

After you have introduced yourself, introduce the course. Tell the participants the name of the course or the equipment that you will be talking about. Make sure that everyone in the room is in the correct class.

Explain how to arrange classroom seating for better interaction between participants and instructors.

Describe the best techniques for instructors to use in introducing themselves and the topic. Show the participants some of these techniques.

SAFETY ORIENTATION AND TRAINING 9.7

Explain how to take care of the administrative aspects of training such as attendance and classroom rules.

6.2.0 Class Administration

Some key points of information should be addressed to help the participants feel more comfortable in the classroom setting. These include the daily start and end times and frequency and length of breaks. The participants will appreciate knowing the course schedule. It is important to adhere to the schedule you present; otherwise, you will quickly lose credibility. You should also review bathroom locations, refreshment availability and location, and smoking locations. Explain the site emergency reporting and response procedures and where to go in the event of an emergency. Gauge how much information you give the students based on their familiarity with the training site.

At this point, explain the ground rules of the training and set the participants' expectations for the course. For example, if you would like to encourage questions and discussions throughout the delivery of the course you should set that ground rule now. If you want to discourage cell phone or other interruptions, request that participants forward their calls or turn off their cell phones. If you plan on spending most of the course time doing hands-on activities with equipment, that should be mentioned at this point. Basically what you're doing is anticipating and answering the questions that most participants have as they walk into a new class so that they can shift their focus from those distractions to the course content.

Describe how icebreakers are an effective technique to help participants feel more comfortable. Give examples of icebreakers that you have used.

6.3.0 Ice-Breaking Activities

A training class can feel like an artificial grouping of people who do not often work or associate with each other. In some cases, participants may not know others in the class. Other times, everyone in the class will be on the same crew or shift. Regardless, they are not typically used to being in a training class together. Ice-breaking activities begin to overcome these barriers.

In its simplest form, the icebreaker can be as simple as going around the room and having everyone introduce themselves and give their titles and their years of experience. While this often suffices, some trainers like to enhance the icebreaker and speed up the process of familiarizing the class members with each other by adding other details into the introduction. An excellent question to ask everyone in the icebreaker is what they hope or expect to get out of the course. If you use this question, you may want to write the responses down on a sheet of flipchart paper and post it in the classroom after the icebreaker is completed.

One other technique that is often used in an ice breaker activity is pairing off the participants, having them ask each other the questions you prescribe, and then going around the room and having everyone take a turn introducing the person with whom they were just paired. This technique is useful in breaking down barriers such as office personnel/maintenance or operations personnel or first shift /second shift. These relationships are very real parts of the everyday workplace. Any steps you can take to remove those barriers from the classroom will make your job as an instructor much easier.

Explain that instructors should use good judgment when using icebreakers so that no one in the audience is alienated.

Be careful with ice-breaking activities. Some participants may find certain ice-breaking activities embarrassing. Make sure that you know your audience well before asking them to do anything unusual or potentially embarrassing. There is no faster way to turn off a participant or an entire class than to force them to do something in an icebreaker that they would rather not do. If someone declines your request in an icebreaker, respect their choice and follow up with them one-on-one later to make sure that everything is all right.

Define the term learning objective. Explain the different types of learning objectives.

Show the participants how to use to the ABCD Method to write learning objectives. Have the participants practice writing learning objectives.

Show Transparency 7 (Figure 4).

6.4.0 Review the Class Objectives

Before jumping into the course content, it is important to review the course objectives with the class. A learning objective is a statement of what the learners will be expected to do once they have completed a specified course of instruction. It explains the conditions, behavior (action), and standard of task performance for the training setting. The objective is sometimes referred to as a performance or behavioral objective. For example, knowledge is a state of mind that cannot be directly measured. This requires an indirect method of evaluation, that of observing behavior or performance.

Objectives should be very simple sentences that state what the participants will be able to do once they have completed the training. This foreknowledge makes the training much more effective because it helps the participants to prepare themselves for what they are about to learn. An effective tool to use when writing objectives is the **A-B-C-D method**, illustrated in *Figure 4*. The A-B-C-D method helps you to make sure that each objective states the audience, behavior, conditions, and degree of acceptance.

6.5.0 Classroom Management

As the instructor, you are responsible for making sure that the class runs smoothly, on schedule, and meets its objectives. This task, known as classroom management, will often take care of itself if

Instructor's Notes:

Describe the elements of classroom management that keep training organized and on-track: pace of instruction, adherence to the schedule, and handing difficult situations.

Explain how to monitor participant comfort. Give examples of the questions you use.

A = **A**UDIENCE	NAME WHO YOU WILL BE TRAINING.	NEW EMPLOYEES, FORKLIFT DRIVERS
B = **B**EHAVIOR	DESCRIBE WHAT THE TRAINEES WILL BE ABLE TO DO.	PUT ON PERSONAL PROTECTIVE EQUIPMENT CORRECTLY, SAFELY REMOVE PALLETIZED LOADS FROM A TRUCK, SET THE GUIDES AND SAFETIES ON A TABLE SAW.
C = **C**ONDITION	DESCRIBE THE CONDITIONS UNDER WHICH THE TRAINEES WILL BE EXPECTED TO PERFORM.	WITHOUT GUIDANCE, GIVEN THE OPERATION MANUAL
D = **D**EGREE	DESCRIBE THE MINIMAL ACCEPTABLE CRITERIA FOR SUCCESS.	100% OF THE TIME

NEW EMPLOYEES WILL BE ABLE TO PUT ON PERSONAL PROTECTIVE EQUIPMENT CORRECTLY, WITHOUT GUIDANCE, 100% OF THE TIME.

FORKLIFT DRIVERS WILL BE ABLE TO SAFELY REMOVE PALLETIZED LOADS FROM A TRUCK, WITHOUT GUIDANCE, 100% OF THE TIME.

CARPENTERS WILL BE ABLE TO SET THE GUIDES AND SAFETIES ON A TABLE SAW, GIVEN THE OPERATOR'S MANUAL, 100% OF THE TIME.

209F04.EPS

Figure 4 ◆ A-B-C-D method.

you have adequately prepared, organized, and outlined your class. Classroom management typically covers three elements:

- The pace of instruction
- Adherence to the schedule
- Handling difficult situations

6.5.1 The Pace of Instruction

It is the instructor's responsibility to make sure that the classroom pace is neither too fast nor too slow. If content is covered too quickly, the participants will feel frustrated and they will not get as much value out of the training as they could. If content is covered too slowly, the participants will lose patience and their minds will begin to wander. Monitor the participants' comfort with the pacing by asking them questions. Use a mix of direct questions about the pacing of the course, for example, "Are we going too fast?" or "Did everyone get a chance to try that?" and questions that test for understanding like, "Who can tell me why we use a hearing protection?" or "Where did we say you could find that procedure in the manual?"

6.5.2 Adherence to the Schedule

It is very important to stick to the schedule you have set up. Begin the class at the designated time and do not wait for stragglers when class reconvenes after a break or lunch. Likewise, give breaks

SAFETY ORIENTATION AND TRAINING 9.9

Explain the importance of adhering to the class schedule. Give the participants tips and ideas for doing so.

Explain the techniques that can be used to handle rude participants or awkward situations.

Describe the reasons why some participants are disruptive.

Explain how to bring closure to the course.

Describe how to assess the performance of class participants.

Explain that remedial training may be needed for some participants. Discuss how to work closely with first-line supervisors to determine if remedial training is necessary.

Describe how to use course objectives and icebreakers to evaluate training.

Explain how participant evaluation forms are used to evaluate both the instructor and training content.

Show Transparency 8 (Figure 5).

and lunch at the scheduled times and end class for the day at the scheduled time. If you find that you need to move too quickly through the content to accomplish everything you want to in the scheduled time, consider revising the way you conduct the training. Perhaps you are going into too much detail in the explanation and not leaving enough time for demonstration or practice. Also consider adding more time to the course.

6.5.3 Handling Difficult Situations

Occasionally, you may have to deal with difficult or rude people or awkward situations. The general rule of thumb is to defuse the situation as quickly and subtly as possible during the class, then address it directly with the individual or individuals in a private conversation during a break. You will find that most of the time people do not intend to be rude or disruptive. After learning that their behavior is detracting from the class, they are likely to improve their performance. Unfortunately, you will occasionally have people in a class who are going to be a problem regardless of what you say or do. Some reasons for these problems include:

- The participant was forced to attend a class they did not want to go to.
- The participant does not get along with others in the class.
- The participant feels that he or she knows as much or more than you do about the topic.
- The participant is so overwhelmed by the content that he or she begins to feel frustrated and angry.

To resolve these cases, first try a private conversation with the participant or participants involved. Then, if you do not see any improvement, you may ask the participant to leave the class. Since your primary responsibility to is to make sure that the class benefits each participant, you may be forced to ask someone to leave so that class may continue in a positive manner. If this does happen, make the request in private during a break so that the person may gather their materials and leave without further disrupting the class.

6.6.0 Course Closure

When you have completed delivering the course content and everyone has had enough time to practice the procedures and tasks covered during the course, there are a few things that need to be done to close out the course. They include:

- Assessing the participants' ability to perform the required tasks
- Reviewing the course objectives and participant goals
- Having the participants evaluate the course
- Verifying attendance and handing out certificates of completion

6.6.1 Performance Assessment

Throughout the demonstrations and practice exercises of the course, you must monitor the participants' performance and offer coaching and feedback to help ensure that each participant ends the class able to perform tasks at least a minimally acceptable level. By the end of the class, you should feel confident that everyone meets a minimum standard of proficiency. If some participants do not meet this minimum acceptable level, you should let their supervisor know this. Their supervisor may want to spend some time with them afterward to help them out with the content, or you may be asked to come back and tutor the participants to help them attain proficiency. The participants may need to attend **remedial training** and then re-attend the course in question. If none of these strategies works, the employee may need to be reassigned or dismissed.

6.6.2 Reviewing Course Objectives and Participant Goals

One of the best ways to end the class is to review the class objectives that you listed in the course opening and to briefly review when and how each was taught. It is also useful for the instructor to review the participants' objectives to verify the course objectives were met. This enables the instructor to evaluate the training method through feedback.

Another way to evaluate course objectives and goals is by using the icebreaker from the beginning of the course. If in the icebreaker activity you asked the participants to list something that they hoped to get out of the class, it is also a very good idea to review those goals and discuss how they were met during the class. These activities help participants feel satisfied with what they have accomplished by the end of the class.

6.6.3 Participant Evaluation of the Course

At the conclusion of the course, you should ask the participants for their opinions of the course. This can be done informally as a large group activity in which you simply ask the participants how they felt about the course content, pacing, and level of activity. Typically, this is handled by passing out an evaluation form. *Figure 5* shows a sample of an evaluation form. The form is helpful because it allows

Instructor's Notes:

PRESENTATION EVALUATION GUIDE – PAGE 1

Instructor ____________________ Name (Optional) ____________________

Date ____________________

	Poor 1	2	Average 3	4	Excellent 5	Not Applicable N/A
Introduction						
Title of lesson stated or displayed?						
Objectives stated?						
Motivation established?						
Lesson overviewed?						
Presentation						
Appropriate information level?						
Objectives covered?						
Presentation follows a logical sequence?						
Visual Aids						
Properly used?						
Illustrate the point?						
Visible by all?						
Out of sight when not in use?						
Used methods suggested in Guide?						
Questioning Techniques						
Different types asked?						
Sufficient number asked?						
Focused attention?						
Created discussion?						

209F05A.EPS

Figure 5 ◆ Evaluation form. (1 of 2)

Go to the following Web site to get more information about designing and developing training programs: www.nwlink.com/~donclark/.

SAFETY ORIENTATION AND TRAINING 9.11

The Video Arts video, *You'll Soon Get The Hang Of It*, can be used to enhance the material presented in this module.

PRESENTATION EVALUATION GUIDE – PAGE 2

Instructor ____________________ Name (Optional) ____________________

Date ____________________

	Poor 1	2	Average 3	4	Excellent 5	Not Applicable N/A
Questioning *(continued)*						
Related information to applications?						
Adapted to level of participants?						
Summary						
Reviewed key points?						
Pointed out benefits to participants?						
Instructor Qualities						
Gestures & Mannerisms						
Eye contact						
Knowledge						
Voice						
Professional attitude						
Enthusiasm for subject						
Overall Comments						
Completed all material?						
Used Task Module properly?						
Followed all safety procedures?						

Additional Comments

What did you like best in the way this lesson was presented?

What one aspect of the presentation would you change and how would you present it differently?

209F05B.EPS

Figure 5 ◆ Evaluation form. (2 of 2)

Instructor's Notes:

____

____

____

each participant to anonymously voice his or her opinions on the course's content, flow, sequence, pacing, and level of activity. The evaluation is useful because it will provide valuable feedback so you can adjust and revise the training for the next class.

6.6.4 Concluding the Training

When conducting safety training, you will have to verify satisfactory course completion for each participant on the sign-in sheet or attendance list. It is important that the record of course completion be entered into their human resource records to prove compliance with OSHA and company standards and policies. You may want to have personalized certificates of completion printed for each class participant. Handing out certificates symbolizes an accomplishment and is a great way to conclude the class.

Summary

In order to work safely and follow correct procedures, workers need to have the necessary skills and knowledge. The most effective way to make sure that everyone has the skills and knowledge necessary is to conduct safety training. To conduct effective safety training courses, you must prepare yourself, prepare your classroom and materials, and then follow your preparation and planning to provide the best training course. If the participants leave the training with a sense of confidence that they can perform safely on the job, then the training was successful.

Review Questions

1. Training is one of the basic elements of a safety program.

a. True
b. False

2. Formal training is _____.

a. less effective than informal training
b. a planned and organized activity
c. may occur spontaneously when the opportunity arises
d. only required for new employees

3. New employee orientation training _____.

a. should cover job-specific safety procedures
b. is not necessary if the new employee is experienced
c. should include a walk-through tour of the work site
d. should be optional

4. Job-specific safety training _____.

a. should be given before site-specific orientation
b. can be very general
c. should include many opportunities for hands-on practice
d. should be used as punishment for workers who cause accidents

5. Supervisory training _____.

a. is mandated by OSHA
b. should reinforce the company's commitment to safety
c. should focus on writing accident reports
d. has little impact on worksite safety

6. The best way to teach people how to perform a new task is to _____.

a. show them a video of the task being performed
b. explain what the task is, why it is done, and how to do it
c. allow for as much hands-on practice as possible
d. review the process in the operation manual

7. Audiovisual equipment should be used to _____.

a. improve the effectiveness of the class
b. keep the course on track
c. organize the presentation
d. present all of the course content

8. When preparing to teach a course, you should _____.

a. create a course based solely on the supervisor's requests
b. rely totally on manuals and documentation
c. practice any procedures and tasks that you will be teaching
d. quickly review drawings and prints right before class starts

9. While organizing the classroom is a nice thing to do, it has little to no effect on the outcome of the training.

a. True
b. False

10. The first thing instructors should do at the opening of the class is _____.

a. run an ice-breaker activity
b. take attendance
c. ask everyone to turn off their cell phones and pagers
d. introduce the course and themselves

Discuss the importance of documenting training. Explain that the documentation may be used either by management and/or OSHA to verify training.

Review the material covered in the module.

Have the participants complete the Review Questions and go over the answers prior to administering the Module Examination.

Administer the Module Examination. Record the results on Craft Training Report Form 200 and submit the results to the Training Program Sponsor.

PROFILE IN SUCCESS

Steven P. Pereira, CSP, Professional Safety Associates, Inc.

President/Consultant

How did you choose a career in the Safety Industry?
Having worked as a petrochemical plant operator, I had an opportunity to work in the insurance industry as a safety and loss-control representative. The company provided me with a considerable amount of in-house safety training, which ultimately lead me to a career in safety.

What types of training have you been through?
I have a B.S. Degree in Industrial Technology and a Master's Degree in Construction Science and Management. I have completed the NCCER Master Trainer Course and the OSHA 500 Course. I am an MSHA Instructor Trainer. Throughout my career I have completed numerous short courses and workshops on such topics as fire protection, crane safety, industrial hygiene, and photography.

What kinds of work have you done in your career?
I have been an insurance company loss-control representative, owner's site-safety representative on a large owner-controlled insurance package (OCIP) project, and a plant safety engineer for a petrochemical facility.

Tell us about your present job.
As a Certified Safety Professional (CSP), I serve as a safety, training, and loss prevention consultant to contractors, industrial manufacturing firms, and governmental agencies. I conduct training for the American Society of Safety Engineers and serve as an Instructor for the NCCER Project Supervisors, Project Managers, and Safety Management Academies held in conjunction with Clemson University.

What factors have contributed most to your success?
A willingness to learn and keep an open mind have helped me greatly. In addition, I worked very hard to develop good people and communication skills and technical writing capabilities.

What advice would you give to those new to safety industry?
Develop your people and management skills. Take advantage of every training opportunity offered. Read and learn as much as you can to stay current and technically competent. Keep an open mind and respect others' opinions. Be flexible, but do not compromise your principles.

Instructor's Notes:

GLOSSARY

Trade Terms Introduced in This Module

Know-show-do: A method of teaching tasks or procedures in which the instructor first explains necessary background information, then demonstrates the task or procedure, and concludes by having the participants practice the task or procedure.

What-why-how: A teaching method in which the instructor describes what an object, task, or procedure is, explains why it is used or done, and finally describes how to use or do it.

Simulators: Teaching aids that are used to simulate a piece of equipment for training purposes rather than conducting training on the actual equipment.

A-B-C-D method: A device used to write training objectives that specifically cover the audience, behavior, conditions, and minimum degree of mastery.

Remedial training: Training that is designed to address specific pieces of a larger course to allow participants to learn more about and practice areas in which they experienced difficulties.

Prerequisite training: Training that provides the skills and knowledge required for a course that is to follow.

SAFETY ORIENTATION AND TRAINING 9.15

Instructor's Notes:

Additional Resources

This module is intended to present thorough resources for task training. The following reference works are suggested for further study. These are optional materials for continued education rather than for task training.

www.osha.gov

www.asse.org

The Psychology of Safety Handbook, 2001. E. Scott Geller,Ph.D. Boca Raton, FL: CRC/Lewis Publishers.

The Participation Factor—How to Increase Involvement in Occupational Safety, 2002. E. Scott Geller, Ph.D. Des Plaines, IL: The American Society of Safety Engineers (ASSE).

Figure Credits

Paul Liberatore	209F01
Phil Elmore	209F02

SAFETY ORIENTATION AND TRAINING 9.17

MODULE 75209-03 — ANSWERS TO REVIEW QUESTIONS

Answer		Section
1.	a	2.0.0
2.	b	2.0.0
3.	c	2.1.0
4.	c	2.2.0
5.	b	2.3.0
6.	c	4.4.0
7.	a	4.5.0
8.	c	5.0.0
9.	b	5.1.0
10.	d	6.1.0

CONTREN™ LEARNING SERIES — USER UPDATES

The NCCER makes every effort to keep these textbooks up-to-date and free of technical errors. We appreciate your help in this process. If you have an idea for improving this textbook, or if you find an error, a typographical mistake, or an inaccuracy in NCCER's Contren™ textbooks, please write us, using this form or a photocopy. Be sure to include the exact module number, page number, a detailed description, and the correction, if applicable. Your input will be brought to the attention of the Technical Review Committee. Thank you for your assistance.

Instructors – If you found that additional materials were necessary in order to teach this module effectively, please let us know so that we may include them in the Equipment/Materials list in the Instructor's Guide.

Write: Curriculum Revision and Development Department
National Center for Construction Education and Research
P.O. Box 141104, Gainesville, FL 32614-1104

Fax: 352-334-0932

E-mail: curriculum@nccer.org

Craft ______________ Module Name ______________

Copyright Date ______________ Module Number ______________ Page Number(s) ______________

Description ______________

____

____

____

(Optional) Correction ______________

____

____

(Optional) Your Name and Address ______________

____

____

Module 75210-03

Work Permit Policies

NCCER STANDARDIZED CRAFT TRAINING PROGRAM

The National Center for Construction Education and Research (NCCER) provides a standardized national program of accredited craft training. Key features of the program include instructor certification, competency-based training, and performance testing. The program provides trainees, instructors, and companies with a standard form of recognition through a National Craft Training Registry. The program is described in full in the *Guidelines for Accreditation*, published by the NCCER. For more information on standardized craft training, contact the NCCER by writing us at P.O. Box 141104, Gainesville, FL 32614-1104; calling 352-334-0911; or e-mailing info@nccer.org. More information may be found at our Web site, www.nccer.org.

HOW TO USE THIS ANNOTATED INSTRUCTOR'S GUIDE

Each page presents two sections of information. The larger section displays each page exactly as it appears in the Participant Module. The narrow column ties suggested participant and instructor actions to each page and provides icons (detailed below) to call your attention to material, safety, audiovisual, or testing requirements. The bottom of each page includes space for your notes.

The **Audiovisual** icon indicates an appropriate time to show a transparency or other audiovisual aid.

The **Classroom** icon prompts you to define a term, stress a point, ask participants to explain a concept, or give examples.

The **Demonstration** icon directs you to show participants how to perform tasks.

The **Examination** icon tells you to administer the written module examination.

The **Homework** icon is placed where you may wish to assign reading for the next class, to assign a project, or to advise participants to prepare for an examination.

The **Laboratory** icon is used when participants are to practice performing tasks.

The **Materials** icon is a reminder for you to gather materials needed for classes, labs, and testing.

The **Performance Testing** icon tells you to administer a performance test or a portion thereof.

The **Safety** icon is used to emphasize safety issues. It is often keyed to *Caution* and *Warning* statements in the Participant Module.

The **Teaching Tip** icon indicates additional guidance is available, such as how to conduct an exercise, get the most educational value from a field trip, or encourage class participation. Teaching Tips may expand on a feature (*Think About It*, *Did You Know?*) or provide Quick Quizzes or similar exercises. You will be referred to the Teaching Tips section at the back of the module if there is additional material.

The **Combination** icon indicates that the laboratory listed corresponds with a performance task. If desired, you can note the proficiency of the participants during the laboratory and use it to satisfy performance testing requirements.

PREPARATION

Before teaching this module, you should review the Objectives, Performance Tasks, Materials and Equipment List, and the Module Outline. Be sure to allow ample time to prepare your own training or lesson plan and gather all required materials and equipment.

MODULE OVERVIEW

In this module, participants will learn the requirements and procedures related to work permits.

PREREQUISITES

Please refer to the Course Map in the Participant Module. Prior to training with this module, it is recommended that the participant shall have successfully completed the following:

Field Safety; Safety Technology, Modules 75201-03 through 75209-03

OBJECTIVES

Upon completion of this module, the participant will be able to:

1. Describe the role and responsibility of the safety technician in relation to work permit policies.
2. State the purpose of work permit policies.
3. Explain the need for hot work permits.
4. Describe the safety technician's role during the performance of hot work.
5. Explain the need for a lockout/tagout program.
6. Describe steps needed during the performance of lockout/tagout procedures.
7. Explain the need for confined-space permits.
8. Describe the safety technician's role regarding confined-space work areas.
9. Describe some of the hazards involved when an excavation work permit is needed.
10. Describe the safety technician's role during the performance of electrical hot work.

PERFORMANCE TASKS

Under the supervision of the instructor, the participant should be able to:

1. Perform a pre-inspection.
2. Properly complete a hot work permit.
3. Demonstrate how to use lockout and tagout devices.

MATERIALS AND EQUIPMENT LIST

Overhead projector and screen
Transparencies
Whiteboard/chalkboard
Markers/chalk
Blank acetate sheets
Transparency pens
Pencils and scratch paper
Appropriate personal protective equipment
Copy of *29 CFR 1926, OSHA Construction Industry Regulations*
Blank hot work permit forms
Blank pre-inspection forms
Lockout/tagout devices
Module Examinations*
Performance Profile Sheets*

* Located in the Test Booklet.

SAFETY CONSIDERATIONS

Ensure that participants are equipped with appropriate personal protective equipment.

ADDITIONAL RESOURCES

This module is intended to present thorough resources for task training. The following reference works are suggested for both instructors and motivated participants interested in further study. These are optional materials for continued education rather than for task training.

Web site: www.osha.gov

Web site: www.asse.org

Guidelines for Hot Work in Confined Spaces: Recommended Practices for Industrial Hygienists and Safety Professionals, 2000. Martin H. Finkel, CIH, CMC. Des Plaines, IL: The American Society of Safety Engineers (ASSE).

TEACHING TIME FOR THIS MODULE

An outline for use in developing your lesson plan is presented below. Approximately 2½ hours are suggested to cover *Work Permit Policies*. This includes 10 minutes at the beginning of the session for administrative tasks and one 10-minute break. You may need to adjust the time required for hands-on activities and testing based on your class size and resources. Because laboratories often correspond to Performance Tasks, the proficiency of the participants may be noted during these exercises for Performance Testing purposes.

Topic	Planned Time
A. Introduction	____________
B. Roles and Responsibilities of the Safety Technician	
1. Laboratory	____________
Have the participants conduct a pre-inspection and complete related paperwork. Note the proficiency of each participant. This laboratory corresponds with Performance Task 1.	____________
C. Work Permits	____________
1. Hot Work Permits	
a. Laboratory	____________
Provide the participants with a hot work scenario and the needed tools and resources. Ask them to complete a hot work permit to familiarize themselves with its contents. Note the proficiency of each participant. This laboratory corresponds with Performance Task 2	
2. Lockout/Tagout Work Policies and Procedures	
a. Laboratory Provide participants with samples of lockout/tagout devices and have them practice using them. Note the proficiency of each participant. This laboratory corresponds with Performance Task 3.	
3. Confined-Space Work Permits	____________
4. Excavation Permits	____________
5. Chemical Hazards	____________
6. Additional Types of Work Permits	____________
D. Review	____________
E. Module Examination	____________
1. Participants must score 70% or higher to receive recognition from the NCCER.	
2. Record the testing results on Craft Training Report Form 200 and submit the results to the Training Program Sponsor.	
F. Performance Testing	
1. Participants must perform each task to the satisfaction of the instructor to receive recognition from the NCCER. If applicable, proficiency noted during laboratory exercises can be used to satisfy the Performance Testing requirements.	

Module 75210-03

Work Permit Policies

Instructor's Notes:

COURSE MAP

This course map shows all of the modules in Safety Technology. The suggested training order begins at the bottom and proceeds up. The local Training Program Sponsor may adjust the training order.

Assign reading of Module 75210-03.

SAFETY TECHNOLOGY

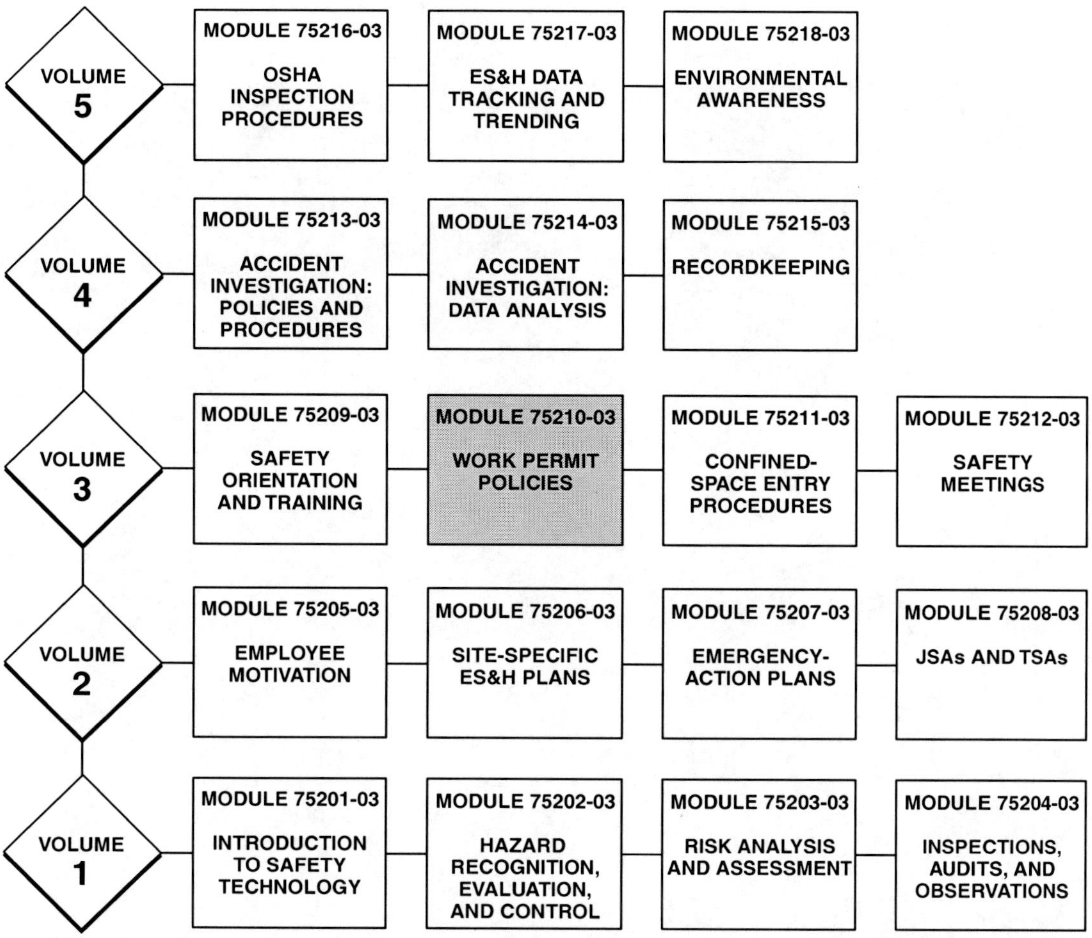

210CMAP.EPS

WORK PERMIT POLICIES 10.iii

Instructor's Notes:

MODULE 75210-03 CONTENTS

Figures

WORK PERMIT POLICIES 10.v

Instructor's Notes:

MODULE 75210-03

Work Permit Policies

Objectives

Upon completion of this module, you will be able to do the following:

1. Describe the role and responsibility of the safety technician in relation to work permit policies.
2. State the purpose of work permit policies.
3. Explain the need for hot work permits.
4. Describe the safety technician's role during the performance of hot work.
5. Explain the need for a lockout/tagout program.
6. Describe steps needed during the performance of lockout/tagout procedures.
7. Explain the need for confined-space permits.
8. Describe the safety technician's role regarding confined-space work areas.
9. Describe some of the hazards involved when an excavation work permit is needed.
10. Describe the safety technician's role during the performance of electrical hot work.

Prerequisites

Before you begin this module, it is recommended that you successfully complete the following: Field Safety; Safety Technology, Modules 75201-03 through 75209-03.

Required Materials

1. Pencil and paper
2. Appropriate personal protective equipment
3. Copy of *29 CFR 1926, OSHA Construction Industry Regulations*

1.0.0 ◆ INTRODUCTION

Work permit policies are needed because some normal work conditions can become a safety hazard when other work is being performed nearby. For example, using a band saw that creates a large amount of wood shavings and fine dust can become a severe fire and explosion hazard when work that requires an open flame is being done in the same area. As the safety technician, you will be responsible for auditing work permit compliance.

A number of work permit systems are used in the construction industry. They all have the same purpose. Permit systems are in place to reduce the chances of accidents and injures to workers. They are also used to prevent damage and loss of property. Work permits also provide a method of communication between the space owner and personnel entering the space.

As the site safety technician, you'll be responsible for advising senior management on all safety issues on a project. Therefore, you must be aware of site requirements for work permits and have a good working knowledge of the hazards and safeguards associated with the various types of permits used on site.

There are a number of different types of work permits and related systems. They include:

- Hot work permits
- Cold work permits
- Hazardous work permits
- **Lockout/tagout devices**
- Electrical work permits
- **Excavation** and **trenching** permits
- Line entry permits
- Confined-space permits

Ensure that you have everything you need to teach the course. Check the Materials and Equipment list in the front of this module.

Explain that terms shown in bold are defined in the Glossary at the back of this module.

Review the goals of the course and explain what will be expected of the participant.

Show Transparency 1, Objectives.

Show Transparency 2, Performance Tasks.

Explain why work permit policies are needed.

Describe the types of work that require work permits.

Discuss the general process for obtaining a work permit, including the roles of controlling employers and contractors.

Explain the roles and responsibilities of the safety technician in the work permit process. Explain that the safety technician's most important role on a site during the work permit process is as a consultant.

Describe how pre-inspections aid in the discovery of potential hazards.

Show Transparency 3 (Figure 1).

Describe who typically issues the work permit and how it is issued.

Show Transparency 4 (Figure 2).

- Crane/critical lift work permits
- Aerial lift permits
- Vehicle permits

When work permits are issued, they are usually for a specific task. It is a very poor practice to issue a blanket work permit for the performance of a variety of tasks. Further, permits are usually issued for a short, specified period of time. Some permits may be valid for the entire workday or shift. Other permits may be valid only between the hours specified on the permit.

The company in charge of the overall operation of the site usually issues permits. This could be the owner of a site, but most likely, it will be a contractor or site manager. In this module, this department will be referred to as the controlling employer.

The controlling employer will be responsible for selecting qualified workers and subcontractors. You may work directly for the controlling employer or for a subcontractor.

Whether you work for the controlling employer or for a subcontractor, your job as safety technician is the same. You need to be sure that your company follows the site's work permit policies and procedures from an auditing standpoint. You also need to be sure that workers in your area(s) of responsibility follow the policies and procedures, too.

The controlling employer needs to explain site work permit requirements to contractors and subcontractors on every tier. They are also responsible for monitoring the work site, auditing the subcontractor's work, and ensuring that all work performed at the site is completed safely.

Subcontractors are responsible for their workers. Subcontractors need to be sure that their crews are capable of safely performing the work. They also need to be sure that they understand the job that needs to be done and the job's safety requirements.

If you work for a subcontractor, you need to be sure that your company's work permit policies are at least as safe as the host site's work permit. If your company's are safer, use them. If the site's are safer, use those. It is critical, however, that everyone on site agrees to which procedures are being used.

Both the controlling employer and subcontractors need to ensure that the workers they hire are trained to do the job. Both need to be sure that the equipment they use is safe and operating properly.

2.0.0 ◆ ROLES AND RESPONSIBILITIES OF THE SAFETY TECHNICIAN

On some sites, you'll be working under the direction of the site safety manager. If you're working with the site safety manager, he or she will determine your role. More often, you'll be working on your own.

There are a number of tasks you will commonly perform as a safety technician. They are:

- Training workers
- Pre-inspection of the work area
- Specifying job conditions and practices
- Completion of any needed paperwork
- Observation of the work area while work requiring permits is being performed

For workers to be allowed to perform certain tasks, they may need to undergo additional training. Sometimes they'll need to attend a class. Other times, you'll need to perform the training. In any case, be sure that you record that the workers received additional training in their employee records. Include the date of the training and how the workers demonstrated to you that they understood their roles.

For almost all work permits, a pre-inspection of the work area is needed. You may need to complete a checklist or form. The purpose of the pre-inspection is to locate and correct any conditions that could become a hazard while performing work that requires permits. *Figure 1* shows a sample of pre-task planning form. Sometimes this type of paperwork needs to be shown to the controlling employer's workers before the work permit is issued.

Once you've prepared the work area and completed any necessary paperwork, you're ready to have a permit issued. Work permit policies vary between sites. At some work sites, you may receive a work permit from the controlling employer after you've completed a written pre-task checklist, inspection, or other item. Work permits can be written (*Figure 2*) or verbal.

Regardless of how permits are issued at the site, the worker and first-line supervisor are responsible for the safe completion of the task. They need to be sure that all of the necessary paperwork is completed and that the area and equipment are safe. Your role as safety technician during these operations is as a site consultant.

While the work is being performed, you will need to observe the work area. You should also be available if the crew leader needs advice on assigning safety observer duties to other workers who have been trained for this role. Safety observers identify and correct any potential hazards on the spot.

In some cases, such as after welding operations, you or another trained employee may need to observe the work area after the work has been done. This is to ensure that a problem doesn't develop. For example, a fire may break out due to an unnoticed spark.

Instructor's Notes:

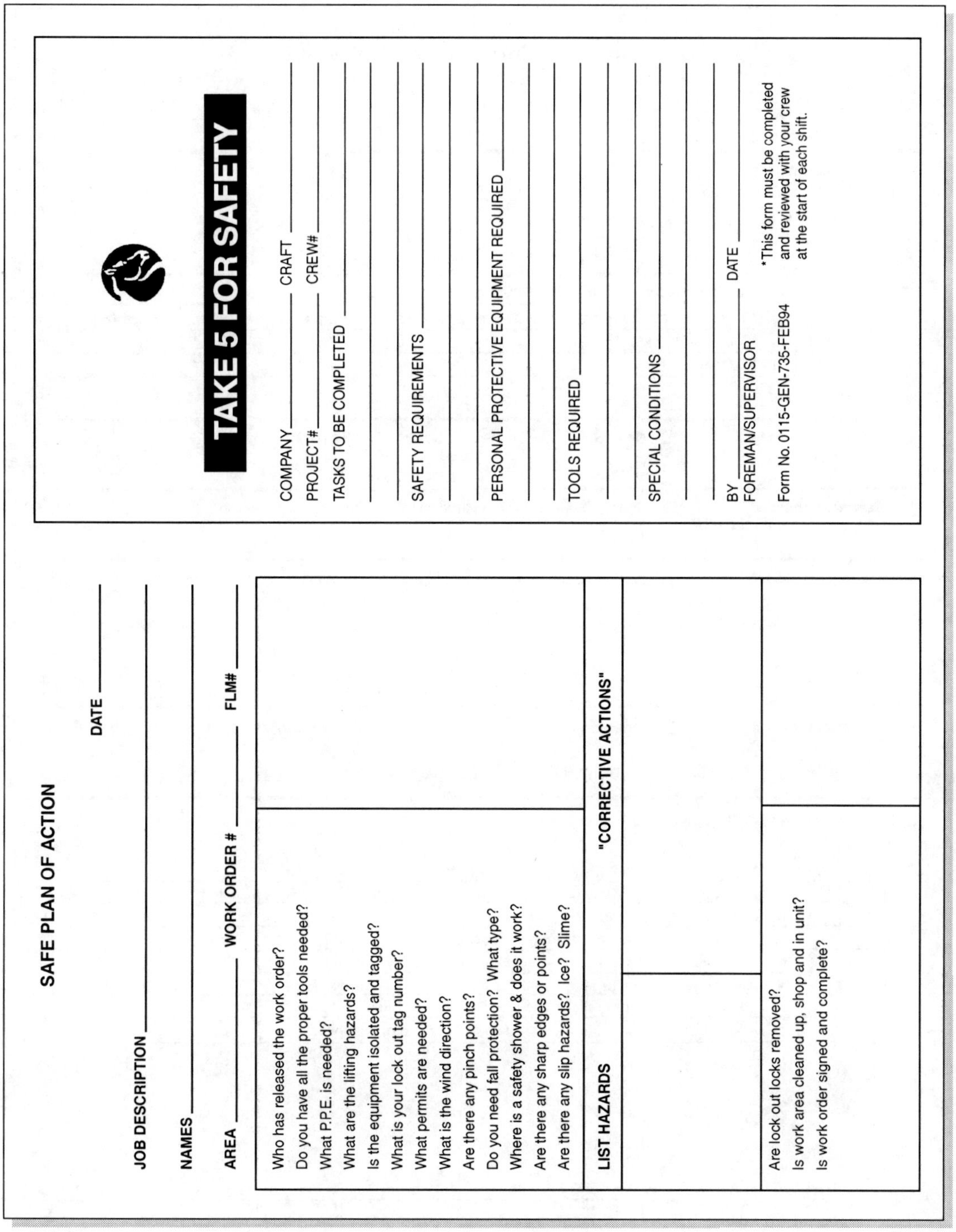

TAKE 5 FOR SAFETY

COMPANY ________ CRAFT ________

PROJECT# ________ CREW# ________

TASKS TO BE COMPLETED ________

SAFETY REQUIREMENTS ________

PERSONAL PROTECTIVE EQUIPMENT REQUIRED ________

TOOLS REQUIRED ________

SPECIAL CONDITIONS ________

BY ________ DATE ________
FOREMAN/SUPERVISOR

Form No. 0115-GEN-735-FEB94

*This form must be completed and reviewed with your crew at the start of each shift.

SAFE PLAN OF ACTION

DATE ________

JOB DESCRIPTION ________

NAMES ________

AREA ________ WORK ORDER # ________ FLM# ________

Who has released the work order?
Do you have all the proper tools needed?
What P.P.E. is needed?
What are the lifting hazards?
Is the equipment isolated and tagged?
What is your lock out tag number?
What permits are needed?
What is the wind direction?
Are there any pinch points?
Do you need fall protection? What type?
Where is a safety shower & does it work?
Are there any sharp edges or points?
Are there any slip hazards? Ice? Slime?

LIST HAZARDS

"CORRECTIVE ACTIONS"

Are lock out locks removed?
Is work area cleaned up, shop and in unit?
Is work order signed and complete?

210F01.EPS

Figure 1 ◆ Take 5 for safety.

Arrange for a visit to a work site. Show the participants how to conduct a pre-inspection.

Under your supervision, have the participants conduct a pre-inspection and complete related paperwork. Note the proficiency of each participant. This laboratory corresponds with Performance Task 1.

Show the participants how to complete a hot work permit.

Provide the participants with a hot work scenario and the needed tools and resources. Ask them to complete a hot work permit to familiarize themselves with its contents. Note the proficiency of each participant. This laboratory corresponds with Performance Task 2.

Burning - Welding - Hot Work Permit

Valid from _____ to _____, __________ Master Card No.________________
(am/pm) (am/pm) DATE

1. Work Description
Equipment Location or Area ______________________________
Work to be done:

2. Gas Test

		Test Results	Other Tests	Test Results
☐ None Required				
☐ Instrument Check				
☐ Oxygen 20.8% Min				
☐ Combustilble % LFL				

Gas Tester Signature Date Time

3. Special Instructions ☐ None ☐ Check with issuer before beginning work

4. Hazardous Materials ☐ None What did the line / equipment last contain?

5. Personal Protection ☐ Standard Equipment: welders hood with long sleeves; cutting goggles

☐ Goggles or Face Shield ☐ Respirator ☐ Forced Air Ventilation

☐ Standby Man ☐ Other, specify: ______________

6. Fire Protection ☐ None Required ☐ Portable Fire Extinguisher

☐ Fire Watch ☐ Fire Blanket ☐ Other, specify: ______________

7. Condition of Area and Equipment

Required Yes	No	THESE KEY POINTS MUST BE CHECKED
		a. Lines disconnected & blanked or if disconnecting is not possible, blinds installed?
		b. Lines steamed, purged, or otherwise properly cleared of combustibles?
		c. Area and equipment satisfactorily clean of oil or combustibles?
		d. Trenches, catch basins & sewer connections properly covered or sealed?
		e. Immediate area and/or area under the work barricaded or roped off?
		f. Adjoining equip. & operations checked to have any effect on the job?
		g. Area fire suppression (fire water and sprinkler system) in service?

Comments

210F02A.EPS

Figure 2 ◆ Work permits. (1 of 2)

10.4 SAFETY TECHNOLOGY — PARTICIPANT MODULE 75210-03

Instructor's Notes:

Burning - Welding - Hot Work Permit

8. Approval	Permit Authorization			Permit Acceptance		
	Area Supv.	Date	Time	Maint.Supv./Engineer Contractor Supv.	Date	Time
Issued by						
Endorsed by						
Endorsed by						

9. Individual Review

I have been instructed in the proper Hot Work Procedures

	Signed	Signed
Persons Authorized to Perform Hot Work	__________	__________
	__________	__________
	__________	__________
	__________	__________
	__________	__________
Fire Watch	__________	__________

10. Job Completion

☐ Yes ☐ No Is the work on the equipment completed?

☐ Yes ☐ No Has the worksite been cleaned and made safe?

Workman answering above questions ____________________

Issuer's Acceptance ____________________

Forward to Production Superintendent within 7 days of job completion

210F02B.EPS

Figure 2 ◆ Work permits. (2 of 2)

Ask the participants if they have ever done work that required the use of a work permit. Ask them to share their experiences.

Discuss the benefits of using work permits.

Describe the hazards associated with hot work.

Describe the type of work that requires a hot work permit.

Identify the safety considerations in a hot work area.

Describe areas and equipment that must be inspected in the pre-inspection process typically performed on a hot-work permit required site.

3.0.0 ◆ WORK PERMITS

Work permits are used to reduce the risk of accidents and injuries to workers while specific tasks are being performed. They are also designed to reduce the chance of property damage. But just having a work permit doesn't mean that the job will be safety completed. A work permit is a tool. Like any tool, it works best when properly used.

Although there are a number of types of work permits, this module will mostly discuss hot work permits, lockout/tagout procedures, confined-space entry permits, and excavation and trenching permits. Confined-space permits are discussed in detail in a later module, *Confined-Space Entry Procedures*.

As the safety technician, you'll need to know when one or more work permits are required to complete a task. You'll need to be able to spot and know how to correct conditions that could become hazards while performing work that requires permits.

> **NOTE**
> Work permits are not normally required for areas that have been specially designed for the work. For example, an area with concrete floor, fire-resistant walls, and special ventilation could be set aside for welding. This area may not need a work permit for welding.

3.1.0 Hot Work Permits

A hot work permit is required whenever an open flame or other type of ignition source is required to complete a task. Hazards associated with hot works include:

- Gases
- Fumes
- Sparks
- Hot metal
- Heat

Some examples of tasks that could require a hot work permit are:

- Welding
- Grinding
- Burning
- Motorized equipment
- Battery-powered drills
- Radios

On a typical site, the supervisor will notify you that a hot work permit is required. You and the crew leader should discuss the following:

- Is there another way to do the job?
- In what area does the job need to be performed?
- Can the area easily be made safe for hot work?
- Can the work be moved to another, safer area?

Once you decide that the hot work permit is needed, the crew leader will need to complete a pre-task planning form similar to the one shown in *Figure 1*. This type of form may be used to apply for the hot work permit.

Along with the crew leader, you will need to inspect the work area. Flammable materials such as wood shavings must be removed or safeguarded. When an item needs to be moved, 35' is considered a safe distance, but you'll need to use your judgment to decide whether a distance is safe.

Items that can't be moved must be covered with a fire-resistive material. This could include specially designed fire-resistant blankets, curtains, or covers. In some cases, items may be wet down with water.

Assess floors and nearby walls to determine whether cracks are present. Sparks could roll into a crack and smolder for hours before a fire is obvious. In addition, you'll need to ensure that all open sewers in the area are covered and sealed, since these can be a source of flammable methane gases.

While you're inspecting the area, assess the need for ventilation. This type of assessment is needed even if the area is not considered a confined area. Because welding *(Figure 3)*, burning, and other hot work can form fumes and toxic gases, adequate airflow is need. The types of fumes generated by welding depend on the type of metal being welded and the type of welding rods used. Some welding fumes include lead, cadmium, and chromium. Toxic gases include ozone and carbon monoxide.

Figure 3 ◆ Welding. 210F03.EPS

Instructor's Notes:

Before any hot work can begin, a flammable gas test must be done. A gas test ensures that no flammable gases or vapors are present. These types of gases and vapors not only represent a fire/explosion risk, but they also represent a potential hazard to the lives of workers who breathe them.

Once the physical area has been made safe, the equipment needs to be inspected. Some of the things that must be checked are:

- The equipment that will be used during the operation
- Personal protective equipment that workers will wear
- Availability of safety equipment such as fire extinguishers

When you and the crew leader are satisfied that the area is safe, safety policies and procedures may require that safety signs be hung in the area (*Figure 4*).

210F04.EPS

Figure 4 ◆ Safety sign.

Now the hot work permit is issued. Before the actual work begins, make a final check to ensure that all needed permits, inspection checklists, and other paperwork have been properly filled out. Then ensure that all hazards and safeguards have been reviewed with the work crew. If you need additional safety observers, you'll need to establish a fire watch. During a fire watch, a person other than the welder or cutting operator must constantly scan the work area for fires. Fire watch personnel must have ready access to fire extinguishers and alarms and know how to use them. They must also know how to shut off oxy-acetylene rigs and welding machines.

During the time that hot work is being done, a trained employee will need to observe the work. The most obvious hazard during hot work is the potential for fire. Post a fire watch nearby with a fire extinguisher.

During hot work, hazards could include a change in conditions that cause the work to become unsafe. For example, rain can leak in and puddle in an area where arc welding is taking place. Unsafe conditions could also include worker issues, such as unauthorized persons in the area. In addition, you should perform air-sampling tests during long work periods to assess the area for gases.

CAUTION

During hot work, the area temperature can increase rapidly. Always observe the workers to assess for heat-related problems.

After the hot work has been completed, observe the area. Most fires begin after the work has been done. A spark that lands in an obscure location can smolder for hours before flames and smoke are detectable. The observation time period will vary by site. Many companies require the fire watch to remain at their post for 30 minutes after the hot work is completed.

After the observation period has been completed, the crew leader must be certain that all information is properly recorded even if the work was done safely. As a safety technician, you should be available to advise the crew leader as needed during this step.

NOTE

For more information about hot work, please refer to the *Welding Safety* module in Field Safety.

3.2.0 Lockout/Tagout Work Policies and Procedures

Lockout/tagout (LOTO) policies and procedures are part of an energy control program. This program is used to prevent the unintended release of stored energy in machines or equipment. The program consists of written **energy control procedures**, a worker-training program, and periodic inspections and audits.

Energy control procedures are written documents that contain information that authorized workers need to know to safely control energy while servicing or maintaining equipment or machinery.

Energy control procedures use **energy-isolating devices** that physically prevent the transmission or release of energy. These devices include manually

Explain the importance of re-checking equipment, checklists, and permits in the hot work area after the hot work permit has been issued to ensure that everything is in order and the area is ready for work.

Explain the role of a fire watch.

Discuss the importance of monitoring the hot work area. Explain why it is important to continue monitoring after the work is complete.

Explain the safety technician's role in advising the crew leader about completing the required paperwork.

Explain the purpose of lockout/tagout policies and procedures.

WORK PERMIT POLICIES 10.7

Explain the safety technician's role in advising management about the development of a lockout/tagout program.

Demonstrate how lockout/tagout devices are used. Show the participants an example of a lockout/tagout device.

Provide participants with samples of lockout/tagout devices and have them practice using them. Note the proficiency of each participant. This laboratory corresponds with Performance Task 3.

Describe the procedures used in a typical lockout/tagout program.

Explain the information that must be included in the written procedures for lockout/tagout.

Explain that the safety technician may be responsible for training workers about lockout/tagout.

Explain the OSHA standards regarding the types of workers who must be trained: authorized, affected, and other employees.

operated circuit breakers, disconnect switches, valves, blocks, blinds, and blank flanges.

Lockout/tagout policies and procedures are designed to prevent the unplanned startup of machines or equipment during servicing or maintenance by using lockout or tagout devices (*Figure 5*) that isolate or identify energy sources. These energy sources could be:

- Electrical
- Mechanical
- Hydraulic
- Pneumatic
- Chemical
- Thermal
- Stored or residual
- Gravitational (when working with suspended or elevated equipment)

210F05.EPS

Figure 5 ◆ Lockout/tagout devices.

Your safety manager has probably already developed written LOTO procedures. If these procedures are in not in place, you may need to help develop them. You'll need to include exactly when the procedure needs to be used. The procedure will need to clearly state when a tagout device may be used in place of a lockout device. In addition, you will need to define exactly what types of devices are to be used.

Both lockout and tagout devices must be provided by the employer and must be standard in color, size, or shape. LOTO devices may be used for their intended purpose only. They must be easily identifiable and durable enough for the work environment. There must be an easy way to identify the name of the person who applied the device. For example, the lock may be engraved with the worker's name or it may be written on the accompanying tag.

Lockout devices must never be used to lock toolboxes, lockers, or other items. Lockout devices are always preferred over tagout devices. Compare the use of a lockout device to prevent an operator from starting a crane while you're servicing it versus a "Do Not Operate/Servicing in Progress" tag that's located near the ignition switch. Which of the two would you prefer?

The LOTO procedure needs to include:

- Steps for shutting down the equipment and releasing stored energy
- Steps to lock out or tag out the energy source
- Steps for testing that the LOTO device(s) are controlling the energy
- Requirements for releasing the LOTO device(s)

> **NOTE**
> Usually, the person who applied the LOTO device is the only person allowed to remove it.

In addition to the steps require for applying and removing the LOTO devices, the written procedures will need to include training requirements, periodic audit programs, and how to handle special situations. For example, the procedure should cover what to do when a LOTO period overlaps work shifts.

3.2.1 Training

As the safety technician, you may be responsible for training workers about LOTO procedures. OSHA standards (*CFR 1910.147*) define three types of employees who may be involved in LOTO procedures. They are:

- **Authorized employees**
- **Affected employees**
- **Other employees**

Authorized employees are those who actually perform the service or maintenance on the involved equipment. These workers need a high degree of training. They need to know about the type and extent of the hazardous energy. They also need to know how to safely isolate and control the energy source.

Instructor's Notes:

Affected employees are usually the equipment users or operators. These workers need to identify when a control procedure is in place. They also need to know that they are not to remove the LOTO device or try to start the equipment during the procedure.

Other employees are those workers who are in the area where the control procedure is taking place. All workers at the site must know the purpose and function of LOTO procedures. They must be able to identify a LOTO device. More importantly, all workers must know that under no conditions should they remove the device unless they applied it.

As stated earlier, lockout devices are better than tagout devices. If an energy source is lockable, lockout devices should be used. Tagout devices may be used if the employer can prove that the tagout device provides at least the same level of safety as a lockout device. However, using a tagout device when a lockout device can be used is a very poor and potentially dangerous practice. As the site safety technician, you should strongly discourage this practice.

During worker training, you must explain the uses and limits of tagout devices. Tagout devices are to be used when an energy source cannot be locked. However, as the safety technician, you should encourage employers to modify or replace such sources so that they can be locked.

Tagout devices may be needed at your work site. You'll need to train workers about their use and limits. Remember—tags may give workers a false sense of security. Tags are warning devices (*Figure 6*). They do not provide physical control of an energy source.

210F06.EPS

Figure 6 ◆ Example of lockout tag.

Tagout devices must be standardized and clearly identifiable. They must be securely attached to the energy source. Tags must be durable enough to withstand the work area and/or the environment. The tag must identify the name of the person who applied it. Once a tag is attached, it may not be removed except by the person who attached it.

Tag attachments usually require at least 50 pounds of pressure to remove. This requirement is based on the strength of a standard ¼" nylon cable tie that has a rated unlocking strength of about 75 pounds.

When tags are used, workers may need to use extra safety means to make the conditions equal to using lockout devices. This could include removing a circuit element, blocking a control switch, opening an extra disconnecting device, or removing a valve handle to reduce the chances that the equipment could be turned on.

You'll need to ensure that all workers receive periodic refresher training about LOTO policies and procedures.

3.2.2 LOTO Procedure

During the performance of a LOTO procedure, the authorized employee will need to be trained to perform the following steps:

Step 1 Before beginning, notify all affected and other workers of the upcoming maintenance.

Step 2 Identify and isolate all energy sources.

Step 3 Securely apply the LOTO device. Be certain to mark the device to identify who applied it.

Step 4 Test the energy source to determine that power has been isolated.

Step 5 Perform servicing or maintenance of the equipment.

Step 6 Release the LOTO device.

Step 7 Test the equipment to ensure that it is safe and operational.

Step 8 Complete the necessary paperwork. *Figure 7* shows a sample checklist for lockout/tagout procedures. You can find this checklist and many others on the OSHA Web site (www.osha.gov).

> **NOTE**
> Always use a safety checklist to be sure that a step isn't omitted.

Explain the difference between a lockout device and a tagout device. Identify specific uses of each.

Show Transparency 5 (Figure 6).

Explain the steps the authorized employee will need to be trained to perform when following a lockout/tagout procedure. Show the procedure while explaining it.

Show Transparency 6, Lockout/Tagout Procedure.

WORK PERMIT POLICIES 10.9

Explain that a lockout/tagout checklist can be used to ensure that steps aren't skipped.

Show Transparency 7 (Figure 7).

Explain the OSHA exceptions related to lockout/tagout policies and procedures.

	LOCKOUT/TAGOUT PROCEDURES
	Is all machinery or equipment that is capable of movement, required to be de-energized or disengaged and locked out during cleaning, servicing, adjusting, or setting up operations, whenever required?
	Where the power disconnecting means for equipment does not also disconnect the electrical control circuit: Are the appropriate electrical enclosures identified? Are means provided to assure the control circuit can also be disconnected and locked out?
	Is the locking-out of control circuits in lieu of locking-out main power disconnects prohibited?
	Are all equipment control valve handles provided with a means for locking-out?
	Does the lockout procedure require that stored energy (mechanical, hydraulic, air, etc.) be released or blocked before equipment is locked-out for repairs?
	Are appropriate employees provided with individually keyed personal safety locks?
	Are employees required to keep personal control of their key(s) while they have safety locks in use?
	Is it required that only the employee exposed to the hazard, place or remove the safety lock?
	Is it required that employees check the safety of the lockout by attempting a startup after making sure no one is exposed?
	Are employees instructed to always push the control circuit stop button immediately after checking the safety of the lockout?
	Is there a means provided to identify any or all employees who are working on locked-out equipment by their locks or accompanying tags?
	Are a sufficient number of accident preventive signs or tags and safety padlocks provided for any reasonably foreseeable repair emergency?
	When machine operations, configuration, or size require the operator to leave his or her control station to install tools or perform other operations, and that part of the machine could move if accidentally activated, is such element required to be separately locked or blocked out?
	In the event that equipment or lines cannot be shut down, locked out and tagged, is a safe job procedure established and rigidly followed?

210F07.EPS

Figure 7 ◆ Sample lockout/tagout procedure checklist.

3.2.3 LOTO Exceptions

Under some conditions, OSHA requirements for LOTO policies and procedures will not apply. However, under these conditions, workers must be provided with other protection that is just as effective as LOTO procedures.

One condition where LOTO procedures are not needed is if the worker is not at risk for injury due to the unexpected release of energy. For example, when the equipment requires the use of a cord and plug, the authorized employee can control the plug during servicing.

LOTO policies and procedures may not apply during hot tap operations that involve transmission and distribution systems for gas, steam, water, or petroleum products. LOTO policies and procedures will not apply when:

- Servicing must be performed on pressurized pipelines.
- Continuation of the service is essential.
- Shutdown of the system is impractical.

CAUTION

Even though LOTO policies and procedures do not apply to hot tap operations, these operations must still be considered potentially hazardous activities that require special precautions and procedures.

Further, there are times when equipment must be powered on during servicing. Making fine adjustments, such as centering a conveyor belt, is one example. Troubleshooting equipment to identify the cause of a problem is another. Although OSHA recognizes that these situations exist, they still require that the employer provide protection to workers performing these duties.

Employees performing minor tool changes, adjustments, and servicing during normal equipment operation may be exempt from LOTO policies. These types of servicing must be routine, repetitive, and integral to the use of the equipment.

Instructor's Notes:

However, other safety protection that provides effective protection must still be used.

Even if the servicing is normally exempt from LOTO procedures, there are certain conditions that will require the use of LOTO policies and procedures. They include some of the following situations:

- The worker must remove or bypass machine guards or other safety devices, therefore exposing the worker to hazards at some time in the operation.
- The worker is required to place any part of his or her body in contact with the operational portion of the equipment.
- The worker needs to place any part of his or her body in the danger zone of the equipment.

3.2.4 Additional LOTO Safety Requirements

There are a number of special conditions that can exist during LOTO procedures. They include:

- *Temporary removal of LOTO devices* – OSHA permits LOTO devices to be temporarily removed under special conditions. For example, LOTO devices can be removed when power is needed to test or reposition equipment. In this case, the power may be applied under strictly controlled conditions. The following steps must be performed in the order given:

  Step 1 Clear the equipment of tools and materials.
  Step 2 Ensure that all workers are out of the area.
  Step 3 Remove all LOTO devices as specified by the site's policy.
  Step 4 Energize the equipment and proceed with testing or repositioning.
  Step 5 De-energize the equipment and isolate the energy source.
  Step 6 Apply the LOTO device as specified by the site's policy.
  Step 7 Test to confirm that the energy source has been isolated.
  Step 8 Complete servicing or maintenance tasks.

- *Outside workers* – When outside workers such as subcontractors are working on your job site, you must ensure that on-site workers and the outside workers are aware of each other's LOTO policies and procedures. Outside workers need to use policies and procedures that at least meet the standards of the host work site.
- *Group workers* – When service and maintenance need to be performed by a group of workers rather than a single worker, each authorized worker must be protected by his or her own LOTO device (*Figure 8*).

In complex lockouts with large numbers of exposed personnel and/or large numbers of isolation points, it is impractical to have each worker affix a personal lock to each isolating device. In situations such as this, the controlling employer will prepare an isolation log or master card listing all isolation points. A controlling employer's lock and accompanying tag will usually be attached to each isolation point. The keys to the locks will then be placed in a job lock box or cabinet that is fitted with one or more multiple lock adapters. Each exposed employee can then review the isolation log and affix his or her personal lock to the multiple lock adapters on the job lock box. In such cases the controlling employer's lock will be the first on and last off. The contractor's lock will be the last on and first off. OSHA mandates that under situations such as this, each exposed employee must have the right to inspect the application of the controlling employer's locks before affixing his/her lock to the job lock box.

> **NOTE**
> There are numerous variations to such systems. It is critically important for you to have an intimate working knowledge of the site-specific lockout/tagout and work permit systems.

210F08.EPS

Figure 8 ◆ Group lockout.

Identify and describe in detail the special conditions that exist during lockout/tagout procedures, including the temporary removal of LOTO devices, outside workers, group workers, and shift changes.

Show Transparency 8, Steps for Temporarily Removing a LOTO Device.

Show Transparency 9 (Figure 8).

WORK PERMIT POLICIES 10.11

Describe the conditions for which a confined-space work permit is required.

Show Transparency 10 (Figure 9).

Identify common confined spaces.

Explain the difference between a permit-required and non-permit-required confined space.

Explain OSHA's definition of a confined space.

Describe the hazards found in confined spaces.

Discuss the basic elements of confined-space entry procedures. Explain that the next module contains more information on this topic.

Discuss the safety technician's role in confined-space jobs.

Show Transparency 11 (Figure 10).

- *Shift changes* – Finally, specific LOTO policies and procedures must be in place to ensure that LOTO protection will continue during shift or personnel changes.

> **NOTE**
> For more information about LOTO procedures, please refer to the *Lockout/Tagout* module in Field Safety.

3.3.0 Confined-Space Work Permits

Confined-space work permits are needed whenever the size and shape of the work area is a hazard to anyone who must enter, work in, and exit the area. These spaces usually have poor natural airflow. In addition, they can contain real or possible hazards and need to be treated with caution and respect *(Figure 9)*.

Some **confined spaces** that may be at your work site include:

- Manholes
- Boiler tanks
- Trenches
- Sewers
- Vats
- Air ducts
- Process vessels

Not all confined spaces require permits. In a non-permit required space, workers can enter and work in the area using the minimum personal protection equipment that is required.

A **permit-required confined space** is one that contains real or possible hazards. These hazards can be atmospheric, physical, electrical, or mechanical. OSHA *(CFR 1910.146)* defines a permit-required confined space as a space that:

- Contains or has the potential to contain a hazardous **atmosphere**
- Contains a material that has the potential to engulf the worker
- Has an internal shape that could trap or asphyxiate the worker
- Contains other serious health hazards

Hazards in a confined space could include:

- Poor atmospheric conditions
- Electric shock
- Purging
- Falling objects
- **Engulfment**
- Extreme temperatures
- Noise
- Slick or wet surfaces
- Moving parts

In the later module, *Confined-Space Entry Procedures,* you will receive in-depth information about working in confined spaces. Briefly, the following apply to confined spaces:

- All permit-required confined spaces must be clearly identified and posted.
- Some confined spaces need permits to both enter and to do work in the space.
- Workers must check the confined space before entering.
- No worker is allowed to enter a confined space unless he or she has been authorized to enter.
- Hazards of the confined space and the work must be identified and eliminated or controlled to an acceptable level.
- The atmosphere in the confined space must be tested and deemed acceptable before entry.
- Proper personal protective equipment must be used in a confined space.
- The entry permit must have a material safety data sheet (MSDS) attached to it that provides information about any toxins the worker may encounter.
- Rescue plans must be developed in advance and communicated to all workers involved in the confined-space operation.
- All confined-space entrants must be thoroughly trained to perform their jobs before entering the confined space.

As the site safety technician, part of your job is to work with the crew leader or first-line supervisor to:

- Identify permit-required and **non-permit required confined spaces**.
- Ensure that all permit-required spaces are clearly posted (*Figure 10*).
- Help develop safety policies and procedures for entering and working in confined spaces.
- Ensure that all workers who need to enter and work in confined spaces are properly trained so that they know what to do if an emergency develops.
- Ensure that confined-spaces policies and procedures include methods to retrieve workers if an emergency develops.
- Ensure that workers are trained and use the proper protective equipment for the confined space.

Instructor's Notes:

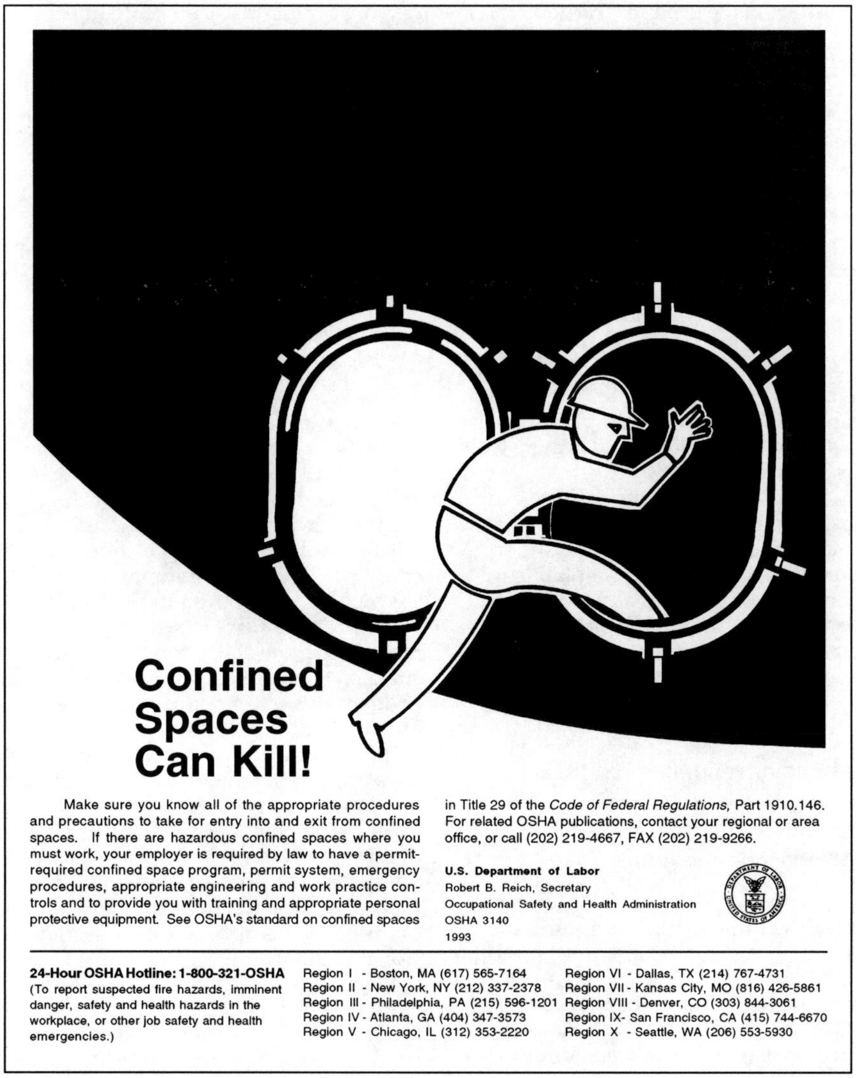

210F09.EPS

Figure 9 ◈ Confined spaces can kill.

- Ensure that periodic safety audits are done to make sure that confined-space policies and procedure are in place, in use, and provide workers with adequate protection to known hazards.

3.4.0 Excavation Permits

OSHA construction safety and health standards (*CFR 1926.650, .651,* and *.652*) recognize that excavation and trenching are very hazardous operations. A trench is defined as a narrow excavation

CAUTION
Confined Space Safe Work Permit Required

210F10.EPS

Figure 10 ◈ Confined-space posting.

Explain that excavation and trenching are very hazardous operations.

Ask the participants if they have ever worked a trenching or excavation job. Have them share their previous work experiences.

WORK PERMIT POLICIES 10.13

Explain that cave-ins are the main hazard of trenching and excavation jobs.

Explain the safety technician's role in safety planning and work permits for trenching and excavation.

Identify some of the areas that will need to be addressed in the plan.

Explain OSHA's requirement for identifying the location of utility lines. Explain that services, such as One-Call, provide information about utilities in the area where trenching or excavation will take place.

Discuss the importance of checking equipment, especially personal protective equipment, before trenching or excavating.

Explain OSHA's requirement for a competent person in regard to trenching.

Describe the methods used to prevent cave-ins.

Show Transparency 12 (Figure 11).

made below the surface of the ground. The depth of a trench must be greater than the width. The width must not exceed 15'. An excavation is defined as any man-made cut, cavity, trench, or depression in the earth's surface formed by earth removal.

Although workers are exposed to many hazards during excavation and trenching, the main hazard is cave-ins. Many accidents occurring during this type of work are the result of poor planning. Trying to correct mistakes after the work has begun increases the chance of cave-ins. It also slows down the work and adds to its cost.

Before beginning the job, you need to review your company's policies, procedures, and practices related to excavation and/or trenching. You need to ensure that they are adequate to protect workers from hazards. In addition, you must ensure that all workers are aware of and adhere to these policies and procedures.

Work permits for excavations and trenching can be very complex. It will be part of your job as the site safety technician to advise the first-line supervisor when he or she is verifying that a plan is in place before work begins. It is a good idea to use safety checklists to be sure that information about the job site and all needed items are on hand.

Some of the items that may need to be addressed are:

- Traffic
- Surrounding buildings and their conditions
- Soil type
- Surface and ground water
- Overhead and underground utilities
- Weather

These and other conditions may require site studies and tests such as geological surveys and soil type tests.

Before any excavation actually begins, OSHA standards require the employer to estimate the location of utility lines that may be encountered during digging. These include:

- Sewer lines
- Telephone cables
- Fuel lines, such as natural or propane gas
- Electric cables
- Water lines
- Other underground installations

In most states, your company will be required to contact local utility companies and the owners of the involved property before starting the excavation. Services such as One-Call provide all of the information need on buried utilities in the area. Contractors must call this service at least 72 hours in advance of doing any work. Adequate response time must be built into plan phase to allow for responses.

When all the needed information about the job site is collected, a review of the equipment and materials needed should be made. This includes personal protective equipment (see *CFR 1926.100* and *102*) and other safety items such as:

- Safety shoes
- Safety glasses
- Hardhats
- Reflective vests
- Breathing equipment
- Safety harness and line
- Atmospheric testing devices

No matter how many of these types of jobs have been done in the past, each new job should be carefully planned. Once the plan has been made and all needed local permits have been issued, the controlling employer will be able to issue work permits at the site. Keep in mind that more than one type of work permit may be required.

Once excavation or trenching begins, OSHA standards require that a **competent person** inspect the site daily for potential cave-ins. The site also needs to be inspected daily for failure of protective systems and equipment, hazardous atmospheres, or other hazardous conditions. Inspections are required after natural events such as heavy rains, or man-made events such as blasting that may add to the potential for hazards.

Since the chief hazard to excavation workers is the danger of cave-ins, OSHA requires that methods be used to prevent such accidents. Most of the methods used to prevent cave-ins need to be planned and approved by a registered professional engineer.

One method to prevent cave-ins is to slope the sides of the site (*Figure 11*).

The actual rise-to-run ratio of the slope will need to account for the soil type, water accumulation, and other factors. Another method to prevent cave-ins involves both sloping and shoring of the side of the site. Trench boxes or shields made of wood, aluminum, or other suitable material may also be used (*Figure 12*).

3.5.0 Chemical Hazards

Chemical use is a special type of risk at work sites. Not only do some chemicals represent a serious direct safety and health hazard, but they can also be a long-term health risk. The effect of exposure to some chemicals may not be visible until years later.

OSHA estimates that annually more than 32 million workers are exposed to 650,000 hazardous

Instructor's Notes:

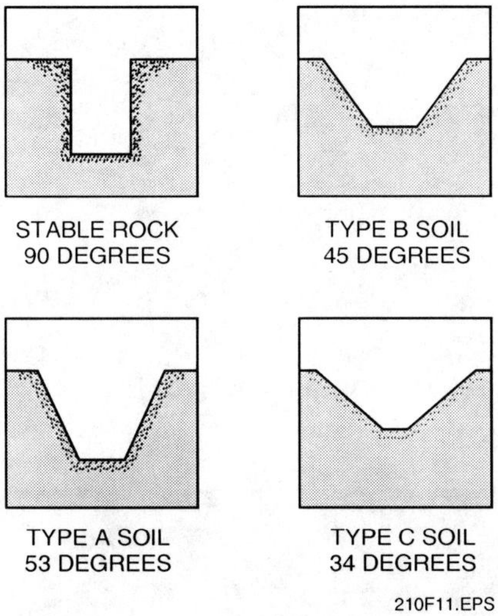

Figure 11 ◆ Sloping the sides of an excavation site.

chemical products in more than 3 million American workplaces. Many host employers will list chemical hazards and precautions on their work permits. Some companies use a safe work permit; others use a hazardous work permit.

> **NOTE**
> For more information about working safely with chemicals, please refer to the *Hazard Communication* module in Field Safety.

OSHA requires that all employers with hazardous chemicals at their sites prepare and use a written hazard communication program (*CFR 1910.1200*). The goal of this program is to be sure employers and employees know about work hazards and how to protect themselves. This should help reduce the chemical source accidents, illnesses, and injuries.

Some of the health hazards of chemicals include:

- Irritation
- Sensitivity
- Carcinogenicity (ability to cause cancer)

In addition, chemicals pose many physical hazards. They include:

- Flammability
- Corrosion
- Reactivity

The site must have a written plan that describes how the OSHA standard will be used at the site. The following are requirements for the written plan:

- The plan must label hazardous chemicals with the identity of the material and appropriate hazard warnings. See *Figure 13* for an example of a hazardous material posting.
- The plan must incorporate the use of material safety data sheets (MSDSs) for each hazardous chemical at the site. These may be obtained from the chemical manufacturers or distributors.
- The plan must include a training program that provides information to all employees who may be exposed to hazardous chemicals. This training must occur before the first work assignment involving hazardous chemicals. Further, more training is needed whenever the hazard changes.

Workers should be trained to handle chemicals safely and use approved policies and procedures, including the proper personal protective equipment.

In addition, when an accident occurs during the use of hazardous chemicals, it must be properly documented. The accident report must include the following:

- Name of the chemical
- Circumstances under which the accident occurred
- Names of all involved workers
- Length of the exposure
- Whether other chemicals were involved
- Immediate action taken

Further, the medical records of the workers involved will need similar entries. For additional information about medical records, see the *Recordkeeping* module in Volume 4.

3.6.0 Additional Types of Work Permits

Some types of jobs require additional types of work permits. The most common types of work that require permits include working on aerial lifts or around vehicles, and while doing some types of electrical work.

Aerial lifts are vehicle-mounted devices that are used to raise workers above the ground. The following are examples of how aerial lifts are used:

- Extensible boom platforms
- Aerial ladders
- Articulating boom platforms
- Vertical towers
- Any combination of the above

Explain the hazards of working with chemicals.

Describe OSHA's standards and requirements for working with chemicals.

Explain that sites with chemical hazards must have a written plan that describes how OSHA standards will be used on site. Identify the elements of the written plan.

Explain the training, reporting, and recording requirements for employees working with chemicals.

Identify the additional types of work permits that may be required for working on or around aerial lifts and vehicles.

WORK PERMIT POLICIES 10.15

Identify the different types of shoring devices that are shown in Figure 12. Share any work experiences involving these types of devices.

www.SHORING.com

www.SHORING.com

www.SHORING.com

www.SHORING.com

www.SHORING.com

210F12.EPS

Figure 12 ◆ Trench shields.

Instructor's Notes:

CAUTION

Hazardous Materials!

Safe Work Permit Required!

210F13.EPS

Figure 13 ◆ Hazardous material sign.

> **NOTE**
> There is no specific requirement for an aerial work permit; however, safety technicians should review the company safety policies regarding requirements for aerial lifts because they are so commonly used.

A number of safety precautions are involved with lifts. They include:

- Testing the lift each day prior to use
- Allowing only authorized persons to operate the lift
- Ensuring that while the lift is in operation all involved workers follow approved safety policies and procedures

> **NOTE**
> Fall protection is an important part of aerial lift safety. For more information on fall protection, please refer to *Fall Protection* in Field Safety. You can also visit this OSHA Web site (www.osha.gov/SLTC/constructionfallprotection/index.html) for more information on fall protection.

Electrical work is safest when performed on de-energized circuits using the lockout/tagout methods just described. When working on energized electrical circuits, special procedures are needed. An electrical hot work permit is required when working on or near energized equipment with a potential of 50 volts or greater above ground.

Only qualified workers may perform hot electrical work. They must be thoroughly familiar with the equipment and use proper safety precautions to avoid accidents.

Workers need to take special precautions such as electrically isolating themselves from the equipment and ground by using electrical safety shoes and proper mats. In addition, workers must de-energize all parts of the equipment that may be de-energized, and should use the lockout/tagout procedures previously described if possible.

During the performance of the work, a competent person should observe the work. You will need to ensure that correct protective devices and procedures are used and that all safety requirements are met. For hot electrical work, the competent person must have:

- Current CPR certification
- Immediate access to a telephone or radio to call for aid in case of emergency
- The ability to immediately cut off all sources of electrical power

A vehicle is defined as a car, bus, truck, trailer or semi-trailer that is used to carry employees or move materials. It may be owned, leased, or rented by the employer. Vehicles must be maintained in a serviceable condition and must be used for their intended purpose. Operators of these vehicles must have a valid permit or license in order to operate them.

Summary

Work permit policies are needed because some normal work conditions can become safety hazards when other work is being performed nearby. There are a number of work permit systems used throughout the industry, but they all have the same purpose. Permit systems are in place to reduce the chances of accidents and injures to workers and to prevent damage to or loss of property.

As the safety technician, you will need to know when work permits must be issued and the process for doing so. Lockout/tagout procedures and work permit systems are the corner stone of an effective safety program. You should audit compliance with these systems on a regular basis. This is important because when properly used, work permits help to identify and control hazards, thus helping to complete the job safely.

Explain the safety precautions that must be considered when operating a lift.

Explain the hazards and safeguards associated with electrical work.

Explain the hazards and safeguards of operating a work vehicle. Identify the types of licenses required to operate company vehicles.

Review the material covered in the module.

Have the participants complete the Review Questions and go over the answers prior to administering the Module Examination.

Administer the Module Examination. Record the results on Craft Training Report Form 200 and submit the results to the Training Program Sponsor.

Administer the Performance Test and fill out Performance Profile Sheets for each participant. If desired, participant proficiency noted during laboratory sessions may be used to complete the Performance Test. Record the results on Craft Training Report Form 200 and submit the results to the Training Program Sponsor.

WORK PERMIT POLICIES 10.17

Review Questions

1. Work permit policies are required because _____.
   a. state and local governments need to be able to control the types of work being performed in their area
   b. ordinary work conditions can become safety hazards when other work is being performed in the area
   c. workers need to have written instructions about who can be on a work site when welding is being performed
   d. sometimes a worker needs to perform a job for the first time and needs written instructions

2. As the site safety technician, you are responsible for observation of the work area while work requiring permits is being performed.
   a. True
   b. False

3. Hot work permits are needed whenever _____.
   a. the outside temperature is above 80°F
   b. welding is being done in an area specially designed for welding
   c. an open flame or ignition source is required in a specific area
   d. work on de-energized circuits needs to be done

4. After the performance of hot work, _____.
   a. all flammable material should be removed from the area
   b. the work area needs to be observed in case a fire breaks out
   c. signs need to be posted indicating that hot work has been performed
   d. all workers should leave the area in case gases have accumulated

5. Lockout/tagout policies and procedures are designed to _____.
   a. control the loss of property and equipment due to employee theft
   b. prevent unauthorized workers from using equipment
   c. keep unauthorized people away from certain parts of a work site
   d. prevent the unplanned startup of equipment during servicing

6. Little training is needed in order to perform lockout/tagout procedures.
   a. True
   b. False

7. Confined-space entry permits are _____.
   a. always required when an area is difficult to enter
   b. needed to enter a posted confined-space work area
   c. usually not needed even when the work area is small
   d. not needed to enter a posted confined-work area

8. As the site safety technician, you need to perform periodic audits to ensure that confined-space work policies and procedures are adequate and being followed.
   a. True
   b. False

9. Excavation sites need to be inspected at least _____.
   a. daily and more frequently if needed
   b. as needed
   c. twice weekly
   d. hourly

10. An electrical hot work permit is needed whenever work is to be performed on _____ electrical equipment.
    a. any
    b. de-energized
    c. energized
    d. broken

Instructor's Notes:

GLOSSARY

Trade Terms Introduced in This Module

Affected employees: Used in reference to LOTO policies and procedures. Affected employees are usually the equipment users or operators. These workers need to recognize when a control procedure is in place. They also need to understand that they are not to remove the LOTO device or try to start the equipment during a control procedure.

Authorized employees: Used in reference to LOTO policies and procedures. It refers to those employees who actually perform the servicing or maintenance on the involved equipment. These workers need a high degree of training. They also need to know about the type and magnitude of the hazardous energy and how to safety isolate and control the energy source.

Atmosphere: The air or climate inside a specific place.

Competent person: A person who is capable of identifying existing and predictable hazards in the surroundings or working conditions that are unsanitary, hazardous, or dangerous to employees, and who has the authority to take prompt corrective measures to fix the problem.

Confined space: A workplace that has a configuration that hinders the activities of employees who must enter, work in, and exit the space.

Energy control procedures: Written documents containing information that authorized workers need to know to safely control energy while servicing or maintaining equipment or machinery. These procedures are also called lockout/tagout procedures.

Energy-isolating devices: Devices that physically prevent the transmission or release of energy.

Engulfment: To be in an environment in which you are covered with a material such as sand, gravel, or grain. The end result is often death due asphyxiation, strangulation, or crushing.

Excavation: Any man-made cut, cavity, trench, or depression in the earth's surface formed by earth removal.

Lockout: The placement of lockout devices on an energy-isolating device, in accordance with established procedures. The lockout device is intended to secure the energy-isolating device in the off or safe position.

Lockout device: Any device that uses positive means, such as a lock, to hold an energy-isolating device in place. It is intended to ensure that the energy-isolating device and the equipment being controlled cannot be operated until the lockout device is removed.

Non-permit required confined space: A workspace free of atmospheric, physical, electrical, and mechanical hazards that could cause death or injury.

Other employees: All employees who are or may be in an area where energy control procedures may be utilized.

Permit-required confined space: A confined space that has real or potential hazards including atmospheric, physical, electrical, or mechanical hazards defined by OSHA.

Tagout: The placement of tags or tagout devices on an energy-isolating device, in accordance with established procedures. It is intended to ensure that the energy-isolating device and the equipment being controlled cannot be operated until the tagout device is removed.

WORK PERMIT POLICIES 10.19

Tagout device: Any prominent warning device, such as a tag, and a means of secure attachment that is placed on an energy-isolating device. The tag indicates that the equipment or machinery to which that the tag is attached should not be started or used until the authorized worker removes the tag.

Trench: A narrow excavation made below the surface of the ground. The depth of a trench must be greater than the width. The width must not exceed 15'.

Instructor's Notes:

Additional Resources

This module is intended to be a thorough resource for task training. The following reference works are suggested for further study. These are optional materials for continued education rather than for task training.

www.osha.gov

www.asse.org

Guidelines for Hot Work in Confined Spaces: Recommended Practices for Industrial Hygienists and Safety Professionals, 2000. Martin H. Finkel, CIH, CMC. Des Plaines, IL: The American Society of Safety Engineers (ASSE).

Figure Credits

Professional Safety Associates, Inc.	210F02
Charles Rogers	210F03
Veronica Westfall	210F05
Trench Shoring Services, Inc.	210F12

WORK PERMIT POLICIES 10.21

MODULE 75210-03 — ANSWERS TO REVIEW QUESTIONS

Answer	Section
1. b	1.0.0
2. a	2.0.0
3. c	3.1.0
4. b	3.1.0
5. d	3.2.0
6. b	3.2.1
7. b	3.3.0
8. a	3.3.0
9. a	3.4.0
10. c	3.6.0

CONTREN™ LEARNING SERIES — USER UPDATES

The NCCER makes every effort to keep these textbooks up-to-date and free of technical errors. We appreciate your help in this process. If you have an idea for improving this textbook, or if you find an error, a typographical mistake, or an inaccuracy in NCCER's Contren™ textbooks, please write us, using this form or a photocopy. Be sure to include the exact module number, page number, a detailed description, and the correction, if applicable. Your input will be brought to the attention of the Technical Review Committee. Thank you for your assistance.

Instructors – If you found that additional materials were necessary in order to teach this module effectively, please let us know so that we may include them in the Equipment/Materials list in the Instructor's Guide.

Write: Curriculum Revision and Development Department
National Center for Construction Education and Research
P.O. Box 141104, Gainesville, FL 32614-1104

Fax: 352-334-0932

E-mail: curriculum@nccer.org

Craft ______________________ Module Name ______________________

Copyright Date ____________ Module Number ____________ Page Number(s) ____________

Description

____

____

____

____

(Optional) Correction

____

____

____

(Optional) Your Name and Address

____

____

____

Module 75211-03

Confined-Space Entry Procedures

NCCER STANDARDIZED CRAFT TRAINING PROGRAM

The National Center for Construction Education and Research (NCCER) provides a standardized national program of accredited craft training. Key features of the program include instructor certification, competency-based training, and performance testing. The program provides trainees, instructors, and companies with a standard form of recognition through a National Craft Training Registry. The program is described in full in the *Guidelines for Accreditation*, published by the NCCER. For more information on standardized craft training, contact the NCCER by writing us at P.O. Box 141104, Gainesville, FL 32614-1104; calling 352-334-0911; or e-mailing info@nccer.org. More information may be found at our Web site, www.nccer.org.

HOW TO USE THIS ANNOTATED INSTRUCTOR'S GUIDE

Each page presents two sections of information. The larger section displays each page exactly as it appears in the Participant Module. The narrow column ties suggested participant and instructor actions to each page and provides icons (detailed below) to call your attention to material, safety, audiovisual, or testing requirements. The bottom of each page includes space for your notes.

The **Audiovisual** icon indicates an appropriate time to show a transparency or other audiovisual aid.

The **Classroom** icon prompts you to define a term, stress a point, ask participants to explain a concept, or give examples.

The **Demonstration** icon directs you to show participants how to perform tasks.

The **Examination** icon tells you to administer the written module examination.

The **Homework** icon is placed where you may wish to assign reading for the next class, to assign a project, or to advise participants to prepare for an examination.

The **Laboratory** icon is used when participants are to practice performing tasks.

The **Materials** icon is a reminder for you to gather materials needed for classes, labs, and testing.

The **Performance Testing** icon tells you to administer a performance test or a portion thereof.

The **Safety** icon is used to emphasize safety issues. It is often keyed to *Caution* and *Warning* statements in the Participant Module.

The **Teaching Tip** icon indicates additional guidance is available, such as how to conduct an exercise, get the most educational value from a field trip, or encourage class participation. Teaching Tips may expand on a feature (*Think About It*, *Did You Know?*) or provide Quick Quizzes or similar exercises. You will be referred to the Teaching Tips section at the back of the module if there is additional material.

The **Combination** icon indicates that the laboratory listed corresponds with a performance task. If desired, you can note the proficiency of the participants during the laboratory and use it to satisfy performance testing requirements.

PREPARATION

Before teaching this module, you should review the Objectives, Performance Tasks, Materials and Equipment List, and the Module Outline. Be sure to allow ample time to prepare your own training or lesson plan and gather all required materials and equipment.

MODULE OVERVIEW

In this module, participants will learn how atmospheric testing is conducted, as well as the roles and duties of all personnel involved in confined-space entry.

PREREQUISITES

Please refer to the Course Map in the Participant Module. Prior to training with this module, it is recommended that the participant shall have successfully completed the following:

Field Safety; Safety Technology, Modules 75201-03 through 75210-03

OBJECTIVES

Upon completion of this module, the participant will be able to:

1. Describe three types of atmospheric hazards in confined spaces.
2. Demonstrate and explain atmospheric testing procedures for confined spaces.
3. Explain the confined-space entry permit system.
4. Explain the different roles and duties for people working in confined spaces.
5. Explain emergency procedures in confined spaces.
6. Explain rescue procedures for confined-space entry.

PERFORMANCE TASKS

There are no performance tasks for this module.

MATERIALS AND EQUIPMENT LIST

Overhead projector and screen

Transparencies

Whiteboard/chalkboard

Markers/chalk

Blank acetate sheets

Transparency pens

Pencils and scratch paper

Appropriate personal protective equipment

Copy of *29 CFR 1926, OSHA Construction Industry Regulations*

Copy of *29 CFR 1910.146, OSHA Permit-Required Confined Spaces Regulations*

Single-gas configuration gas detection meter

Multiple-gas configuration gas detection meter

Flammability detection meters

Gas detector tubes

Module Examinations*

* Located in the Test Booklet.

SAFETY CONSIDERATIONS

Ensure that participants are equipped with appropriate personal protective equipment.

ADDITIONAL RESOURCE

This module is intended to present thorough resources for task training. The following reference work is suggested for both instructors and motivated participants interested in further study. This is optional material for continued education rather than for task training.

Web site: www.osha.gov

TEACHING TIME FOR THIS MODULE

An outline for use in developing your lesson plan is presented below. Approximately 2½ hours are suggested to cover *Confined-Space Entry Procedures*. This includes 10 minutes at the beginning of the session for administrative tasks and one 10-minute break. You may need to adjust the time required for hands-on activities and testing based on your class size and resources.

Topic	Planned Time
A. Introduction	______
B. Atmospheric Hazards in Confined Spaces	______
1. Oxygen	______
2. Flammable Atmospheres	______
3. Toxic Atmospheres	______
C. Atmospheric Testing Procedures for Confined Spaces	______
1. Oxygen Meters and Monitors	______
2. Testing Equipment – Flammability Meters	______
3. Toxic Air Contamination Testers	______
4. Monitoring	______
D. Confined-Space Entry Permits	______
E. Roles and Duties of Personnel in Confined Spaces	______
1. Entrant	______
2. Attendant	______
3. Supervisor	______
F. Ventilation and Equipment	______
1. Ventilation	______
2. Electrical Equipment	______
3. Pneumatic Tools and Equipment	______
G. Emergency Training	______
1. Loss of Air Line	______
2. Loss of Communications	______
3. Loss of Light	______
H. Rescue Procedures	______
1. Non-Entry Rescue	______
2. Entry Rescue	______
I. Review	______
J. Module Examination	______

1. Participants must score 70% or higher to receive recognition from the NCCER.
2. Record the testing results on Craft Training Report Form 200 and submit the results to the Training Program Sponsor.

Module 75211-03

Confined-Space Entry Procedures

Instructor's Notes:

COURSE MAP

This course map shows all of the modules in Safety Technology. The suggested training order begins at the bottom and proceeds up. The local Training Program Sponsor may adjust the training order.

Assign reading of Module 75211-03.

SAFETY TECHNOLOGY

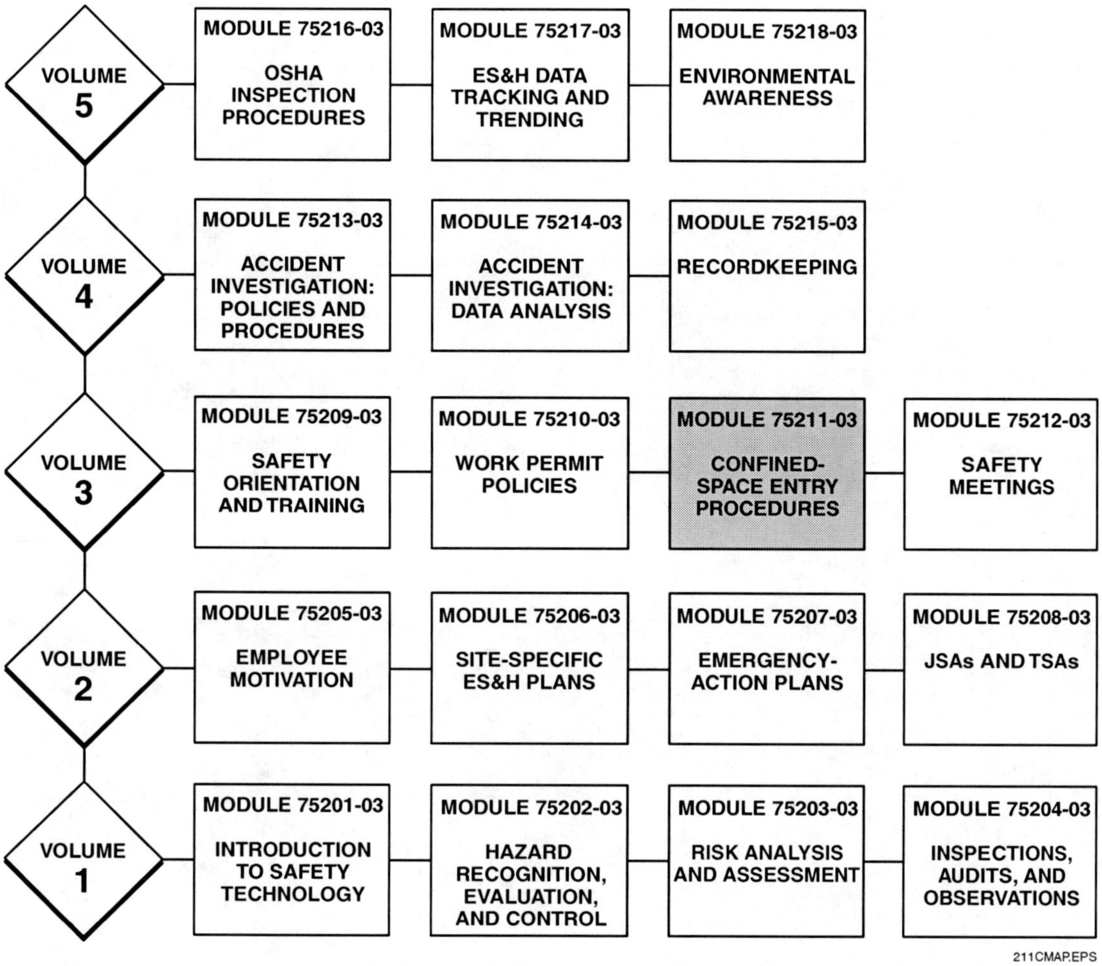

CONFINED-SPACE ENTRY PROCEDURES 11.iii

Instructor's Notes:

MODULE 75211-03 CONTENTS

CONFINED-SPACE ENTRY PROCEDURES 11.v

Figures

Table

Instructor's Notes:

MODULE 75211-03

Confined-Space Entry Procedures

Objectives

When you have completed this module, you will be able to do the following:

1. Describe three types of atmospheric hazards in confined spaces.
2. Demonstrate and explain atmospheric testing procedures for confined spaces.
3. Explain the confined-space entry permit system.
4. Explain the different roles and duties for people working in confined spaces.
5. Explain emergency procedures in confined spaces.
6. Explain rescue procedures for confined-space entry.

Prerequisites

Before you begin this module, it is recommended that you successfully complete the following: Field Safety; Safety Technology, Modules 75201-03 through 75210-03.

Required Materials

1. Pencil and paper
2. Appropriate personal protective equipment
3. A copy of *29 CFR 1926 OSHA Construction Industry Regulations*
4. A copy of *29 CFR 1910.146 OSHA Permit-Required Confined Spaces Regulations*

1.0.0 ◆ INTRODUCTION

Confined spaces hold many hazards for workers. For that reason, management of work activities in confined spaces is one of a safety technician's most important duties. As a safety technician, you may be the person who signs or audits confined-space work permits.

The term *confined space* means any space having a limited means of egress (entry and exit). These spaces are subject to the accumulation of toxic or flammable contaminants or have an oxygen-deficient atmosphere. Confined or enclosed spaces include, but are not limited to, storage tanks, process vessels, bins, boilers, ventilation or exhaust ducts, sewers, underground utility vaults, tunnels, pipelines, and open top spaces more than 4' in depth such as pits, tubs, vaults, and vessels. An enclosed space is a space, other than a confined space, which is enclosed by bulkheads and overhead. It includes cargo holds, tanks, quarters, and machinery and boiler spaces.

The most common hazards, and often the most deadly, are invisible, atmospheric hazards. These hazards can take various forms. A lack of oxygen can lead to **asphyxiation** in a matter of seconds. Flammable gas can turn a confined space into a deadly explosion hazard. Toxic gases can creep up on the worker with little or no signs or warnings.

Confined spaces must be tested before workers enter them. Gas detection meters provide the necessary information to ensure worker safety. Additional ventilation and/or personal protective equipment may be needed. Many types of detection equipment are available to test oxygen depletion or flammable- and toxic-gas concentrations. Some meters provide several tests in a single multi-use unit.

The permit system is a method of identifying confined-space hazards and verifying that all possible safety measures are taken in confined-space work. Each of the team members plays a part in ensuring the success of the system. This includes those working in the space, controlling access, conducting gas tests, or supervising the operation.

Ensure that you have everything you need to teach the course. Check the Materials and Equipment list in the front of this module.

Explain that terms shown in bold are defined in the Glossary at the back of this module.

Review the goals of the course and explain what will be expected of the participant.

Show Transparency 1, Objectives.

Define the term confined space. Describe the hazards associated with confined-space entry as well as the basic testing, permit, and emergency rescue requirements.

Explain the atmospheric conditions that make confined spaces hazardous.

Explain the physical forms of atmospheric contaminants.

In an emergency, team members must know how to react. Rescue personnel must be prepared to deal with any situation. Training and practice provide the keys to a safe, productive work environment.

2.0.0 ◆ ATMOSPHERIC HAZARDS IN CONFINED SPACES

Atmospheric hazards are the most common hazards found in confined spaces. They are also the most deadly. A hazardous atmosphere may contain:

- Too little or too much oxygen
- Explosive or flammable gases, vapors or dusts
- Toxic gases

Show Transparency 2, Atmospheric Contaminants.

Do not allow anyone to enter a confined space before complete atmospheric testing is done. If it is necessary to enter the space to complete testing, then the person entering the space must wear the appropriate respiratory protection. You must use respiratory protection until you are positive, based on test results, that it is not needed.

Atmospheric contaminants take many forms. The physical forms include:

- *Dusts* – Small particles generated by crushing solids
- *Fumes* – Small particles created by condensation from vapor state, especially volatized metals
- *Mists* – Suspended liquid particles formed by condensation from gases or by liquid dispersion
- *Smoke* – An aerosol mixture from incomplete combustion
- *Vapors* – Gaseous forms of materials that are liquids or solids at room temperature. Many solvents generate vapors.
- *Gases* – Materials that do not exist as solids or liquids at room temperature, such as carbon monoxide and ammonia

Describe how the percentage of oxygen in a confined space is critical for safety. Identify safe and unsafe levels of oxygen.

Describe the physical symptoms that occur at various levels of oxygen deficiency.

2.1.0 Oxygen

The percentage of oxygen in the confined space is critical for safety. In an **oxygen-deficient atmosphere** the concentration of oxygen is too low, and can quickly lead to asphyxiation. An **oxygen-enriched atmosphere**, where the concentration of oxygen is too high, presents a flammability/explosion hazard. The normal range for oxygen concentration is between 19.5% and 23.5%. Below 19.5% is oxygen-deficient, above 23.5% is oxygen-enriched. An ideal concentration is 21% oxygen.

Show Transparency 3, Symptoms of Oxygen Deficiency.

An oxygen-deficient atmosphere in the confined-space environment will have a rapid effect on workers. Symptoms of an oxygen-deficient atmosphere are as follows:

- *12–16% oxygen* – Deep breathing, fast heartbeat, poor attention, poor thinking, poor coordination
- *10–14% oxygen* – Faulty judgment, intermittent breathing, rapid fatigue (possibly causing heart damage), very poor coordination, lips turning blue
- *Less than 10% oxygen* – Nausea, vomiting, loss of movement, loss of consciousness, death
- *Less than 6% oxygen* – Spasmodic breathing, convulsive movement, and death in approximately eight minutes
- *4–6% oxygen* – Coma in 40 seconds, death

Explain how hot work affects oxygen levels in a confined space.

Even if the concentration of oxygen in a confined space is initially safe, many processes can cause oxygen deficiency. **Hot work** such as welding, cutting, or brazing consumes oxygen. Oxygen can also be depleted due to bacterial action such as fermentation or chemical reactions such as rust. Each additional person working in a confined space, coupled with the amount of physical activity, reduces the concentration of oxygen. The concentration of oxygen can also be reduced through displacement by other gases.

List the flammable gases found in confined spaces.

2.2.0 Flammable Atmospheres

A flammable atmosphere in a confined space is one that contains sufficient oxygen and contains other flammable gases, vapors, or dust. Common flammable gases found in confined spaces include:

- Acetylene
- Butane
- Propane
- Hydrogen
- Methane
- Natural gas

Many flammable gases are heavier than air, and will sink to the lower levels of storage tanks, pits, and vessels. Other gases are lighter than air, and may rise and develop a flammable concentration near the top of a closed top tank. It is critical that you always test for flammable gases at several depths in a confined space. You can find out if the gas is lighter or heavier than air by checking its **vapor density**. The vapor density is listed on the material safety data sheet (MSDS) in Section 9: Physical and Chemical Properties. Check the *NIOSH Pocket Guide to Hazardous Chemicals* for naturally occurring gases.

Instructor's Notes:

Asphyxiation in Oklahoma

CASE HISTORY

In Oklahoma, a 3-person work crew ruptured a water line while boring through a street to extend water service. The workers were instructed to close three valves to cut off water flow to the damaged pipe. The valves were in three separate pits. They were aware of a company policy that required atmospheric testing before entry, but they decided that shutting off the water was more important. The workers had no personal protective equipment or training for confined-space entry.

They had no trouble with the first valve. The employee who entered the second pit, which had not been opened in three years, soon called for help. The crew leader entered the pit to assist the initial entrant but was overcome. The third crewman realized that entering the pit was unsafe and went for help. Firefighters equipped with self-contained breathing apparatus were on the scene within a few minutes. They entered the second valve pit and retrieved the victims. Both victims died shortly afterward, asphyxiated due to oxygen deprivation. The accident report noted that the oxygen level at a valve pit two miles downstream from the scene of the accident was only 3 percent.

The Bottom Line: The workers were not overcome by toxic fumes or air contaminants, but by a lack of oxygen. Water can displace oxygen in a confined space. Always test confined spaces before entry.

Source: The Oregon Occupational Safety and Health Administration (OR-OSHA)

Discuss the Case History. Ask the participants to share any personal experiences they may have had with confined spaces.

NOTE

Work in other areas can affect the atmosphere in a confined space. Gases created or released above or below a confined space can rise or sink and become trapped in the confined space. You need to know the hazards around a confined space as well as hazards in the space.

Flammable or explosive conditions can also be created by working in a confined space. For example, spray painting can create explosive gases or vapors. Transferring coal, grain products, nitrated fertilizers, or finely ground chemical products can lead to a concentration of combustible dust.

Chemicals can combine to form flammable gases. For example, sulfuric acid reacts with iron to form hydrogen. Calcium carbide reacts with water to form acetylene. Both of these gases are highly flammable. Simply opening the door to a confined space and allowing the outside air in can also cause chemical reactions that form flammable gases.

Hot work in a confined space is a major cause of explosions in areas that contain combustible gas. However, flammable atmospheres can just as easily be ignited through a single spark generated from static electricity.

Explain that flammable or explosive conditions can be caused not just by pre-existing conditions but also by the work done in the confined space, such as welding or spray painting.

2.2.1 Measuring Flammable Atmospheres

In order to check the flammability of an atmosphere, you'll need to know the limits associated with particular gases. These limits are available in Section 5 of the MSDS. You can also find them in the *NIOSH Pocket Guide to Chemical Hazards* or other reference sources.

The **lower flammable limit (LFL)**, also known as the **lower explosive limit (LEL)**, is the lowest concentration of air-fuel mixture at which a gas can ignite. The **upper flammable limit (UFL)**, or **upper explosive limit (UEL)**, is the highest concentration that can be ignited. Explosion can occur at any point between the LFL and UFL. A confined space should not be considered safe unless the concentration of gas is well below the LFL.

The potential for a flammable atmosphere increases as the temperature rises (*Figure 1*). A rise in temperature, such as from morning to afternoon, will require additional testing. For example, assume a hazardous material is present and its flash point is 50°F. A test run with a combustible gas indicator in the morning when the temperature is 34°F may not show a hazardous atmosphere. However, in the afternoon, when the temperature increases to 65°F, you would likely get a reading at or above the LFL. Confined spaces must be tested periodically to confirm that they are still safe to enter.

Identify Section 5 on an MSDS as the place to look for information about the flammable and explosive limits of particular gases.

Define the following terms: lower flammable limit (LFL), lower explosive limit (LEL), upper flammable limit (UFL), and upper explosive limit (UEL).

Explain how the potential for a flammable atmosphere increases as temperature rises.

Show Transparency 4 (Figure 1).

CONFINED-SPACE ENTRY PROCEDURES 11.3

Explain OSHA's standards for permissible exposure limits (PELs).

Identify the sources and/or causes of toxic atmospheres.

Identify Section 2 on an MSDS as the place to look for PELs for toxic substances.

Explain that PELs can also be found in the *NIOSH Pocket Guide to Chemical Hazards*.

Identify the PELs for common toxic gases.

Show Transparency 5 (Table 1).

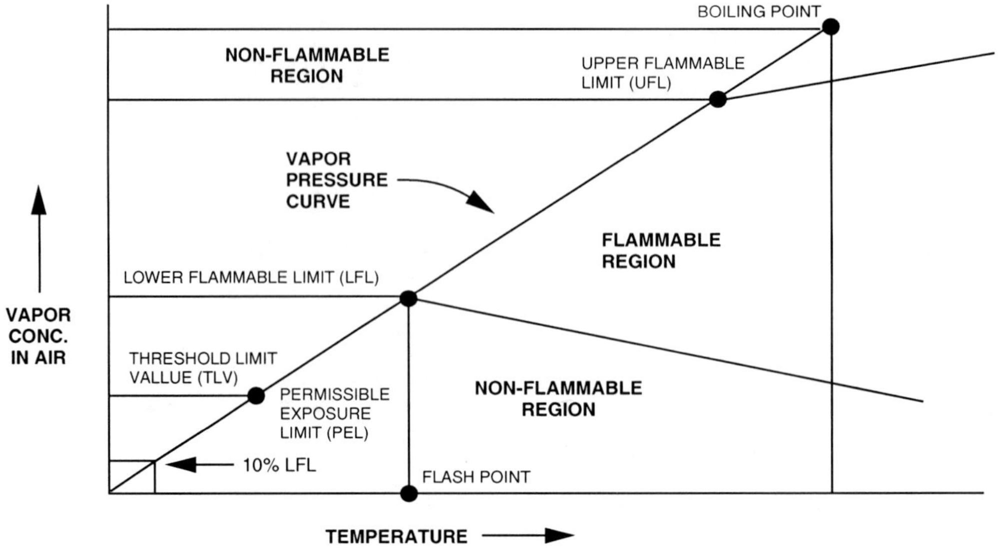

Vapor pressure, temperature diagram shows the relationship among upper and lower flammable (explosive) limits, flammable and non-flammable regions, threshold limit value, boiling point, flashpoint, and vapor pressure curve. This diagram shows what happens to a vapor / air mixture as concentrations and temperature vary.

211F01.EPS

Figure 1 ◆ Flammable vapor thresholds.

The **permissible exposure limit (PEL)** is set by OSHA standards at 10% of the LFL. If the concentration is greater than 10% of the LFL, you may not enter, according to OSHA. You must ventilate the area to reduce the concentration of flammable vapors before you enter. Technically, the air would not be flammable at 90% of the LFL. However, the air in a confined space often has pockets of concentrated vapors. It is impossible to test every part of the air in a space, so OSHA set 10% of the LFL as a safe limit for entry.

Technically, a concentration above the UFL would not be flammable. But it is not safe to enter a space above the UFL. The air can easily become diluted when you enter. The concentration will sink below the UFL into the flammable range. The atmosphere may also contain an oxygen-deficient or toxic atmosphere.

Flammable vapors may be diluted through ventilation, or through the process of **inerting**. Inerting is when a noncombustible gas such as nitrogen is released into the confined space so that the atmosphere is no longer combustible. Inerting will also drive out oxygen, creating an oxygen-deficient atmosphere.

2.3.0 Toxic Atmospheres

There is a wide range of substances in gas or particulate (dust) form that can cause or contribute to a toxic atmosphere in a confined space. Toxic atmospheres result from several sources, including manufacturing processes, storage procedures, and operations such as welding or paint application. Human error and mechanical malfunctions may also contribute to the production of toxic gases.

PELs for toxic substances can be found in Section 2 of the MSDS. A PEL is listed in **parts per million (ppm)**. You can also find them in the *NIOSH Pocket Guide to Chemical Hazards* or other reference sources. The PEL for a particular substance is the maximum concentration of a substance that a worker may be exposed to in an 8-hour shift without any adverse effects. MSDSs also provide chemical-specific toxicity characteristics, health hazards, and reactivity hazards. PELs for several common toxic gases are listed in *Table 1*.

Instructor's Notes:

DID YOU KNOW?

Hydrogen Sulfide Gas

Hydrogen sulfide gas is a toxic gas that smells like rotten eggs. It is formed when waste containing sulfur is broken down by bacteria. Sewers, septic tanks, manholes, and well pits often contain hydrogen sulfide gas. You can smell hydrogen sulfide at a level of about 10 parts per billion (ppb) or 0.01 ppm. This is about the same as a thimbleful of hydrogen sulfide gas in a house full of air.

If you are exposed for a while, you will not be able to smell the gas. At 50 to 100 ppm of hydrogen sulfide gas in air, your sense of smell begins to break down. You must not rely on your sense of smell to decide if hydrogen sulfide gas is present. Always use an instrument to detect hydrogen sulfide gas.

Carbon Monoxide

Carbon monoxide (CO) is a hazardous gas that is colorless, odorless, and has roughly the same density as air. It can be fatal at 1,000 ppm. It is dangerous at 200 ppm because it quickly replaces oxygen in the bloodstream. Early symptoms of CO intoxication are nausea and headache.

You should know that a safe reading on a combustible gas indicator does not mean that CO is not present. You must test specifically for carbon monoxide. CO is the result of incomplete combustion of materials such as wood, coal, gas, oil, and gasoline. It can also be formed from decomposition of organic matter in sewers, silos, and fermentation tanks. Fatalities due to CO poisoning are not limited to any particular industry.

Explain the reasons for atmospheric testing. Describe the basic process for atmospheric testing in confined spaces.

Show the participants both a single- and multiple-gas configured gas detection meter. Explain how they are used to test the atmosphere.

Explain that testing equipment must be calibrated before initial atmospheric testing begins.

Table 1 PELs of Common Toxic Gases

SUBSTANCE	PEL(PPM)
CARBON DIOXIDE	5,000
CARBON MONOXIDE	50
HYDROGEN SULFIDE	20
METHANE	1,000
NITRIC OXIDE	25
OXYGEN DIFLOURIDE	0.05
PHOSGENE (CARBONYL CHLORIDE)	0.1
SULFUR DIOXIDE	5
STODDARD SOLVENT	500

211T01.EPS

3.0.0 ◆ ATMOSPHERIC TESTING PROCEDURES FOR CONFINED SPACES

You must test the internal atmosphere of a confined space before anyone enters and periodically thereafter if conditions change. It must be tested for oxygen content, flammable gases and vapors, and toxic air contaminants, in that order. Anyone entering the space must be allowed to observe the pre-entry testing. Testing provides both an evaluation of the hazards and verification of acceptable entry conditions.

Test equipment that measures both oxygen and a range of flammables is the best choice for initial testing. In a rapidly changeable atmosphere such as a sewer, the test equipment will be carried and used by the entrant. He or she will monitor the atmosphere. The entrant must also test in the direction of movement before advancing. Tests must be repeated periodically to warn the entrant of any deterioration in atmospheric conditions.

In some instances, the attendant may monitor the gas testing instrument from the outside. The attendant will alert the entrants if there is a problem.

3.1.0 Oxygen Meters and Monitors

Gas detection meters are available in both single- and multiple-gas configurations (*Figure 2*). Regardless of the type of meter, the oxygen concentration must be tested first. Test equipment may give misleading results in an oxygen-deficient atmosphere.

For example, an atmosphere might consist of 90% methane and 10% air. If you did not test for oxygen and went straight to a flammability test, the meter would show that this atmosphere is not explosive because there isn't enough oxygen for combustion. As outside air moves into the space, the atmosphere becomes highly explosive. A safe oxygen level (between 19.5 and 21.5%) must be the first step.

3.1.1 Testing Procedures for Oxygen

Before beginning any atmospheric testing, the test equipment must be calibrated according to the manufacturer's instructions. Calibration ensures that the equipment provides accurate, dependable

CONFINED-SPACE ENTRY PROCEDURES 11.5

Explain the procedure for initial confined-space testing for oxygen concentration. Explain that initial testing results must be recorded on the permit.

Explain that the confined space must be periodically monitored for oxygen concentration. Identify the conditions after which retesting must be done. Describe the procedure for periodic monitoring.

Explain the purpose and use of single-gas disposable test meters.

Explain that meters used to test flammability are often used to test oxygen levels. Show the participants examples of flammability detection meters.

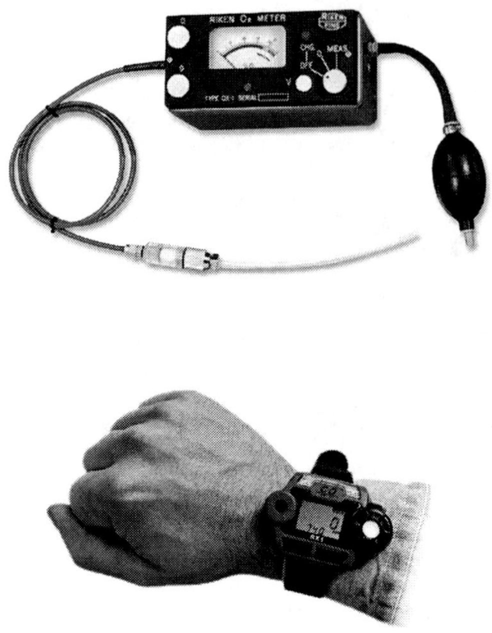

211F02.EPS

Figure 2 ◆ Gas detection meters.

readings of oxygen concentrations. Oxygen meters must be adjusted to read 20.9%. Combustible gas detectors and toxic gas monitors should read zero.

Begin testing at the entrance to the space, moving the meter probe further into the space at four-foot intervals. Do not enter the space to test without appropriate respiratory protection.

3.1.2 Monitoring

The initial testing for oxygen concentration must be noted on the permit. Ventilation may be necessary. You must retest after ventilation and have normal oxygen levels before entering the space.

The space must be periodically monitored for oxygen. There is no guarantee that the space is still safe the next day, or even an hour later. Retesting must occur after any hot work in the space. Activities that create combustion consume oxygen.

Retesting must be performed at regular intervals. This depends on several factors, including the size of the space, the number of entrants and the work being performed. At a minimum, retesting must occur once each shift for the duration of the work permit.

Continual testing can be achieved by using single-gas disposable test meters. These units are similar to household smoke detectors and do not require calibration. They have an audible alarm when oxygen concentration drops below normal levels. They may be placed in the confined space for the duration of the work permit.

3.2.0 Testing Equipment – Flammability Meters

The meters used for flammability testing are often also used for oxygen testing (*Figure 3*). Meters that test for flammability can be set to detect a combustible gas as a percentage of LEL.

211F03.EPS

Figure 3 ◆ Flammability detection meters.

3.2.1 Testing Procedures for Flammable Atmospheres

Some flammable gases, such as propane and methane, are heavier than air, and will collect at the bottom of a confined space. Others, such as hydrogen, are lighter, and will rise to the top. It is

Instructor's Notes:

not sufficient to test the atmosphere only near the entrance of a confined space. The air may be very different at different levels inside the space. Tests should be done at a minimum of 4' intervals from top to bottom and bottom to top.

As with oxygen meters, the equipment must be calibrated according to the manufacturer's instructions before testing.

Flammable detection equipment must have an alarm before the concentration reaches the LEL. OSHA standards define the highest permissible level of flammable gas in a confined space at 10% of the LEL for that particular gas. This does not mean that the atmosphere can be composed of 10% flammable gas.

Testing for concentrations of flammable dust is not so straightforward. A good rule of thumb is that the concentration of dust that meets or exceeds its LEL may be approximated as a condition in which the dust obscures vision at a distance of 5' (1.52 m) or less. Do not attempt to work in an environment with such a high concentration of airborne dust.

3.2.2 Monitoring

As with oxygen testing, conditions may change. The initial test for flammable gases may not be valid over time. You must periodically retest to make sure the work area is safe. This is particularly important if the ambient temperatures vary widely throughout the day.

Certain work activities can contribute to the amount of flammable gas in the atmosphere. Operations within the space, such as purging welding lines, will contribute to the concentration of flammable gases. The application of protective coatings and paints can also create hazardous atmospheres. Operations outside the space can also change the air in a confined space. For example, a release of methane, a dense gas, on the ground may not immediately affect that floor, but can quickly fill a connected below-ground space.

Testing must be redone after any operation that might increase gas concentrations. It should also be carried out on at least once a shift for the duration of the work permit.

3.3.0 Toxic Air Contamination Testers

Toxic gases may be discovered using gas detection meters (*Figure 2 and 3*). There are many toxic gases that are not typically included on multi-use or single-use meters. You may also need gas detector tubes to check for toxic atmospheres (*Figure 4*).

Each tube tests for the presence of a different toxin. It indicates if the toxic agent is present, usually by a change in color. The tubes are a single-use product, so you'll need a new tube for each test.

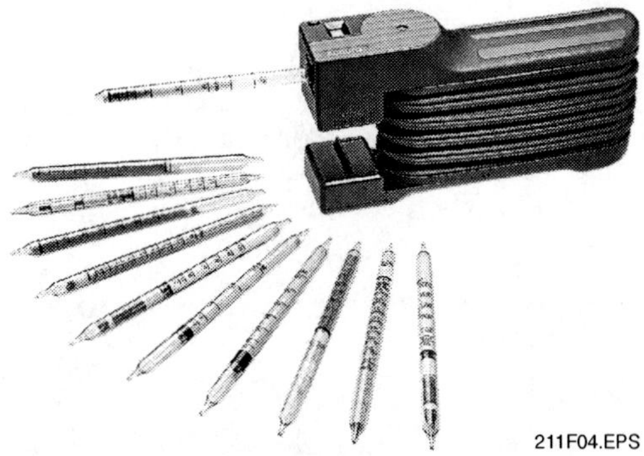

211F04.EPS

Figure 4 ◆ Gas detector tubes.

3.3.1 Toxic Air Contamination Testing

Before testing for toxic chemicals, make sure that oxygen and flammability tests have been completed and the atmosphere is safe. Calibrate the test meter (if it allows for toxic gas testing) and check for the presence of toxins. Note that toxins, like flammable gases, may be heavier or lighter than air. You must test at a minimum of 4' intervals. All areas of the space will need to be tested.

If air toxins have been identified, or are suspected, you must determine the specific concentration of contaminant. This is done by using a gas detector tube. First, select the appropriate tube for the suspected hazard. After calibrating the pump, open the tube and insert it in the pump or the sample line. The pump pulls the air sample through the tube. The toxin changes the color of the material in the tube. The concentration of toxic gas can be read directly from the markings on the tube. It is often difficult to tell where the color changes on the tube. This gives the tubes limited accuracy.

Single-use gas detection tubes have limitations. You must fully understand these limitations when you use them. These include:

- Other gases may cause an inaccurate reading.
- Readings may vary among different tube manufacturers.
- You must only use tubes from the same pump manufacturer.

Always read and follow the instructions on the information sheet provided with each package of tubes.

Describe the behavior of flammable gases in a confined space.

Discuss the procedures for flammability testing. Explain how flammability detection meters are used.

Discuss OSHA standards for flammability testing.

Describe the procedure for monitoring flammable gases thoroughout the entire job. Explain the similarities to oxygen monitoring.

Explain how gas detector tubes are used. Show the participants examples of gas detector tubes.

Explain the procedure for toxic air contamination testing. Explain the limitations of single-use gas detection tubes.

CONFINED-SPACE ENTRY PROCEDURES 11.7

Explain the conditions that require retesting for toxic gases. Identify safe intervals for retesting.

Explain the confined-space permit system. Describe the procedure and requirements for obtaining a confined-space entry permit. Identify the information that must be on a confined-space entry permit.

3.4.0 Monitoring

After the initial testing of a confined space, you must purge or dilute any toxic levels of gases. You will need to periodically retest. Retesting must be done after any process in or near the space that might produce toxic gases. At a minimum, it must be done at least daily for the duration of the work permit.

> **WARNING!**
> When wearing respiratory protection to protect against toxic vapors, you must make sure that the filter is rated for that particular toxin.

> **NOTE**
> You must always check your instruments for leaks. A leak in the sampling hose or pump will significantly change the readings. In addition to calibrating the instrument, you must perform a leak check as directed by the manufacturer. This should be done each time you use the equipment.

Show Transparency 6, Entry Permit Contents.

4.0.0 ◆ CONFINED-SPACE ENTRY PERMITS

The **confined-space entry permit system** is the written procedure for preparing and issuing permits for entry, and for returning the permit-required space to service after entry work is completed. Confined spaces requiring permits are those with a high likelihood of atmospheric or other hazards.

Before work can begin, the designated supervisor must approve the entry permit by signing it. The authorized permit will be posted at the entry point or made available to all entrants at the time of entry. The duration of the permit may not exceed the time required to complete the assigned task or job listed on the permit.

The supervisor will cancel the entry permit when operations covered by the permit have been completed. They will also cancel the permit if something occurs in or near the permit-required space that is not allowed by the permit.

The supervisor will note any problems encountered during the entry operation on the permit. The employer will maintain a file of canceled entry permits for at least one year.

The entry permit will include the following:

- Identification of the space to be entered
- Purpose of entry
- Date and authorized duration
- A means of tracking authorized entrants within the permit-required space, for example, a list of names, a roster, or other tracking system
- Names of all authorized attendants
- Name of the current entry supervisor, with a space for the signature or initials of the entry supervisor who originally authorized entry
- Known hazards of the confined space covered by the permit
- Measures used to isolate the space and to eliminate or control hazards before entry, including lockout or tagging of equipment, as well as procedures for purging, inerting, ventilating, and flushing spaces
- Acceptable entry conditions
- Test results (both initial and periodically recurring) will be accompanied by time, date, and the testers' names or initials
- Responsible rescue and emergency services and the means for calling for those services
- Communication procedures used by entrants and attendants to maintain contact
- Equipment to be provided, including PPE, test equipment, communications gear, alarm systems, and rescue equipment
- Other information needed in order to ensure employee safety
- Any additional permits that have been issued to authorize work in the space

Identify the personnel that will be working in or near confined spaces.

Discuss in detail the roles and duties of the confined-space entrant.

5.0.0 ◆ ROLES AND DUTIES OF PERSONNEL IN CONFINED SPACES

Working in permit-required spaces is a team effort involving entrants, attendants, and supervisors. All of these personnel must be trained to do the tasks they are expected to perform before any work in a confined space. Entrants, attendants, gas testers, and entry supervisors should have skills training.

Show Transparency 7, Entrant Roles and Duties.

5.1.0 Entrant

Authorized entrants are those permitted by an employer to enter a permit-required space. Each entrant must be aware of the permit-required space hazards, including the symptoms and consequences of exposure. They must be trained on the proper use of all required test and work equipment.

Entrants must remain in regular communication with the authorized attendant. They must immediately notify the attendant if hazardous conditions, or warning signs of those conditions, arise. Entrants will leave the confined space immediately if hazardous conditions arise.

Instructor's Notes:

Entrants will also leave if ordered to evacuate by the attendant.

5.2.0 Attendant

Authorized attendants are those who monitor the entrants' activities from outside the space. The attendant must know all of the permit-required space hazards, including symptoms and consequences of exposure. He or she must monitor the entrants. This includes knowing at all times how many entrants are in the confined space and making sure that everyone involved is aware of the permit requirements.

The attendant will stay out of the space during entry operations. The attendant's duties are as follows:

- Keep unauthorized personnel away from the space.
- Maintain communication with the entrants.
- Keep track of the status of each entrant.
- Monitor activities inside and outside the confined space at all times.
- Determine if it is safe for entrants to remain in the space.

The attendant will order an evacuation if a hazardous condition arises. This includes detection of behavioral symptoms of hazard exposure or a problem outside the space that could harm the entrants.

If necessary, the attendant will activate approved rescue procedures. A non-entry rescue will be used if possible. Summoning rescue and other emergency services may be required, if the attendant determines that entrants may need assistance to evacuate. The attendant must not attempt an internal rescue, except as part of a trained rescue team. In that case, another person must assume the role of attendant.

5.3.0 Supervisor

The entry supervisor makes sure that attendants and entrants follow entry-permit procedures. The supervisor needs to know all of the permit-space hazards, including the symptoms and consequences of exposure. He or she must verify that:

- The entry permit is accurate and current.
- All permit tests have been conducted.
- All permit procedures and equipment are in place.
- Personnel comply with the provisions of the permit.
- Rescue personnel will be available in an emergency.
- Entry operations are consistent if another authorized person replaces an attendant or an entrant.

The supervisor must monitor the job and stop all entry operations and cancel the entry permit during a hazardous condition or when permit-required confined-space work is done.

6.0.0 ◈ VENTILATION AND EQUIPMENT

To work safely in a confined space, you must have adequate ventilation and some specialized tools. A ventilation system must be set up to circulate fresh air into a space if supplied air is not used. Also most power tools create a spark hazard. If flammable vapors are suspected, special tools will be needed to prevent fires or explosions.

6.1.0 Ventilation

Ventilation is a critical aspect of confined-space entry. Several methods are used to circulate fresh air into a confined space. These methods include:

- Natural ventilation
- Air movers
- Air horns
- Exhaust systems

Natural ventilation is simply opening the hatches to allow airflow into the space. It is not usually enough to circulate fresh air into a confined space. You must be aware of possible dead zones. These are areas where air can become trapped. These areas have the potential to contain a hazardous atmosphere even after the space is ventilated. Continuous air monitoring is recommended when using natural ventilation. As mentioned, working in the space will create hazardous atmospheric conditions.

A positive means of ventilation must be established. It must be operated the entire time work is in progress. The accepted practice and recommended level of ventilation is one complete air change every three minutes.

An air horn is often used to circulate air through a vessel as shown in *Figure 5*. Air horns should be set to exhaust air out of the space. This will draw in clean air from other openings. Pneumatic air movers must use compressed air. Never use other gases such as nitrogen. Air movers or fans must be electrically bonded to the vessel. The vessel must also be grounded.

Air movers should be placed at the top of a vessel when vapors that are lighter than air may be present. This is the most common ventilation configuration. More than one air mover and fresh air

Discuss in detail the roles and duties of the confined-space attendant.

Show Transparency 8, Attendant Roles and Duties.

Discuss in detail the roles and duties of the confined-space entry supervisor.

Show Transparency 9, Entry Supervisor Roles and Duties.

Describe how ventilation systems help eliminate some of the hazards of working in a confined space.

Identify and describe the methods used to circulate fresh air into a confined space: natural ventilation, air movers, air horns, and exhaust systems.

Show Transparencies 10 through 12 (Figures 5 through 7).

CONFINED-SPACE ENTRY PROCEDURES 11.9

Explain that when workers must enter at the same opening that is used for ventilation, the equipment should minimize the restriction that may occur.

Explain the conditions under which retrieval equipment should be used.

Describe the safety precautions that must be considered when using electrical tools in a confined space.

Identify the minimum requirements for electrical tools, lights, and equipment that are used in confined spaces.

Define the term intrinsically safe in relation to power tool operation in a confined space.

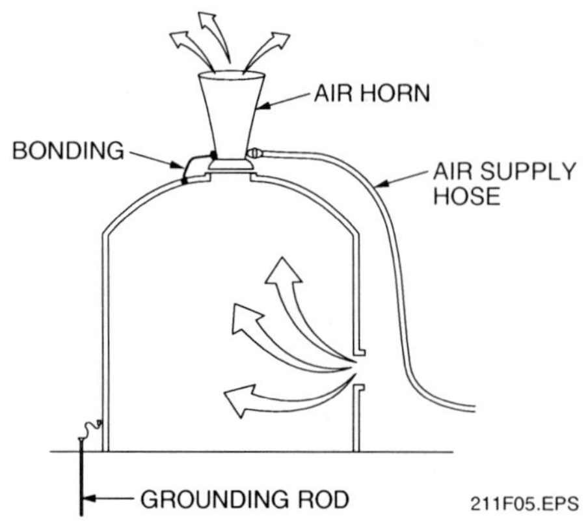

Figure 5 ◆ Air horn ventilation with bonding and grounding.

inlet should be used whenever possible to ensure adequate ventilation of the entire space.

Air movers should be placed at the bottom of a space when vapors that are heavier than air may be present. When this system is used, be certain the exhausted air does not create a hazard for area personnel. Air movers or fans must not be positioned to blow into a space. The use of a compressed air hose to ventilate a vessel or space is prohibited.

Figure 6 depicts a common ventilation system for subsurface spaces. In this instance, fans move clean air into the space. The system must be upwind so only clean, fresh air is drawn into the space. If it is downwind, the fans will blow the bad air back into the space. The system must be so arranged to allow complete purging of the space. The exhausted air must be located so it will not be a hazard to other personnel or equipment.

Additional air may be blown into a space in addition to purging ventilation. This provides clean air for general comfort and breathing. Make-up air must not be blown into the space when there are flammable or toxic materials present or being generated by the work in progress. Only explosion-proof electric fans may be used to ventilate vessels or spaces that may contain flammable gases, vapors, or dust.

Local exhausting may also be required in addition to forced ventilation. A local exhaust system (*Figure 7*) is used when the work generates fumes or vapors. Welding is a common example. The exhaust from this local exhaust system should be located so it does not present a hazard to other personnel or equipment.

When the worker must enter at the same opening that is used for ventilation, the equipment should minimize the restriction that may occur. This system is described later in this module. A flexible hose in a manhole would be an example. Retrieval equipment should be used whenever:

- Rescue may be difficult
- There is a possibility of a fall
- Self contained or supplied breathing air is used

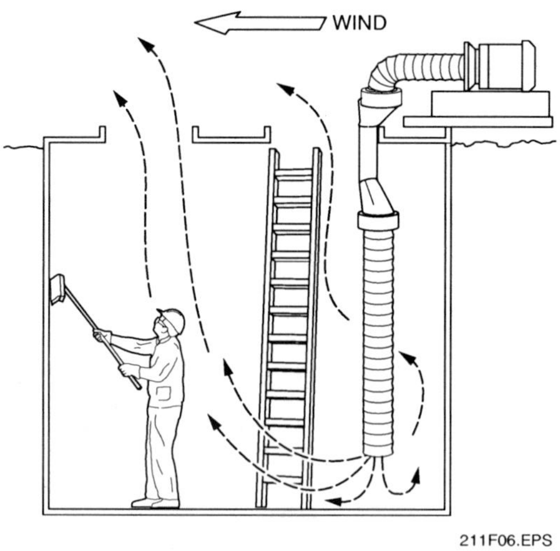

Figure 6 ◆ Positive ventilation system.

6.2.0 Electrical Equipment

Portable electric hand tools pose an additional hazard in a confined space. Use of these tools should be discouraged. If portable electrical tools must be used, they must be protected by a ground fault circuit interrupter (GFCI). Portable electric tools, lights, and equipment used in confined spaces must meet the following minimum requirements:

- Industrial grade
- In good condition
- Operated at a maximum of 110 volts AC and connected to a GFCI or operated at more than 12 volts DC

Additional care must be taken when using power tools where there is a potential for a flammable atmosphere. These tools must be **intrinsically safe**. This means that they do not pose an explosion hazard under normal use. The tools must be certified by a nationally recognized testing laboratory like Underwriters Laboratory (UL) as being approved for use in hazardous locations.

Instructor's Notes:

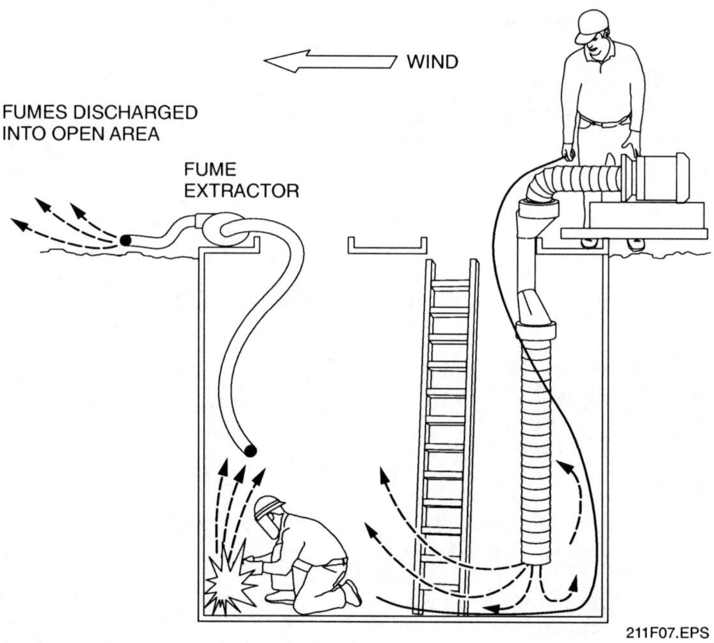

Figure 7 ◆ Fume extractor used for local exhaust.

6.3.0 Pneumatic Tools and Equipment

Pneumatic tools used in confined spaces must be driven by air. Using any other compressed gas to drive tools can create fire or suffocation hazards. Using pneumatic tools does not eliminate the fire hazard posed by using electric tools. It only reduces the hazard by eliminating the electrical energy source.

Compressed gas cylinders must not be brought into confined spaces, unless they are part of a **self-contained breathing apparatus (SCBA)** or an emergency escape unit. The only gas cylinders that should be brought into a confined space are those used for respiratory protection.

7.0.0 ◆ EMERGENCY TRAINING

In addition to hazard analysis training, all entrants, attendants, and supervisors must be trained in emergency escape procedures. Operating procedures for handling emergency situations must be reinforced with training and practice. You should not accept written procedures without testing them in simulated emergency scenarios. Effective procedures and practices can only be established through real-world testing. Training must include scenarios dealing with loss of air line, communications, and/or light.

7.1.0 Loss of Air line

A supplied air respirator includes an air line connected to an outside air supply. Emergency training must include the possibility of air line failure. SCBA can also fail without warning. If this occurs, the entrant must be prepared to evacuate the confined space as quickly as possible.

A limited air supply must be used efficiently. A personal egress bottle will provide a short supply of air for the entrant. You should teach and practice air conservation, such as controlled breathing techniques, for situations in which the egress bottle is either too limited or not available.

It is usually wise to maintain an air line connection during escape procedures, even if the air line has failed. It is possible that the situation that caused the airline to fail has been corrected. This provides a supply of air to the escaping entrant. If rescue efforts are needed, the air line connection may provide the best means of finding the entrant, especially in a complex space.

Entrants using SCBA and working as a team must be equipped with buddy breather hose attachments. If one entrant's SCBA fails, a fellow entrant may then supply them with breathable air until they both escape. Once again, training and practice will provide the difference between a potentially fatal panic and an orderly, safe evacuation.

Explain the hazards and safeguards for using pneumatic tools and equipment in confined spaces.

Discuss the importance of emergency training for all personnel working in or near a confined space. Explain that emergency training must include scenarios dealing with loss of air line, communications, and/or light.

Explain what can happen to entrants when there is a loss of air line. Describe the precautions and procedures that must be used to save and protect entrants, especially the use of SCBA equipment.

CONFINED-SPACE ENTRY PROCEDURES 11.11

Explain the role of communications in confined-space entry work.

Describe the back up devices that may be used when there is a loss of communication.

Explain the type of evacuation training entrants need to safely exit an emergency situation involving the loss of light. Describe the secondary lighting sources that can be used when there is a loss of light.

Explain the purpose and necessity of establishing rescue procedures. Emphasize the importance of training and the use of proper equipment for all rescue personnel.

Describe in detail non-entry rescue procedures. Describe the role of the attendant during a non-entry rescue.

Show Transparency 13 (Figure 8).

Explain each step of a typical rescue plan.

Show Transparency 14, A Typical Rescue Plan.

7.2.0 Loss of Communications

Communications equipment failure must be considered in any emergency training plan. Whatever the primary form of communication, alternatives and backup procedures must be established and practiced. This training will also establish when the attendant should call in rescue team personnel after communication with entrants is lost.

A secondary communications device is one option. Another option is a personal distress device or some other signaling device. Emergency procedures could also be initiated by using a whistle, small air horn, tapping a tool, or a similar non-electronic means. Whatever tools are used must be used in training, and tested regularly in practice. Each participant must understand the signals used.

7.3.0 Loss of Light

An emergency situation may involve a loss of power and loss of primary lighting. Entrants may need training to evacuate the space with little or no light. This includes accounting for team members, managing air supply, maintaining communication, and operating rope systems.

You should consider using secondary light sources. Each entrant can use a combination of flashlights, helmet lights, and chemical light sticks. In an emergency, the attendant should set up approved lighting at the entrance. This will mark the escape portal for entrants still in the space.

8.0.0 ◆ RESCUE PROCEDURES

Rescue procedures are necessary if one or more entrants are incapacitated or unable to exit without assistance. Effective rescue operations require adequate planning and training.

Rescue procedures must be set up in advance and will be specific to the space. Rescue personnel may be team members with specialized training, or a separate group dedicated to rescue operations.

Sixty percent of all confined-space fatalities are would-be rescuers. Case histories of confined-space accidents often mention one or more rescuers who died with the people they were trying to save. While time is certainly an important element in rescuing a co-worker, you should never attempt any rescue without proper training and equipment.

8.1.0 Non-Entry Rescue

Non-entry rescue (*Figure 8*) is the preferred method for extracting an entrant from a confined space. No one should enter a confined space to respond to an emergency unless they have been properly trained and equipped.

If a confined-space rescue is necessary, the attendant will summon the emergency rescue team. The attendant can also attempt to rescue entrants using only non-entry rescue equipment, unless it would increase the entrant's risk of injury. During an emergency, the attendant will monitor the situation. He or she must inform the rescue team of the:

- Number of victims
- Victims' condition
- Hazards in the space

Retrieval systems should be used every time an entrant enters a confined space, unless it would increase the overall risk of entry or would not aid a rescue. Each entrant must wear a chest or full-body harness with an attached retrieval line. The line should be attached at the center of the entrant's back near shoulder level or at some other point that presents a profile small enough for the successful removal of the entrant. Wristlets are another option if the use of a harness is not feasible or might create a hazard to the entrant.

Attach the other end of the retrieval line to a mechanical device or fixed point outside the space so that rescue can begin as soon as possible. A mechanical device must be available to retrieve personnel from vertical spaces more than 5' (1.52 m) deep.

A typical rescue plan would include the following steps:

Step 1 As soon as a problem is discovered, the attendant gives the order to evacuate.

Step 2 If an entrant cannot vacate the space under her or his own power, the attendant will contact the rescue team. The attendant must not enter.

THINK ABOUT IT

Personal Distress Units

The entrants on your team use wireless personal distress units, which provide an instant signal to the attendant if anything goes wrong. Will these devices provide an appropriate level of safety in a space lined with steel? What about reinforced concrete? Will line of sight affect either of these scenarios?

Instructor's Notes:

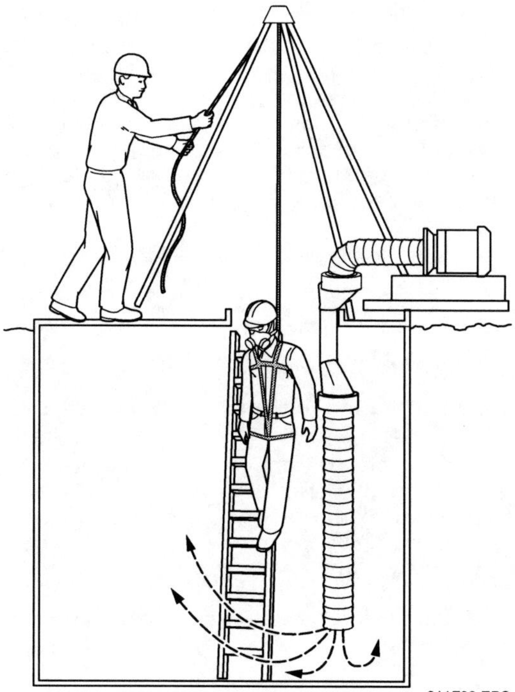

211F08.EPS

Figure 8 ◆ Non-entry rescue.

Step 3 If the attendant is certain that the entrant is incapacitated due to non-environmental causes, the entrant should not be moved until the rescue team arrives and directs otherwise.

Step 4 If the entrant is incapacitated due to environmental causes, and is wearing a safety harness, the attendant and standby entrants must use the lifting device to remove the employee from the space.

8.2.0 Entry Rescue

Only designated emergency responders can enter a confined space during an emergency. Each emergency responder must have full confined-space entrant training, in addition to emergency training. An emergency response team responding to a confined-space emergency is shown in *Figure 9*.

The rescue team may include employees or an outside emergency response team. The supervisor must ensure that they have the ability to perform rescue services and are able to reach a victim in time depending on the hazards present.

Rescue-team members must practice confined-space rescue operations at least once annually. Practice should include:

- Hazard analysis
- Escape procedures
- Rescue team preparedness
- Use of all retrieval equipment and personal protective equipment

Practice escape procedures must have real challenges. They should include rapidly changing environments and problems with protective gear, especially failures in breathing apparatus. Consider all possibilities, including loss of air line, loss of communications, and loss of lighting. Practice should be in simulated spaces similar to actual permit-required spaces, or in the spaces themselves.

The safety of entrants is the main objective of any rescue operation in a hazardous environment. Before entering the confined space, rescuers must understand the physical layout and be aware of any hazards inside. All rescue personnel must be equipped with appropriate personal protective equipment and the training necessary to use it. They must also be trained in basic first aid and cardiopulmonary resuscitation (CPR).

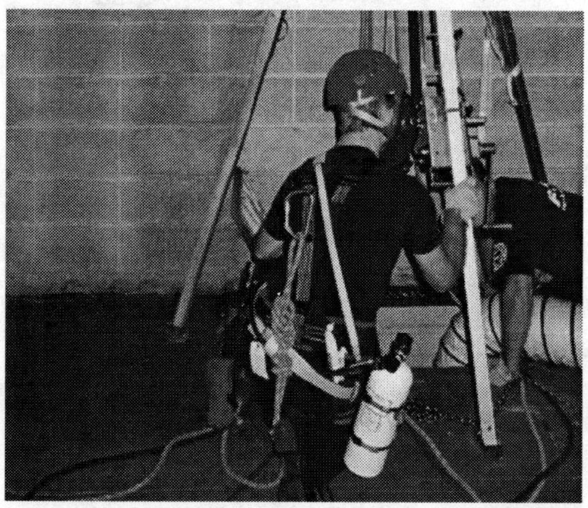

211F09.EPS

Figure 9 ◆ Emergency response team responding to a confined-space emergency.

Refer to the Think About It. Open a discussion about the effectiveness of wireless personal distress units.

Explain that only designated emergency responders can enter a confined space during an emergency. Explain that responders can be company employees or outside emergency response personnel.

Identify the skills that emergency responders must practice at least annually. Explain that emergency responders must know the layout of the confined space and that they must be equipped with appropriate personal protective equipment.

CONFINED-SPACE ENTRY PROCEDURES 11.13

Review the material covered in the module.

Have the participants complete the Review Questions and go over the answers prior to administering the Module Examination.

Summary

All personnel working in and around a confined-space environment must take an active role in their own safety, and that of their co-workers. Every member of the team needs to be aware of potential atmospheric hazards and their symptoms. They need to know how the work processes may contribute to those hazards. Workers need to be trained before using test equipment for hazard identification.

The air in a space must be tested before entry. Adequate ventilation must be established. Training must include emergency preparedness for evacuation and rescue procedures. Non-entry rescue should be used if it is available. An entry rescue must only be done by those with the proper training and equipment.

Review Questions

Administer the Module Examination. Record the results on Craft Training Report Form 200 and submit the results to the Training Program Sponsor.

1. Atmospheric contaminants include all of these *except* _____.

a. gas
b. vapor
c. dust
d. liquid

2. An oxygen-enriched atmosphere is one in which the concentration of oxygen is greater than _____.

a. 19.5%
b. 21%
c. 23.5%
d. 25%

3. An unprotected worker in an atmosphere of 5% oxygen can fall into a coma in _____.

a. less than one minute
b. 5 minutes
c. 8 minutes
d. 15 minutes

4. A flammable gas does not pose an explosion threat as long as the concentration of the gas in the atmosphere is below 10%.

a. True
b. False

5. Permissible exposure limits to toxic substances can be determined from the _____.

a. authorized work permit
b. MSDS
c. signed UFL
d. gas detector tube

6. Atmospheric testing should be done in the following order:

a. Oxygen content, toxic contaminants, flammable gases
b. Toxic contaminants, oxygen content, flammable gases
c. Oxygen content, flammable gases, toxic contaminants
d. Flammable gases, oxygen content, toxic contaminants

7. Which of the following does *not* appear on the authorized confined-space entry permit?

a. Purpose of entry
b. Authorized duration
c. Communication procedures
d. Internal temperatures

8. The person responsible for ordering a space evacuation is the _____.

a. entrant
b. attendant
c. supervisor
d. rescue team leader

9. An entrant who experiences a malfunction with an air line should _____.

a. maintain the air-line connection and evacuate the space
b. immediately disconnect the air line and evacuate the space
c. contact the attendant and ask for instructions
d. sit down to avoid expending oxygen

10. Of all confined-space fatalities, _____% are would-be rescuers.

a. 40
b. 50
c. 60
d. 70

Instructor's Notes:

GLOSSARY

Trade Terms Introduced in This Module

Asphyxiation: Death due to lack of oxygen.

Atmospheric Contaminants: Impurities in the air. Any natural or artificial matter capable of being airborne, other than water vapor or natural air, which, if in high enough concentration, harm man, other animals, vegetation, or material. They include many physical forms, liquid, solids, and gases as listed below.

Confined-space entry permit system: The written procedure for preparing and issuing permits for entry, and for returning the permit-required space to service after entry work is completed.

Hot work: Any work function that involves ignition or combustion. Examples include welding, burning, cutting, and riveting.

Inerting: The use of an inert gas, such as nitrogen, to supplant a flammable gas in a confined space.

Intrinsically safe: An electric tool or device that is UL-rated to be explosion-proof under normal uses. For example, a fuel pump must be intrinsically safe to prevent explosions sparked by static electricity.

Lower flammable limit (LFL) or Lower explosive limit (LEL): The lowest concentration of air-fuel mixture at which a gas can ignite.

Oxygen-deficient atmosphere: A body of air that does not have enough oxygen to sustain normal breathing. This is usually considered less than 19.5% by volume.

Oxygen-enriched atmosphere: A body of air that contains enough oxygen to be flammable or explosive. This is usually considered more than 23.5% by volume.

Parts per million (ppm): A measure of concentration of a substance in a solution like air or water. In a volume of air, it is the number of milliliters of a substance in a cubic meter of air: 10 ppm = 10 ml/m^3. One part per million is about the same as two soda cans full of gas in a house full of air.

Permissible exposure limit (PEL): The limit set by OSHA as the maximum concentration of a substance that a worker can be exposed to in an 8-hour work shift. Most flammable gases have a PEL defined as 10% of the LFL, while toxic gases have individual PELs.

Self-contained breathing apparatus (SCBA): A device allowing an individual to breath in a toxic or oxygen-deficient atmosphere.

Upper flammable limit (UFL) or Upper explosive limit (UEL): The highest concentration of air-fuel mixture at which a gas can ignite.

Vapor density: The relative weight of gases and vapors as compared with some specific standard, usually hydrogen, but sometimes air.

CONFINED-SPACE ENTRY PROCEDURES 11.15

REFERENCES & ACKNOWLEDGMENTS

Additional Resources

This module is intended to be a thorough resource for task training. The following reference works are suggested for further study. These are optional materials for continued education rather than for task training.

www.osha.gov

www.asse.org

Complete Confined Spaces Handbook, 1994. John F. Rekus, MS, CIH, CSP. Boca Raton, FL: CRC/Lewis Publishers.

Guidelines for Hot Work in Confined Spaces: Recommended Practices for Industrial Hygienists and Safety Professionals, 2000. Martin H. Finkel, CIH, CMC. Des Plaines, IL: The American Society of Safety Engineers (ASSE).

Figure Credits

RKI Instruments, Inc.	211F02, 211F03
Draeger Safety, Inc.	211F04
Brad Krauel	211F06

Instructor's Notes:

MODULE 75211-03 — ANSWERS TO REVIEW QUESTIONS

Answer		Section
1.	d	2.0.0
2.	c	2.1.0
3.	a	2.1.0
4.	b	2.2.1
5.	b	2.3.0
6.	c	3.0.0
7.	d	4.0.0
8.	b	5.2.0
9.	a	7.1.0
10.	c	8.0.0

CONTREN™ LEARNING SERIES — USER UPDATES

The NCCER makes every effort to keep these textbooks up-to-date and free of technical errors. We appreciate your help in this process. If you have an idea for improving this textbook, or if you find an error, a typographical mistake, or an inaccuracy in NCCER's Contren™ textbooks, please write us, using this form or a photocopy. Be sure to include the exact module number, page number, a detailed description, and the correction, if applicable. Your input will be brought to the attention of the Technical Review Committee. Thank you for your assistance.

Instructors – If you found that additional materials were necessary in order to teach this module effectively, please let us know so that we may include them in the Equipment/Materials list in the Instructor's Guide.

Write: Curriculum Revision and Development Department
National Center for Construction Education and Research
P.O. Box 141104, Gainesville, FL 32614-1104

Fax: 352-334-0932

E-mail: curriculum@nccer.org

Craft ______________________ Module Name ______________________

Copyright Date __________ Module Number __________ Page Number(s) __________

Description

____

____

____

____

(Optional) Correction

____

____

____

(Optional) Your Name and Address

____

____

____

Module 75212-03

Safety Meetings

NCCER STANDARDIZED CRAFT TRAINING PROGRAM

The National Center for Construction Education and Research (NCCER) provides a standardized national program of accredited craft training. Key features of the program include instructor certification, competency-based training, and performance testing. The program provides trainees, instructors, and companies with a standard form of recognition through a National Craft Training Registry. The program is described in full in the *Guidelines for Accreditation*, published by the NCCER. For more information on standardized craft training, contact the NCCER by writing us at P.O. Box 141104, Gainesville, FL 32614-1104; calling 352-334-0911; or e-mailing info@nccer.org. More information may be found at our Web site, www.nccer.org.

HOW TO USE THIS ANNOTATED INSTRUCTOR'S GUIDE

Each page presents two sections of information. The larger section displays each page exactly as it appears in the Participant Module. The narrow column ties suggested participant and instructor actions to each page and provides icons (detailed below) to call your attention to material, safety, audiovisual, or testing requirements. The bottom of each page includes space for your notes.

The **Audiovisual** icon indicates an appropriate time to show a transparency or other audiovisual aid.

The **Classroom** icon prompts you to define a term, stress a point, ask participants to explain a concept, or give examples.

The **Demonstration** icon directs you to show participants how to perform tasks.

The **Examination** icon tells you to administer the written module examination.

The **Homework** icon is placed where you may wish to assign reading for the next class, to assign a project, or to advise participants to prepare for an examination.

The **Laboratory** icon is used when participants are to practice performing tasks.

The **Materials** icon is a reminder for you to gather materials needed for classes, labs, and testing.

The **Performance Testing** icon tells you to administer a performance test or a portion thereof.

The **Safety** icon is used to emphasize safety issues. It is often keyed to *Caution* and *Warning* statements in the Participant Module.

The **Teaching Tip** icon indicates additional guidance is available, such as how to conduct an exercise, get the most educational value from a field trip, or encourage class participation. Teaching Tips may expand on a feature (*Think About It, Did You Know?*) or provide Quick Quizzes or similar exercises. You will be referred to the Teaching Tips section at the back of the module if there is additional material.

The **Combination** icon indicates that the laboratory listed corresponds with a performance task. If desired, you can note the proficiency of the participants during the laboratory and use it to satisfy performance testing requirements.

PREPARATION

Before teaching this module, you should review the Objectives, Performance Tasks, Materials and Equipment List, and the Module Outline. Be sure to allow ample time to prepare your own training or lesson plan and gather all required materials and equipment.

MODULE OVERVIEW

In this module, participants will learn how to coordinate and conduct informal and formal safety meetings.

PREREQUISITES

Please refer to the Course Map in the Participant Module. Prior to training with this module, it is recommended that the participant shall have successfully completed the following:

Field Safety; Safety Technology, Modules 75201-03 through 75211-03

OBJECTIVES

Upon completion of this module, the participant will be able to:

1. Communicate safety issues and concerns to workers through safety meetings.
2. Prepare for and conduct an effective safety meeting.
3. Evaluate the quality of a safety meeting.

PERFORMANCE TASK

Under the supervision of the instructor, the participant should be able to:

1. Choose a topic and conduct a three- to five-minute informal safety meeting.

MATERIALS AND EQUIPMENT LIST

Overhead projector and screen
Transparencies
Whiteboard/chalkboard
Markers/chalk
Blank acetate sheets
Transparency pens
Pencils and scratch paper
Appropriate personal protective equipment
Copy of *29 CFR 1926, OSHA Construction Industry Regulations*
TV (optional)
VCR (optional)
The CREATTP video, *Tricks to Teach* (optional)
Module Examinations*
Performance Profile Sheets*

* Located in the Test Booklet.

SAFETY CONSIDERATIONS

Ensure that participants are equipped with appropriate personal protective equipment.

ADDITIONAL RESOURCES

This module is intended to present thorough resources for task training. The following reference works are suggested for both instructors and motivated participants interested in further study. These are optional materials for continued education rather than for task training.

Web site: www.osha.gov

Web site: www.asse.org

The Psychology of Safety Handbook, 2001. E. Scott Geller, Ph.D. Boca Raton, FL: CRC/Lewis Publishers.

The Participation Factor—How to Increase Involvement in Occupational Safety, 2002. E. Scott Geller, Ph.D. Des Plaines, IL: The American Society of Safety Engineers (ASSE).

TEACHING TIME FOR THIS MODULE

An outline for use in developing your lesson plan is presented below. Approximately 2½ hours are suggested to cover *Safety Meetings*. This includes 10 minutes at the beginning of the session for administrative tasks and one 10-minute break. You may need to adjust the time required for hands-on activities and testing based on your class size and resources. Because laboratories often correspond to Performance Tasks, the proficiency of the participants may be noted during these exercises for Performance Testing purposes.

Topic	Planned Time
A. Introduction	____________
B. Formal Safety Meetings	____________
1. Coordinating and Conducting Safety Meetings	____________
2. Evaluating Safety Meetings	____________
C. Toolbox/Tailgate Safety Talks	____________
1. Safety Talk Preparation	____________
2. Five Ps for Successful Safety Talks	____________
D. Recordkeeping	____________
E. Laboratory	____________
Have the participants choose a topic and conduct a three- to five-minute informal safety meeting. If you have a particularly large class such that individual presentations are not feasible due to time constraints, participants may work in small groups. Note the proficiency of each participant. This laboratory corresponds with Performance Task 1.	
F. Review	____________
G. Module Examination	____________
1. Participants must score 70% or higher to receive recognition from the NCCER.	
2. Record the testing results on Craft Training Report Form 200 and submit the results to the Training Program Sponsor.	
H. Performance Testing	____________
1. Participants must perform each task to the satisfaction of the instructor to receive recognition from the NCCER. If applicable, proficiency noted during laboratory exercises can be used to satisfy the Performance Testing requirements.	
2. Record the testing results on Craft Training Report Form 200 and submit the results to the Training Program Sponsor.	

Module 75212-03

Safety Meetings

Instructor's Notes:

COURSE MAP

This course map shows all of the modules in Safety Technology. The suggested training order begins at the bottom and proceeds up. The local Training Program Sponsor may adjust the training order.

Assign reading of Module 75212-03.

SAFETY TECHNOLOGY

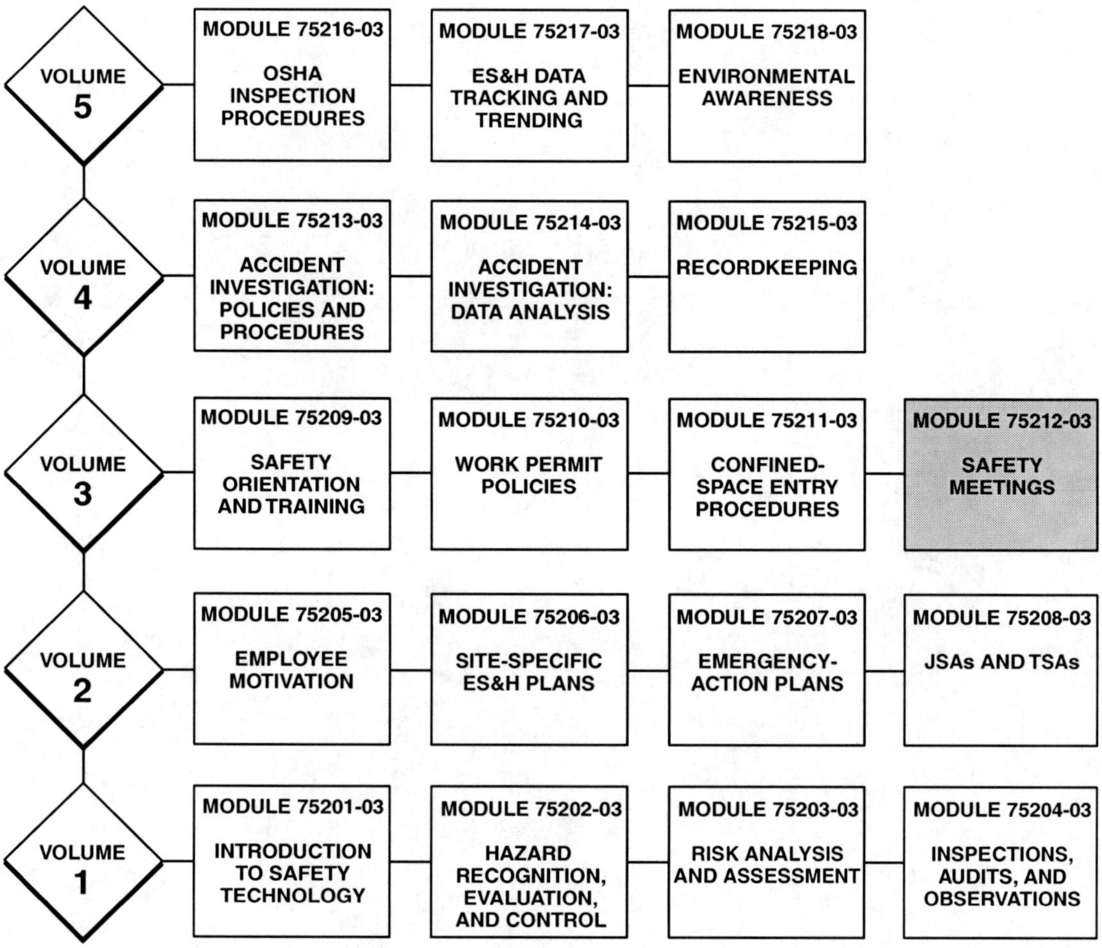

212CMAP.EPS

SAFETY MEETINGS 12.iii

Instructor's Notes:

MODULE 75212-03 CONTENTS

Figures

SAFETY MEETINGS 12.v

Instructor's Notes:

MODULE 75212-03

Safety Meetings

Objectives

When you have completed this module, you will be able to do the following:

1. Communicate safety issues and concerns to workers through safety meetings.
2. Prepare for and conduct an effective safety meeting.
3. Evaluate the quality of a safety meeting.

Prerequisites

Before you begin this module, it is recommended that you successfully complete the following: Field Safety; Safety Technology, Modules 75201-03 through 75211-03.

Required Materials

1. Pencil and paper
2. Appropriate personal protective equipment
3. Copy of *29 CFR 1926, OSHA Construction Industry Regulations*

1.0.0 ◆ INTRODUCTION

Safety meetings can range from short, informal safety talks to longer, more formal meetings. Safety meetings can be used to exchange information regarding specific safety matters, defuse potential job disruptions by providing an outlet for problems, provide a written record of the actions taken to correct a problem, and establish an effective communications link between management and employees.

The length and location of a meeting depends on the safety issues to be discussed. Short, informal meetings cover specific jobs workers are doing. Formal safety meetings are longer and can feature a guest speaker, films, photos, or other **audiovisual materials**. Whether you are conducing a short, informal safety meeting or a formal safety meeting, it is important to provide workers with effective lessons that will help them do their jobs safely.

> **NOTE**
> For more information on safety training and orientation, please refer to the *Safety Orientation and Training* module.

2.0.0 ◆ FORMAL SAFETY MEETINGS

Formal safety meetings are typically held once a month for job-site personnel at a specific time and place. These meetings should last no more than an hour, and the audience should include all employees, including project managers, supervisors, and workers. On some jobs, subcontractors are included in the meetings.

Topics for a safety meeting can vary from hand-tool safety to emergency response. Safety meeting topics should always be practical and relevant to the work that is done on the site. Suggested safety meeting topics and resources for topics include:

- Recent accidents and near misses
- OSHA "Fatal Facts"
- Department of Energy "Lessons Learned"
- MSDSs
- Accident analysis
- Upcoming jobs or tasks
- National-interest topics such as National Fire Prevention Week
- OSHA training requirements
- Results of safety audits, inspections, and observations

> **NOTE**
> OSHA's Fatal Facts can be found at http://www.osha.gov/OshDoc/toc_FatalFacts.html .The Department of Energy's Lessons Learned can be found at http://tis.eh.doe.gov/ll/listdb.html.

Ensure that you have everything you need to teach the course. Check the Materials and Equipment list in the front of this module.

Explain that terms shown in bold are defined in the Glossary at the back of this module.

Review the goals of the course and explain what will be expected of the participant.

Show Transparency 1, Objectives.

Show Transparency 2, Performance Task.

Explain the two types of safety meetings that are typically conducted for workers.

Describe how often formal safety meetings are generally held and the employees that must attended.

Identify common safety meeting topics.

Explain the steps used to coordinate and conduct safety meetings.

2.1.0 Coordinating and Conducting Safety Meetings

Safety meetings should be planned in advance. When they are planned in advance, the presented information will be organized and easy to understand. This helps to ensure that the message of the meeting, which will always be safety, gets through to the audience. The following items should be considered in preparing for and conducting safety meetings:

Show Transparency 3, Safety Meeting Considerations.

- *Meeting site preparation* – Check to make sure all of the necessary projectors, pencils, paper, blackboard/whiteboard, lighting, and sound equipment have been set up, if needed. Make sure equipment is working properly before the meeting starts.
- *Meeting time* – As a general rule, make sure that training is conducted during working hours. It should not be conducted during the employees' breaks or lunches.
- *Meeting location* – Make certain that the meeting area is neither too hot nor too cold. Make sure the lighting is good and that there is no background noise. Confirm that everyone can see and hear what is being presented.
- *Meeting materials* – Provide handouts, visual aids, models, and examples to help your audience see as well hear the information being presented. This will reinforce what is being said in the presentation. Job site tools, equipment, and materials can also be used to reinforce concepts.
- *Presenter preparation* – Make sure you have a thorough understanding of the material you are about to present. This will help you to anticipate audience questions and needs. It will also increase your credibility as a presenter.
- *Group composition* – Ask all employees, supervisors, and upper-level managers to attend safety meetings. The presence of management demonstrates company interest in good safety practices.
- *Audience pet peeves* – Make sure you are aware of the things that will distract your audience and keep them from retaining the presented information. Some pet peeves include:
  - There is no real purpose or agenda for the meeting.
  - The objectives are unclear.
  - The topic doesn't apply to the audience.
  - The presenter is not prepared.
  - The meeting starts late or lasts too long.
  - There is no closure on the topic or a decision hasn't been made on the issue.
  - The meeting is used to complain about issues rather than solve them.
  - The presenter doesn't allow anyone in the audience to speak or participate in the discussion.
  - One person in the audience does all the talking, while everyone else is quiet.
- *Appropriate topic* – Make sure that the topic is related to work in progress, future activities, or recently completed tasks. Limit the number of topics to one or two per meeting.
- *Participant preparation* – Learn how to get the audience involved in the meeting *(see Appendix)*. State the objectives and purpose of the meeting at the beginning. This will help your audience know what they are expected to learn. Leave time at the end of the presentation for questions and answers.

Discuss the qualities that make safety meetings successful.

Refer the participants to the Appendix in the Participant Module. Demonstrate some of the techniques suggested in it.

Good safety meetings incorporate all of these ideas. The key to making the presentation a success is knowing how to coordinate them. For example, before you can set up a time and place for a meeting, you must choose a topic. Next, you need to make sure the meeting location has all of the equipment you may need, such as a chalk or drawing board, a TV and VCR, or a film projector and screen. Once you know that all of the equipment is in place, you can prepare any handouts or visual aids that will be used in the presentation. Next, before the meeting date, you should practice what you are going to say. Practice until you have the presentation memorized. When the meeting date and time arrive, make sure the temperature of the room and the lighting are comfortable for the audience. Next, during your presentation, speak clearly and at a good pace. Learn to watch the audience for signs of boredom or uneasiness and involve them in the presentation when appropriate. At the end of the presentation, leave room for a question and answer period to allow the participants to clarify anything that may not have been understood. Summarize the objectives in order to reinforce them. By incorporating all of these ideas, you are helping to ensure that the message of the meeting is understood and carried over into the audience's work.

Explain how evaluation forms are used to get feedback that will improve the quality of future meetings.

2.2.0 Evaluating Safety Meetings

Show Transparency 4 (Figure 1).

Conducting effective meetings can be a challenge. One way to improve the quality of meetings is to have the audience, including project managers and supervisors, provide feedback. The form shown in *Figure 1* is an example of a speaker evaluation form. This form provides the presenter with feedback about the effectiveness of the meeting. The feedback received from these forms should be taken seriously, especially if the overall score is average or below average.

Instructor's Notes:

PRESENTATION EVALUATION GUIDE – PAGE 1

Presenter ______________________ Name (Optional) ______________________

Date ______________________

	Poor 1	2	Average 3	4	Excellent 5	Not Applicable N/A
Introduction						
Title of lesson stated or displayed?						
Objectives stated?						
Motivation established?						
Lesson overviewed?						
Presentation						
Appropriate information level?						
Objectives covered?						
Presentation follows a logical sequence?						
Visual Aids						
Properly used?						
Illustrate the point?						
Visible by all?						
Out of sight when not in use?						
Used methods suggested in Guide?						
Questioning Techniques						
Different types asked?						
Sufficient number asked?						
Focused attention?						
Created discussion?						

212F01A.EPS

Figure 1 ◆ Typical speaker evaluation form (1 of 2).

CREATTP's video, *Tricks to Teach*, shows how to incorporate brainteasers and energizers into meetings and presentations for the purpose of engaging participants.

SAFETY MEETINGS 12.3

Go to the Web site: www.makesafetyfun.com/effective_safety_meeting_articles.htm to look for articles and ideas on how to make safety meetings more effective.

PRESENTATION EVALUATION GUIDE – PAGE 2

Presenter ____________ Name (Optional) ____________

Date ____________

	Poor 1	2	Average 3	4	Excellent 5	Not Applicable N/A
Questioning *(continued)* Related information to applications?						
Adapted to level of participants?						
Summary Reviewed key points?						
Pointed out benefits to participants?						
Instructor Qualities						
Gestures & Mannerisms Eye contact						
Knowledge						
Voice						
Professional attitude						
Enthusiasm for subject						
Overall Comments Completed all material?						
Followed all safety procedures?						

Additional Comments

What did you like best in the way this lesson was presented?

What one aspect of the presentation would you change and how would you present it differently?

212F01B.EPS

Figure 1 ◆ Typical speaker evaluation form (2 of 2).

12.4 SAFETY TECHNOLOGY — PARTICIPANT MODULE 75212-03

Instructor's Notes:

This form can also be an effective tool in determining if the quality of safety meetings justifies their cost. This is important because coordinating and conducting safety meetings can be expensive. The equation in *Figure 2* shows exactly how expensive safety meetings can be. If the effectiveness of meetings does not justify the cost, the number and length of meetings may be reduced. This would be detrimental to both the workers and the company itself. Using evaluation forms to make changes and improvements to presentations helps to document the effectiveness of meetings and therefore increase the likelihood that safety meetings will be done often and correctly.

Another way to improve the quality of meetings is to make their effectiveness a personal goal. *Figure 3* shows a form that can be used to evaluate yourself as a presenter. By making sure you are doing the best possible job in presenting safety issues, you are ensuring that meetings continue to take place and that the quality of the meetings remains high. This can reduce the number of accidents and missed days of work.

WHERE: E = NUMBER OF EMPLOYEES
N = NUMBER OF MEETINGS PER YEAR
R = AVERAGE HOURLY WAGE RATE INCLUDING FRINGE BENEFITS
C = AVERAGE LENGTH OF MEETING

EXAMPLE: 25 EMPLOYEES ATTEND 12 1-HOUR MEETINGS PER YEAR. THE AVERAGE HOURLY COST PER EMPLOYEE WITH BENEFITS = $27.00

$E \times N \times R \times C$

$25 \times 12 \times 1 \times 27 = \$8,100.00$

212F02.EPS

Figure 2 ◆ Cost-of-safety equation.

3.0.0 ◆ TOOLBOX/TAILGATE SAFETY TALKS

Toolbox/tailgate talks are short, informal safety meetings led by the crew supervisor for his or her crew. The safety technician's role in toolbox/tailgate talks is to act as a resource for the supervisor and attend the meetings when possible or needed. He or she should also maintain the records of toolbox talks and alert management if problems arise with the frequency or quality of the meetings.

Toolbox/tailgate talks are designed to inform workers of specific hazards associated with a job. They can also act as a refresher to workers, reminding them about the hazards and safeguards of their job. Safety talks are usually held on site, near the location of the work.

Toolbox/tailgate safety talks can be held daily or weekly depending on company policy. They should always be conducted, however, when any of the following conditions exist:

- A certain job hasn't been performed for some time.
- New employees join the crew.
- A task and/or location poses specific hazards. For example:
  - Elevated work locations
  - Work over water
  - High-volume, high-speed traffic
  - Very limited sight distance in approach to a work zone
  - Areas known for high accident rates
- Substantial changes in work conditions and procedures have occurred.
- A recent accident in this or another crew needs to be reviewed.
- A near-miss incident has occurred.
- A supervisor feels employees are becoming lax about safety.

The best time to have a safety talk is before the workday begins. This helps to remind workers about safety issues before they begin work rather than during or after work. Sometimes, however, if an incident or accident occurs during the day, it is also appropriate to have a short safety talk after the accident or at the end of the day to discuss it.

NOTE
The Construction Industry Institute (http://construction-institute.org/) has more information about the effectiveness of safety meetings.

3.1.0 Safety Talk Preparation

Even though they are informal and take place on site, safety talks should be planned in advance. The following are some helpful hints for organizing and conducting toolbox/tailgate safety talks.

- Start with a review of recent safe work and practices.
- Present short talks on topics related to current or upcoming work activities.
- Limit prepared talks to 10 to 15 minutes.
- Encourage employee participation.
- Use **open-ended questions**, ask for opinions, invite suggestions, and provide appropriate follow-up.
- Review recent near misses and workplace injuries.

Explain how evaluation forms can be used to determine the cost effectiveness of safety meetings.

Show Transparency 5 (Figure 2).

Explain how evaluation forms can be used to evaluate the performance of the instructor and how the instructor should use the feedback to improve their technique.

Show Transparency 6 (Figure 3).

Explain the purpose, uses, and most effective times to hold toolbox/tailgate safety talks.

Explain the conditions that require the use of safety talks and discuss .

Explain how to prepare for a safety talk. Give participants examples of things you've done to prepare for a safety talk.

SAFETY MEETINGS 12.5

For topics and ideas about toolbox safety talks, go to the following Web site: www.toolboxtopics.com

SELF-EVALUATION CHECKLIST

Ask yourself the following questions to determine your effectiveness as a presenter. If you answer no to any of these questions, stop and think about ways to improve.

	YES	NO	WAYS TO IMPROVE
TOPIC			
Are presented topics related to the work that is being done on site?			
Do you use other resources to look for relevant topics such as OSHA, NIOSH, eLCOSH, and the CDC?			
PREPAREDNESS			
Before each safety meeting do you:			
Inspect the job site for hazards related to your topic?			
Read over the material you plan to cover?			
Look up any terms or concepts you don't understand?			
Make sure you are familiar with any laws, regulations, and company rules related to the meeting's topic?			
Review reports of recent accidents on the site, including "near misses"?			
PARTICIPANT/CREW INVOLVEMENT			
Do you:			
Begin with a real-life example, or with information that will capture interest?			
Encourage full participation by the crew throughout the meeting?			
Invite the crew to ask questions and make suggestions related to the topic?			
Respond to questions that you can answer, and offer to find answers you don't know?			
Allow time at the end of the meeting for questions and suggestions on any safety issue?			
Ask the crew for feedback about the meeting?			
Involve the crew in preparing for and/or leading future safety meetings?			
FOLLOW UP			
Do you:			
Look into complaints, concerns, and suggestions that the crew brought up?			
Report back later to let the crew know what will be done?			
Keep good records of each tailgate meeting and other safety matters?			
CREDIBILITY/RESPECTABILITY			
Do you:			
Set an excellent safety example yourself?			
Invite crew members to come to you anytime with safety problems and suggestions?			
Encourage and reward safe work practices?			

Source: Oregon Occupational Safety and Health Administration (OR-OSHA)

212F03.EPS

Figure 3 ◆ Self-evaluation checklist.

Instructor's Notes:

- Discuss how near misses happened and how they could have been prevented.
- Review and discuss hazards encountered while working with other workers on site.
- Look ahead to potential safety hazards involved in upcoming work and remind workers of the proper use of safety equipment and the procedures to be followed in dealing with those hazards.

3.2.0 Five Ps for Successful Safety Talks

One technique that can help you give better safety talks is a technique called the **Five Ps for Successful Safety Talks**. You can easily apply this technique to safety talk topics that are related to work being done on a site.

The Five Ps for Successful Safety Talks are:

- *Prepare* – Think safety. Write ideas down for future use. Read safety materials thoroughly. Listen to others' ideas and attitudes. Organize and outline your talks. Practice what you are going to say.
- *Pinpoint* – Concentrate on one safety rule, one first-aid hint, one unsafe practice, or one main idea. Too many topics can confuse your audience.
- *Personalize* – Establish common ground with your audience. Make the lesson personal and important. Use as many real-life examples as possible.
- *Picture* – Create clear mental pictures for your listeners. Appeal to both their ears and their eyes. Help them really see what you mean. Use visual aids.
- *Prescribe* – In closing your safety talk, tell workers what you expect them to do. Give them something to think about after the meeting.

The Five Ps technique is a good way to organize your thoughts and prepare a meeting that effectively communicates the purpose of the meeting: safety.

4.0.0 ◆ RECORDKEEPING

Keeping records of safety meetings and toolbox/tailgate safety talks is an important part of any safety program. Recordkeeping provides documentation that proves that safety meetings are taking place, that workers are attending, and that actions are being taken to correct safety problems. Proving that safety meetings take place on a regular basis can help shield a company from legal liability, help control the **recordable incident rate** of the company, and justify adding safety meetings and training into the company's budget. For example, *Figure 4* shows the findings a recent Construction Industry Institute report stating that the companies that included safety in their budgets had a lower recordable incident rate. This means less time and money spent on accidents. The Construction Industry Institute wouldn't have been able to gather this information if accurate records of safety meetings had not been kept. The easiest way to document safety meetings is to record attendance by using a sign-up sheet like the one shown in *Figure 5*. Speaker evaluation forms are also a way to record safety meetings.

> **NOTE**
> You should always review each safety meeting record to be sure that each attendee's name is clearly legible and that no fictitious names have been entered.

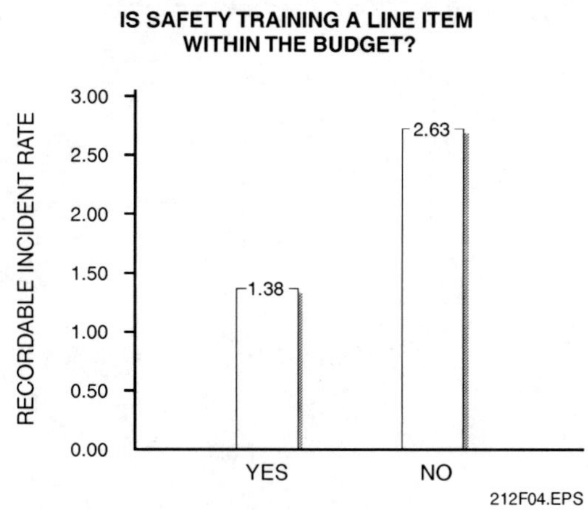

Figure 4 ◆ Recordable incident rate graph.

Explain the Five Ps for Successful Safety Talks technique.

Show Transparency 7, The Five Ps for Successful Safety Talks.

Explain how keeping accurate training records can help justify the cost of safety training to management.

Define the term recordable incident rate.

Show Transparency 8 (Figure 4).

SAFETY MEETINGS 12.7

Have the participants choose a topic and conduct a three- to five-minute informal safety meeting. If you have a particularly large class such that individual presentations are not feasible due to time constraints, participants may work in small groups. Note the proficiency of each participant. This laboratory corresponds with Performance Task 1.

SIGN-IN FORM

TOPIC: ______________________________

Date Presented: ____________________ By: ____________________

Project Name/No.: ____________________ Location: ____________________

NAMES OF THOSE WHO ATTENDED THIS SAFETY MEETING

PRINTED NAME	SIGNATURE

212F05.TIF

Figure 5 ◆ Attendance sheet.

Instructor's Notes:

____

____

____

Summary

Safety meetings, whether formal or informal, are an important part of any safety program because they help train workers on how to work safely. They also provide an outlet for workers to share safety concerns, participate in the solutions that correct work place hazards, and provide feedback about the training they receive. You must be able to learn from these meetings in order to improve as a presenter. This will improve the quality of the meetings, which benefits the workers and the company.

As a safety technician, you are responsible for coordinating and conducting all safety meetings, with the exception of toolbox/tailgate safety talks. These short, informal meetings are usually conducted by a crew's first-line supervisor. Even though it is not likely that you will conduct these safety talks, you will still be a resource for the supervisor who is conducting the meeting. Therefore, it is important for you to not only know to how to prepare, conduct, and record formal safety meetings, it is important for you to know how to provide guidance to the supervisor conducting safety talks.

Review the material covered in the module.

Have the participants complete the Review Questions and go over the answers prior to administering the Module Examination.

Review Questions

1. The types of meetings that are typically held for all job-site personnel are formal safety meetings.
   a. True
   b. False

2. The Five Ps for Successful Safety Talks are prepare, pinpoint, personalize, prescribe, and _____.
   a. practice
   b. plan
   c. picture
   d. participate

3. Toolbox/tailgate meetings are held on site, but *not* near the location of the work.
   a. True
   b. False

4. A presenter only needs to be somewhat knowledgeable about a topic in order to give a good presentation.
   a. True
   b. False

5. As a general rule, formal safety meetings should last no more than _____.
   a. 30 minutes
   b. one hour
   c. one and a half hours
   d. two hours

6. As a general rule, safety meetings should *not* be held during breaks or lunches.
   a. True
   b. False

7. Because they are short and informal, toolbox/tailgate safety talks do *not* have to be planned in advance.
   a. True
   b. False

8. Feedback from an audience can help improve the quality of safety meetings.
   a. True
   b. False

9. OSHA's "Fatal Facts" web page is a good resource for finding safety meeting topics.
   a. True
   b. False

10. Safety training records can help justify adding safety meetings into a company's budget.
    a. True
    b. False

Administer the Module Examination. Record the results on Craft Training Report Form 200 and submit the results to the Training Program Sponsor.

Administer the Performance Test and fill out Performance Profile Sheets for each participant. If desired, participant proficiency noted during laboratory sessions may be used to complete the Performance Test. Record the results on Craft Training Report Form 200 and submit the results to the Training Program Sponsor.

SAFETY MEETINGS 12.9

Instructor's Notes:

GLOSSARY

Trade Terms Introduced in This Module

Audiovisual materials: Materials such as photos, films, charts, and graphs that are designed to aid in learning or teaching by making use of both hearing and sight.

Five Ps For Successful Safety Talks: A technique for conducting effective safety talks involving five key elements: preparing, pinpointing, personalizing, picturing, and prescribing.

Open-ended questions: Questions that require more than a yes or no answer. These types of questions are used to encourage the audience to participate in the discussion.

Recordable incident rate: An equation that calculates the number of job-related injuries and illnesses, or lost workdays per 200,000 hours of exposure on a construction site. For example, Incident Rate = the number of injuries and illnesses × 200,000 ÷ employee hours worked for the year.

SAFETY MEETINGS 12.11

Instructor's Notes:

APPENDIX

Meeting Tips: How to Get the Crew Involved

Safety meetings work best if the whole crew actively participates. Here are some ways to encourage everyone to get involved.

- **ASK QUESTIONS INSTEAD OF LECTURING** – During the meeting, introduce each new point you want to make by asking the crew a question. After you ask each question, wait a short time to let people think. Then call on volunteers to answer. Use the answers as a springboard for discussion. Don't just read the answers.
- **ASK ABOUT PERSONAL EXPERIENCE** – If you ask a question and no one has an answer, rephrase the question. It may be too abstract. Try to make it more direct and personal. Ask if someone has had any personal experience that can help the group figure out an answer. For example, suppose no one can answer the question, "What are the health effects of breathing asbestos?" You could try to make the question more personal by asking, "Have you ever known anyone who got sick from working with asbestos? What kind of illness did they have?"
- **LIMIT THE AMOUNT OF TIME ANY ONE PERSON CAN TALK** – If a crew member is talking too much, invite someone else to speak. Do it tactfully. For example, wait until the person takes a breath, quickly say "thank you," and then move along.
- **ROLE PLAYING** – Provide workers with a scenario involving a safety issue. Have the worker play the role of supervisor or safety technician. Ask them how they would resolve the issue in the scenario if they were the supervisor or safety technician.
- **DON'T FAKE IT** – If someone has a question and you don't know the answer, don't guess or fake an answer. Write the question down. Promise that you will get back to the person, and then make sure you do.
- **STICK TO THE TOPIC** – If the crew's questions and comments move too far from the topic, tell them that their concerns can be addressed later; either in private conversation or in an upcoming safety meeting.
- **COMPETITION** – Add some activities that make workers compete against each other for prizes. For example, use the game show type format using questions about the topic and reward the winner. Winning prizes can range from a t-shirt to a gift certificate.

Source: Electronic Library of Construction Occupational Safety and Health (eLCOSH)

Instructor's Notes:

Additional Resources

This module is intended to be a thorough resource for task training. The following reference works are suggested for further study. These are optional materials for continued education rather than for task training.

www.osha.gov

www.asse.org

The Psychology of Safety Handbook, 2001. E. Scott Geller,Ph.D. Boca Raton, FL: CRC/Lewis Publishers.

The Participation Factor—How to Increase Involvement in Occupational Safety, 2002. E. Scott Geller, Ph.D. Des Plaines, IL: The American Society of Safety Engineers (ASSE).

Figure Credits

Construction Industry Institute	212F04
Labor Occupational Health Program at UC Berkeley	Appendix

MODULE 75212-03 — ANSWERS TO REVIEW QUESTIONS

Answer		Section
1.	a	2.0.0
2.	c	3.2.0
3.	b	3.0.0
4.	b	2.1.0
5.	b	2.0.0
6.	a	2.1.0
7.	b	3.1.0
8.	a	2.2.0
9.	a	2.0.0
10.	a	4.0.0

CONTREN™ LEARNING SERIES — USER UPDATES

The NCCER makes every effort to keep these textbooks up-to-date and free of technical errors. We appreciate your help in this process. If you have an idea for improving this textbook, or if you find an error, a typographical mistake, or an inaccuracy in NCCER's Contren™ textbooks, please write us, using this form or a photocopy. Be sure to include the exact module number, page number, a detailed description, and the correction, if applicable. Your input will be brought to the attention of the Technical Review Committee. Thank you for your assistance.

Instructors – If you found that additional materials were necessary in order to teach this module effectively, please let us know so that we may include them in the Equipment/Materials list in the Instructor's Guide.

Write: Curriculum Revision and Development Department
National Center for Construction Education and Research
P.O. Box 141104, Gainesville, FL 32614-1104

Fax: 352-334-0932

E-mail: curriculum@nccer.org

Craft ______________________ Module Name ______________________

Copyright Date __________ Module Number __________ Page Number(s) __________

Description ______________________

____

____

____

(Optional) Correction ______________________

____

____

(Optional) Your Name and Address ______________________

____

____

Module 75213-03

Accident Investigation: Policies and Procedures

NCCER STANDARDIZED CRAFT TRAINING PROGRAM

The National Center for Construction Education and Research (NCCER) provides a standardized national program of accredited craft training. Key features of the program include instructor certification, competency-based training, and performance testing. The program provides trainees, instructors, and companies with a standard form of recognition through a National Craft Training Registry. The program is described in full in the *Guidelines for Accreditation*, published by the NCCER. For more information on standardized craft training, contact the NCCER by writing us at P.O. Box 141104, Gainesville, FL 32614-1104; calling 352-334-0911; or e-mailing info@nccer.org. More information may be found at our Web site, www.nccer.org.

HOW TO USE THIS ANNOTATED INSTRUCTOR'S GUIDE

Each page presents two sections of information. The larger section displays each page exactly as it appears in the Participant Module. The narrow column ties suggested participant and instructor actions to each page and provides icons (detailed below) to call your attention to material, safety, audiovisual, or testing requirements. The bottom of each page includes space for your notes.

The **Audiovisual** icon indicates an appropriate time to show a transparency or other audiovisual aid.

The **Classroom** icon prompts you to define a term, stress a point, ask participants to explain a concept, or give examples.

The **Demonstration** icon directs you to show participants how to perform tasks.

The **Examination** icon tells you to administer the written module examination.

The **Homework** icon is placed where you may wish to assign reading for the next class, to assign a project, or to advise participants to prepare for an examination.

The **Laboratory** icon is used when participants are to practice performing tasks.

The **Materials** icon is a reminder for you to gather materials needed for classes, labs, and testing.

The **Performance Testing** icon tells you to administer a performance test or a portion thereof.

The **Safety** icon is used to emphasize safety issues. It is often keyed to *Caution* and *Warning* statements in the Participant Module.

The **Teaching Tip** icon indicates additional guidance is available, such as how to conduct an exercise, get the most educational value from a field trip, or encourage class participation. Teaching Tips may expand on a feature (*Think About It*, *Did You Know?*) or provide Quick Quizzes or similar exercises. You will be referred to the Teaching Tips section at the back of the module if there is additional material.

The **Combination** icon indicates that the laboratory listed corresponds with a performance task. If desired, you can note the proficiency of the participants during the laboratory and use it to satisfy performance testing requirements.

PREPARATION

Before teaching this module, you should review the Objectives, Performance Tasks, Materials and Equipment List, and the Module Outline. Be sure to allow ample time to prepare your own training or lesson plan and gather all required materials and equipment.

MODULE OVERVIEW

In this module, participants will learn the policies and procedures for conducting accident investigations.

PREREQUISITES

Please refer to the Course Map in the Participant Module. Prior to training with this module, it is recommended that the participant shall have successfully completed the following:

Field Safety; Safety Technology, Modules 75201-03 through 75212-03

OBJECTIVES

Upon completion of this module, the participant will be able to:

1. Explain the purposes and uses of accident investigations.
2. Identify the person responsible for conducting an accident investigation.
3. Complete an accident investigation form.
4. Explain the procedure for conducting accident investigation interviews.

PERFORMANCE TASKS

Under the supervision of the instructor, the participant should be able to:

1. Conduct an accident investigation interview.
2. Complete an accident investigation form.

MATERIALS AND EQUIPMENT LIST

Overhead projector and screen
Transparencies
Whiteboard/chalkboard
Markers/chalk
Blank acetate sheets
Transparency pens
Pencils and scratch paper
Appropriate personal protective equipment
Copy of *29 CFR 1926, OSHA Construction Industry Regulations*
TV (optional)
VCR (optional)
The Safe Expectations, International video, *Accident Investigation: A Tool for Effective Prevention* (optional)
Accident investigation forms
Module Examinations*
Performance Profile Sheets*

* Located in the Test Booklet.

SAFETY CONSIDERATIONS

Ensure that participants are equipped with appropriate personal protective equipment.

TEACHING TIME FOR THIS MODULE

An outline for use in developing your lesson plan is presented below. Approximately 2½ hours are suggested to cover *Accident Investigation: Policies and Procedures*. This includes 10 minutes at the beginning of the session for administrative tasks and one 10-minute break. You may need to adjust the time required for hands-on activities and testing based on your class size and resources. Because laboratories often correspond to Performance Tasks, the proficiency of the participants may be noted during these exercises for Performance Testing purposes.

Topic	Planned Time
A. Introduction	____________
B. Accidents vs. Incidents	____________
C. Accident Investigation	____________
1. Need for Investigation	____________
2. Roles and Responsibilities	____________
3. Equipment Used in Accident Investigation	____________
4. Investigation Time Line	____________
5. Conducting Interviews	____________
a. Laboratory Given an accident scenario, have participants pair off and practice performing accident investigation interviews. Note the proficiency of each participant. This laboratory corresponds with Performance Task 1.	____________
6. Accident Investigation Reports	
a. Laboratory Given an accident scenario, have participants practice completing accident investigation forms. Use either the OSHA form or one you've created. Note the proficiency of each participant. This laboratory corresponds with Performance Task 2.	____________
D. Review	____________
E. Module Examination	____________
1. Participants must score 70% or higher to receive recognition from the NCCER.	
2. Record the testing results on Craft Training Report Form 200 and submit the results to the Training Program Sponsor.	
F. Performance Testing	____________
1. Participants must perform each task to the satisfaction of the instructor to receive recognition from the NCCER. If applicable, proficiency noted during laboratory exercises can be used to satisfy the Performance Testing requirements.	
2. Record the testing results on Craft Training Report Form 200 and submit the results to the Training Program Sponsor.	

Module 75213-03

Accident Investigation: Policies and Procedures

Instructor's Notes:

COURSE MAP

This course map shows all of the modules in Safety Technology. The suggested training order begins at the bottom and proceeds up. The local Training Program Sponsor may adjust the training order.

Assign reading of Module 75213-03.

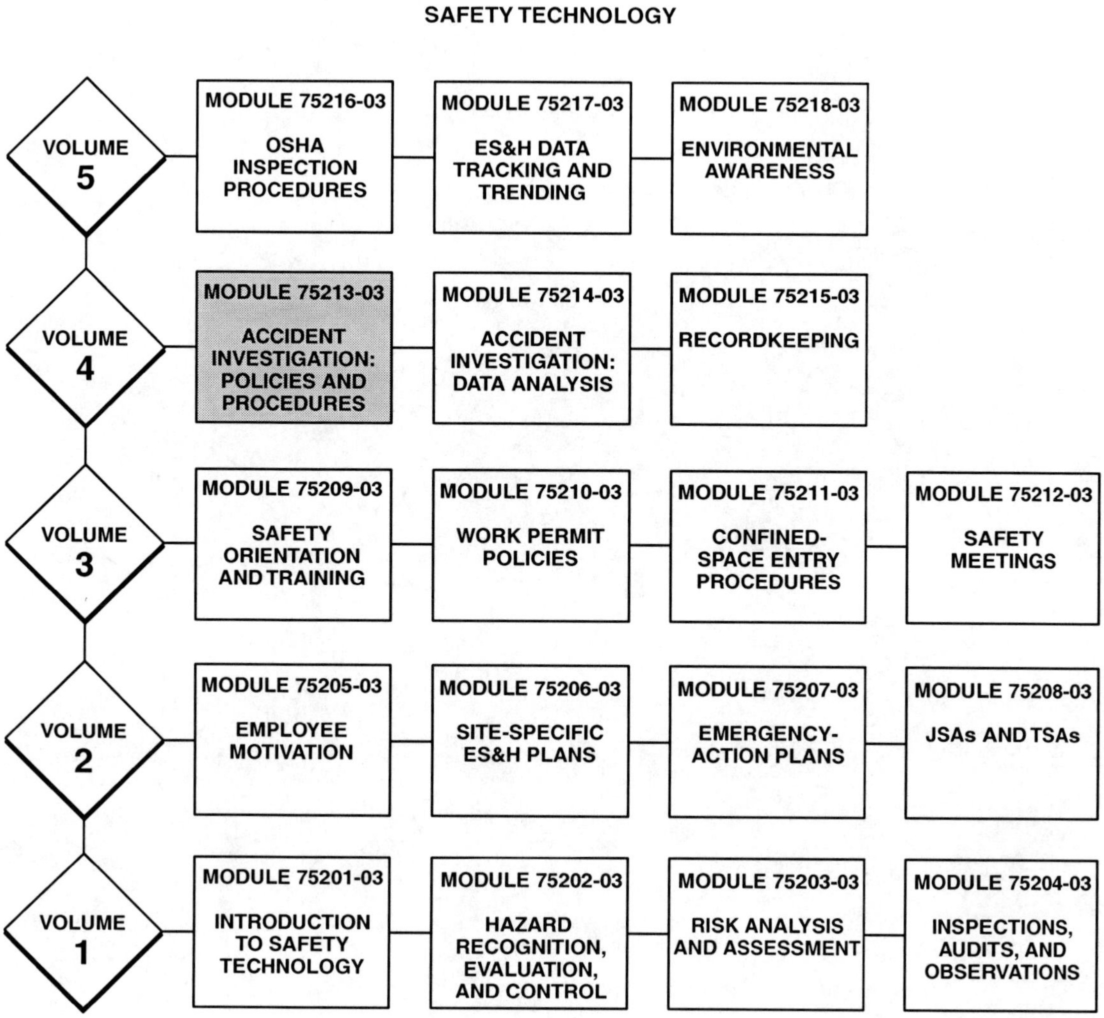

213CMAP.EPS

ACCIDENT INVESTIGATION: POLICIES AND PROCEDURES 13.iii

Instructor's Notes:

MODULE 75213-03 CONTENTS

Figures

ACCIDENT INVESTIGATION: POLICIES AND PROCEDURES 13.v

Instructor's Notes:

MODULE 75213-03

Accident Investigation: Policies and Procedures

Objectives

When you have completed this module, you will be able to do the following:

1. Explain the purposes and uses of accident investigations.
2. Identify the person responsible for conducting an accident investigation.
3. Complete an accident investigation form.
4. Explain the procedure for conducting accident investigation interviews.

Prerequisites

Before you begin this module, it is recommended that you successfully complete the following: Field Safety; Safety Technology, Modules 75201-03 through 75212-03.

Required Materials

1. Pencil and paper
2. Appropriate personal protective equipment
3. Copy of *29 CFR 1926, OSHA Construction Industry Regulations*

1.0.0 ◆ INTRODUCTION

Accident investigations play an important part in preventing future accidents. They help to determine direct and indirect causes of accidents, prevent similar accidents from happening, document facts, satisfy government regulations, detect trends, provide information on costs, and promote safety.

The on-site supervisor, also known as the first-line supervisor, is generally responsible for conducting accident investigations. The safety technician's role in accident investigations is to act as resource for the supervisor. Once the investigation is complete, the safety technician is responsible for reviewing the report for completeness and accuracy. In a serious or potentially serious accident, the safety technician may assist in the investigation. Safety technicians are also responsible for following up with the site supervisor to ensure that the needed safety policies and procedures have been put in place and the control actions have be implemented. It is important, however, for safety technicians to know the entire accident investigation process, not just their role. Gaining a better understanding of the process will help make investigations more efficient and effective.

> **NOTE**
> This module discusses the policies and procedures used to conduct an accident investigation. In the next module, *Accident Investigation: Data Analysis*, the process of analyzing accident investigation data is discussed in detail.

2.0.0 ◆ ACCIDENTS VS. INCIDENTS

The only difference between an accident and an **incident** is the outcome. Incidents are events that do not result in personal injury or property damage. Incidents act as free warnings if they are reported and acted on. Accidents are events that result in property damage, personal injury, or death. Regardless if an event was an incident or an accident, it should be reported and investigated.

Ensure that you have everything you need to teach the course. Check the Materials and Equipment list in the front of this module.

Explain that terms shown in bold are defined in the Glossary at the back of this module.

Review the goals of the course and explain what will be expected of the participant.

Show Transparency 1, Objectives.

Show Transparency 2, Performance Tasks.

Explain the safety technician's role in accident investigations.

Explain the difference between accidents and incidents.

ACCIDENT INVESTIGATION: POLICIES AND PROCEDURES 13.1

Discuss the purpose of an accident investigation. Explain what the focus of the investigation should be.

Explain when and why accident investigations are done.

Explain the roles and responsibilities of first-line supervisors and safety technicians during an accident investigation.

Describe the basic equipment that is used during an investigation.

Show Transparency 3, Accident Investigation Equipment.

3.0.0 ◆ ACCIDENT INVESTIGATION

The purpose of accident investigations is to determine the cause of the accident and recommend the corrective actions needed to prevent accidents from happening again. Accident investigations should be fact-finding rather than fault-finding activities. The emphasis of an investigation should never be on identifying who should be blamed for the accident, but on why the accident happened. Trying to focus the blame on one person can damage an investigator's credibility. It can also reduce the amount and accuracy of information received from workers. This does not mean that personal responsibility should not be determined. It means that the investigation should be concerned with only the facts. Effective accident investigations are not emotional or judgmental. They are objective examinations that determine the reasons an accident happened.

3.1.0 Need for Investigation

Most accidents result in personal injury or property damage. Therefore, all accidents must be investigated to find the cause or causes. The results of the investigation can pinpoint the corrective actions that will prevent future accidents.

Near misses or incidents also need to be investigated because they also affect the work process. Some accidents and incidents are more serious than others. The amount of time and effort that goes into an investigation will depend on the seriousness of the accident or near miss.

3.2.0 Roles and Responsibilities

Investigating accidents requires cooperation from all levels of management and every worker on site. The first-line supervisor's unique position gives him or her special priority and responsibility during an accident investigation. This person has certain qualifications and advantages other workers do not have. The following lists the qualities and advantages that make the first-line supervisor the best person to investigate accidents.

- Supervisors know the most about the situation. They have daily contact and familiarity with the people, machines, and materials involved. They know the standard practices and circumstances in the area, as well as the hazards.
- Supervisors have a personal interest in identifying accident causes because they are often protective of their workers, machines, and materials. They can also give insight into how to correct the problem.
- Supervisors can take the most immediate action to prevent an accident from recurring because they are so familiar with the situation. Being in direct control of the people, procedures, and property in the area gives the supervisor the advantage of taking immediate corrective action and the greatest opportunity for effective follow-up.
- Supervisors can communicate more effectively with workers. A worker may be employed by the company, but he or she works for and with the first-line supervisor. Workers know the supervisor is interested in them and their safety. To a worker, the first-line supervisor is a level of management that he or she should feel comfortable approaching. During an accident investigation, workers should feel free to be honest with their supervisor. This helps to ensure accurate information is obtained.

The disadvantage of a supervisor investigating his or her own accident is that it may be difficult to be unbiased. For example, the supervisor may have to admit some personal failure , such as not making the right work assignment. This is where the role of the safety technician becomes important. The safety technician will be able to provide an unbiased analysis of the accident; offer objective, corrective feedback; and provide quality control.

3.3.0 Equipment Used in Accident Investigation

Accident investigation requires specific equipment *(Figure 1)*. The following is a list of equipment that is essential to an investigation.

- Camera, film, flash, and fresh batteries to take pictures of the accident scene
- Tape measure (preferably 100') to measure evidence and the distance between pieces of evidence
- Clipboard and writing pad to take notes
- Graph paper to draw diagrams of the accident scene
- Ruler to use as a scale reference in photos
- Pens and pencils to write notes and draw diagrams
- Accident investigation forms to properly document the accident
- Flashlight and fresh batteries to see clearly if the accident area is dark

The following additional equipment can be helpful during an investigation.

- Accident investigator's checklists to make sure you are following your company's procedures

Instructor's Notes:

DID YOU KNOW?

Digital Evidence

In some jurisdictions, digital cameras are not allowed to be used to gather information during an accident investigation. This is because digital photos can be easily altered, which can change the outcome of the investigation.

- Meters, detectors, and test equipment to gather any needed data about the accident
- Magnifying glass to see small details in evidence
- Sturdy gloves to protect your hands when handling materials
- High-visibility plastic tapes to mark off the accident area
- First aid kit to treat minor injures
- Cassette recorder and spare cassette tapes to tape information rather than write it down
- Identification tags to identify and mark evidence
- Masking tape to mark the accident area or seal containers that should not be opened
- Specimen containers to collect evidence
- Compass to gather the most accurate location data
- Chalk to identify the location of evidence
- Video camera to make a visual record of the accident scene
- Tarp to protect evidence from weather or contamination of the accident scene
- Bloodborne pathogen kit
- Lockout/tagout set
- Personal protective equipment for the area

3.4.0 Investigation Time Line

Gathering and preserving information during an investigation is critical to finding the cause or causes of an accident. The sooner the investigator is on the scene, the better. This way there is less chance of evidence tampering and workers are more likely to remember details. The following steps provide a sample time line for investigating accidents.

Step 1 The investigator should begin investigating the accident immediately after the injured person has been treated, before the scene can be changed, and before important evidence is removed, destroyed, or cleaned up. The investigator should look and listen for clues.

Step 2 The investigator should discuss the accident with the injured person if possible, after first aid or medical treatment has been administered. Care and tact are important when interviewing injured workers.

213F01.EPS

Figure 1 ◆ Accident investigation equipment.

Step 3 The investigator should talk with witnesses who were at the scene of the accident, and those who arrived shortly thereafter. Talking to a witness should take place at the scene, if possible. If the situation is stressful for the witness, conduct the interview away from the accident scene. Interview witnesses separately, never as a group, because one person's story may influence others.

Step 4 The investigator should ask for all details that can give clues to the cause of the accident. Keep in mind that witnesses sometimes give more detailed information when they explain events in **reverse chronological order**. The investigator should encourage witnesses to share their ideas about what happened but only after the factual questions of who, what, where, when, why, and how have been answered. This may lead to more clues about the accident.

If the accident is fatal, these additional steps should be taken:

Step 1 Cover the body, but do not move it.

Step 2 Take accurate measurements to define the physical relationship between the body and any equipment and materials involved.

Step 3 Photograph the accident scene and surroundings. Photos are important since the information will be lost on the scene has been cleaned up. Photos also allow you to revisit the scene. Do not take photos of the body unless there is a specific purpose for doing so, because they will generally be of limited value. Photos of the body can also cause needless heartache to the victim's family.

Describe the additional equipment that may be helpful during an investigation.

Explain each step of an accident investigation.

Define the term reverse chronological order.

Describe the additional steps that must be taken during an investigation of a fatal accident.

ACCIDENT INVESTIGATION: POLICIES AND PROCEDURES 13.3

Explain the conditions under which a second-level investigation is done.

Explain why interviews are an important part of an accident investigation.

Identify the people who should be considered for interviews. Explain why each is considered a valuable source of information.

Explain the ultimate goal of every accident investigation interview.

Step 4 Collect and identify any and all related material. Mark it in relation to the accident scene. This evidence may be used if the accident scene needs to be reconstructed.

Following these steps gives investigators the best chance at collecting accurate information and providing corrective feedback.

3.4.1 Second-Level Investigations

Many companies use a team or a committee of workers to investigate accidents involving serious injury or property damage. This team or committee may replace the first-line supervisor's investigation or may serve as a second-level investigation. Second-level investigations should be considered when:

- The accident has or could have had very serious results.
- The nature of the accident is very complex.
- The accident involved more than one supervisor's crew.
- The initial investigation did not clearly establish a full range of contributing factors or corrective and preventive actions.

NOTE

When a team or committee investigates an accident, the team leader or chairperson must have enough authority and status in the organization to do whatever is needed to conduct a thorough investigation.

3.5.0 Conducting Interviews

Once you have documented the accident scene, it is important to start gathering data through the interview process. Conducting interviews is perhaps the most difficult part of an investigation. The purpose of the accident investigation interview is to obtain an accurate and comprehensive picture of what happened. This is done by obtaining all pertinent facts, interpretations, and opinions. Your job is to construct a composite story using the various accounts of the accident and other evidence. In order to be effective, you must have a firm understanding of the techniques for interviewing.

Your first task when beginning the interviewing process is to determine who needs to be interviewed. Questions will need to be personalized for each person interviewed. Interviews should occur as soon as possible, but usually do not begin until things have settled down a bit. Some people you may want to consider for an interview include:

- *The injured worker* – Determine the specific events leading up to the accident.
- *Co-workers* – Find out if appropriate procedures were being used at the time of the accident.
- *First-line supervisor* – Get background information on the worker. He or she can also provide procedural information about the task that was being performed.
- *Project manager* – This person can be the main source for information on related systems.
- *Maintenance workers* – Determine background on equipment and machinery maintenance.
- *Emergency responders* – Learn what they saw when they arrived and during the response.
- *Medical personnel* – Get medical information, as allowed by law.
- *Coroner* – This person can be a valuable source to determine the type and the extent of fatal injuries.
- *Police* – If a police report was filed, talk to the reporting officer.
- *The injured worker's spouse and family* – They may have insight into the worker's state of mind or other work issues.

Cooperation, not intimidation, is the key to a successful accident investigation interview. It's counterproductive to give the impression that you are trying to establish blame. The purpose of the accident interview is to uncover additional information about the hazardous conditions, unsafe work practices, and related system weaknesses that contributed to the accident. Therefore, it's very important that the interviewer use effective techniques to establish a cooperative atmosphere.

It's also important to remember that you are conducting an accident investigation, not a criminal investigation. These two interview processes may be similar, but each has a unique purpose. Each process requires different techniques to fulfill their intended role. The following is a list of effective interviewing techniques that will help you find the facts, not assign fault.

- When interviewing, keep the purpose of the investigation in mind. You are attempting to determine the cause of the accident so that similar accidents will not recur. Make sure the **interviewee** understands this.
- Approach the investigation with an open mind. It will be obvious if you have preconceptions about the individuals or the facts.
- Go to the scene. Just because you are familiar with the location or the victim's job, don't assume that things are always the same. If you can't conduct a private interview at the loca-

Instructor's Notes:

tion, find an office or meeting room that the interviewee considers a neutral location.

- Interview everyone involved including the injured workers, witnesses, and people involved with the process. Witnesses should be interviewed separately, never as a group.
- Put the person at ease. Explain the purpose of the investigation and your role. Sincerely express concern regarding the accident and a desire to prevent a similar occurrence.
- Express to the individual that the information given is important. Be friendly, understanding, and open minded. Be calm and unhurried.
- Let the individual talk. Ask background information such as name or job first. Ask the witness to tell you what happened but don't ask leading questions, interrupt answers, or make facial expressions.
- Ask open-ended questions to clarify particular areas. Try to avoid yes and no answer type questions. Don't ask "why," because these types of questions tend to make people respond defensively. For example: Do not ask: "Why did you drive the forklift with under-inflated tires?" Instead, ask: "What are the forklift inspection procedures? What are the forklift hazard reporting procedures?"
- Repeat the facts and sequence of events back to the person to avoid any misunderstandings.
- Take notes as carefully and casually as possible. Let the individual read them if desired. Give the interviewee a copy of the notes you take.
- Don't use a tape recorder unless you get permission. Tape recorders often intimidate witnesses. Tell the interviewee that the purpose of the recorder is to ensure accuracy. Offer to give the interviewee a copy of the tape.
- Ask for suggestions as to how the accident/incident could have been avoided.
- Conclude the interview with a statement of appreciation for the interviewee's contribution. Ask him or her to contact you if he or she thinks of anything else. If possible, advise these people personally of the outcome of the investigation before it becomes public knowledge.

Understanding and applying this information during the interview process will help you establish a cooperative relationship so that you can obtain the facts. Intimidation and blaming will always result in an ineffective interview process.

3.6.0 Accident Investigation Reports

Accident investigation reports should document the full range of facts. The investigation should include thorough interviews of everyone with any knowledge of the event. Six key questions should be answered: who, what, when, where, how, and why. Facts should be presented carefully and clearly.

A good investigation is likely to reveal several contributing factors. It will also enable the investigator to recommend corrective actions that will help prevent future accidents. The report has to be strong enough and clear enough to justify the corrective actions recommended. This means that the corrective action must be strong enough to support spending time and money on correcting the problem.

The accident investigator should avoid blaming the injured worker, even if the worker admits blame. Instead, the accident investigator should be objective and find all contributing causes. The error made by the worker is most probably not the root or basic cause. The worker who did not follow proper procedures may have been encouraged directly or indirectly by a supervisor or another worker to cut corners.

All supervisors and others who investigate accidents should be held accountable for describing causes carefully and clearly. When reviewing accident investigation reports, pay careful attention to phrases such as "the worker did not plan the job properly." While this statement may suggest an underlying problem with this worker, it does not identify all possible causes, preventions, and controls. It is the supervisor's responsibility to identify, anticipate, and report hazardous conditions to keep workers safe. The accident investigation report should list all the ways to correct the hazardous condition or unsafe activity. Some of these corrective actions can be accomplished quickly. Others may take time, planning, and money. Either way, the accident investigation report is the tool for identifying the cause of accidents and providing corrective actions that prevent more accidents.

3.6.1 Other Uses of Accident Investigation Reports

The primary purpose of accident investigations is to prevent future occurrences. The information obtained through the investigation should also be used to update and revise the inventory of hazards and the company-established safety program. Reports should also be available to top management because they are ultimately responsible for safety on the site and must be aware of the results of investigations.

NOTE
In larger firms, the results should be shared with sister organizations.

Describe the most effective interviewing techniques used during an accident investigation interview.

Demonstrate how to conduct an accident investigation interview.

Given an accident scenario, have participants pair off and practice performing accident investigation interviews. Note the proficiency of each participant. This laboratory corresponds with Performance Task 1.

Explain that accident investigation reports provide a method for documenting the facts found during an accident investigation.

Explain why it is important to remain objective and unbiased during an accident investigation.

Describe the primary and secondary uses of accident investigation reports.

ACCIDENT INVESTIGATION: POLICIES AND PROCEDURES 13.5

Explain the importance of completing all applicable fields on an accident investigation form.

Describe the additional documentation that can be included with the accident investigation form.

Show Transparency 4 (Figure 2).

Explain OSHA's requirements regarding accident investigation forms.

Show Transparencies 5 and 6 (Figures 3 and 4).

3.6.2 Filling Out an Accident Investigation Report

Filling out an accident investigation report is an essential part of the investigation process. It helps to document accidents so that they can be properly analyzed and corrective actions can be made to the work process.

> **NOTE**
> Accident investigation reports can also be used to investigate work-related illnesses from single or multiple exposure to materials found on a job site. The illnesses can include conditions such as contact dermatitis or respiratory conditions caused by exposure to toxic gas.

When completing an accident investigation form, it's important to answer all questions on the form. Answers should be complete, specific, and factual. Do not include opinions or speculations in the report about why the accident happened. If the question does not apply to the accident, indicate this on the report with a written note that says D.N.A. (Does Not Apply). The report should also indicate if no answer was available for a question on the form. Any documentation supporting the investigation, such as sketches of the scene (*Figure 2*), photos, or diagrams, should be attached to the report.

The accident investigation report shown in *Figure 3* is an example of a report that is commonly used to report accidents. This report form meets the recordkeeping requirements specified by OSHA Form 301 (*Figure 4*). You'll notice, however, that the sample accident investigation report includes more information than the OSHA form. This is because OSHA does not provide a mandatory format for accident investigation reports but does require that specific information is provided on it. By including additional information on the report, companies can easily access information such as that which would be helpful in reporting workers' compensation claims.

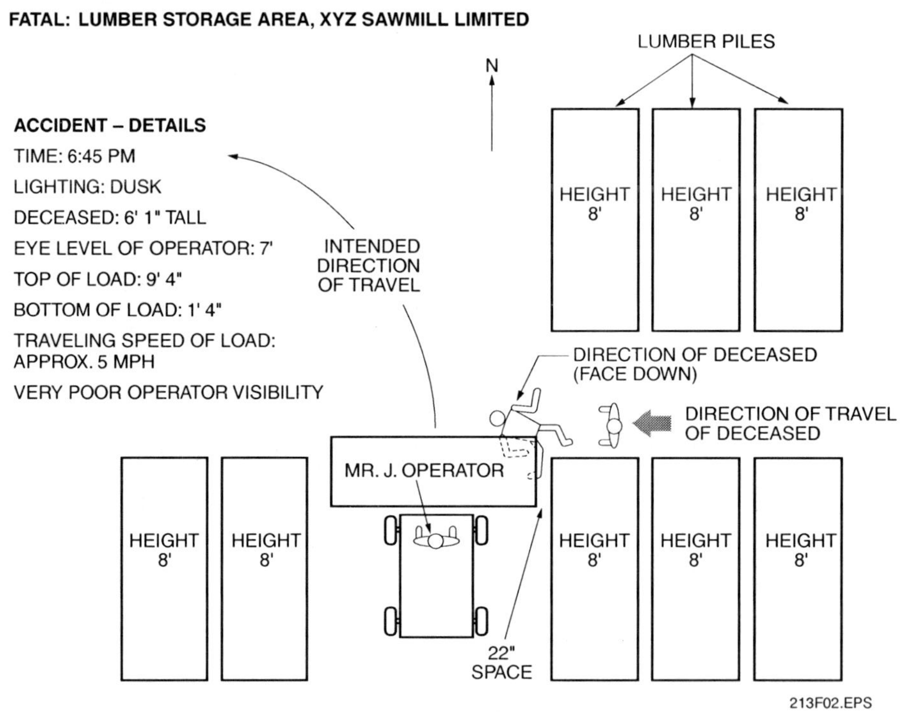

Figure 2 ◆ Accident scene sketch.

Instructor's Notes:

Accident/Incident Investigation Form

▯ Rx/Treatment ▯ First Aid

▯ Property Damage Initial Report ▯ Follow-up

INJURED/INVOLVED EMPLOYEE INFORMATION

Employee #: __________ Name: ______________________________ S.S.# ______________

DOB: ____________ Date of Hire: __________ State Where Hired: __________ Occupation: ______________ Job #: ______________

Pay Rate: $______________ per hr./day/month/year Employees scheduled days off: ______________

Workday begins: __________ am/pm Ends: __________ am/pm All involved Employees Drug tested? Yes No Paid for DOI? Yes No

INJURY/DAMAGE DESCRIPTION: Date of Incident: ______________ Time: __________ am/pm

Describe incident (include specific injury/damage detail) ______________________________

____

____

Complete this section only if the incident resulted in injury.

MEDICAL INFORMATION: Physician Name: ______________________________ Phone: ______________

Physician address: ______________________________ Was Rx Prescribed? Y / N

Trmt. Provided: (Rx, X-ray, Sutures, Exam, Chiro, etc.) ______________________________

Date RTW: ______________ Regular? ❍ Restricted? ❍

RESTRICTED DUTY

LIFTING	❍ CANNOT	❍ CAN	IF "CAN" LIMITATION IN LBS ______________
STOOPING, BENDING, KNEELING	❍ CANNOT	❍ CAN	LIMITATION ______________
CLIMBING	❍ CANNOT	❍ CAN	LIMITATION ______________
REACHING	❍ CANNOT	❍ CAN	LIMITATION ______________
USE OF LEFT OR RIGHT UPPER EXTREMITY	❍ CANNOT	❍ CAN	LIMITATION ______________
USE OF LEFT OR RIGHT LOWER EXTREMITY	❍ CANNOT	❍ CAN	LIMITATION ______________

ANY OTHER RESTRICTIONS (INCLUDE RESTRICTIONS FROM PRESCRIBED MEDS)

ESTIMATED DURATION OF RESTRICTED DUTY

SUPERVISOR INFORMATION: Date reported: __________ Time reported: __________ am/pm

Person reported to: ______________________________ Phone: ______________

Supervisor: ______________________________ Phone: ______________

Signed: ______________________________ Date: ______________

(Person completing report) (Report completed date)

Maintain the original report at the jobsite and transmit copy to the appropriate safety personnel.

DO NOT WRITE BELOW THIS LINE

Timed Received:		Forward to Carrier:	Yes / No	❍ This event resulted in Fatality
Number of incidents for this project:		❍ This event is medical recordable	❍ This event is lost work day	❍ This event is 1st Aid Only

SW-70 Revised 2/03

213F03.EPS

Figure 3 ◆ Sample accident investigation form.

Given an accident scenario, have participants practice completing accident investigation forms. Use either the OSHA form or one you've created. Note the proficiency of each participant. This laboratory corresponds with Performance Task 2.

ACCIDENT INVESTIGATION: POLICIES AND PROCEDURES 13.7

The Safe Expectations, International video, *Accident Investigation: A Tool for Effective Prevention*, can be used to enhance the topics covered in this module.

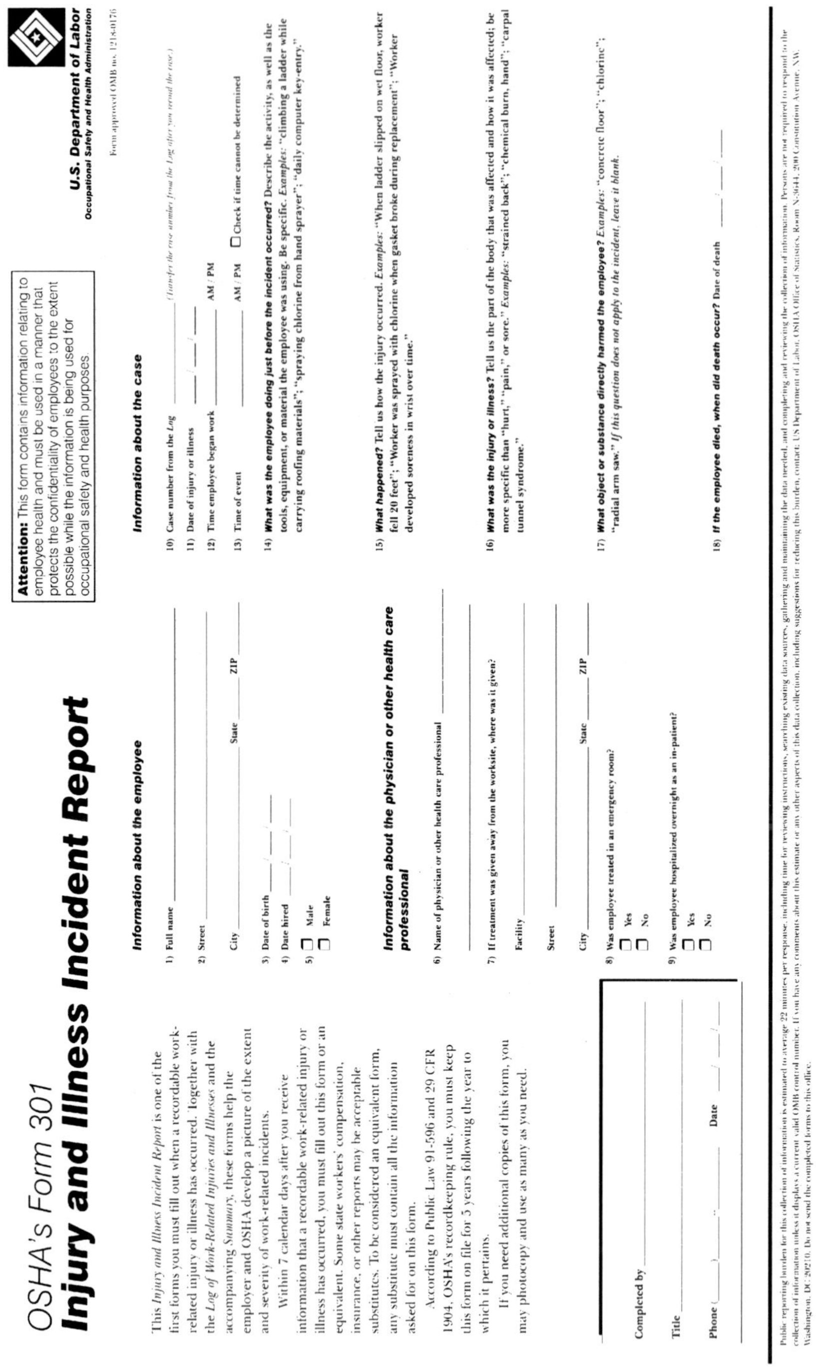

U.S. Department of Labor
Occupational Safety and Health Administration

Form approved OMB no. 1218-0176

Attention: This form contains information relating to employee health and must be used in a manner that protects the confidentiality of employees to the extent possible while the information is being used for occupational safety and health purposes.

OSHA's Form 301
Injury and Illness Incident Report

This *Injury and Illness Incident Report* is one of the first forms you must fill out when a recordable work-related injury or illness has occurred. Together with the *Log of Work-Related Injuries and Illnesses* and the accompanying *Summary*, these forms help the employer and OSHA develop a picture of the extent and severity of work-related incidents.

Within 7 calendar days after you receive information that a recordable work-related injury or illness has occurred, you must fill out this form or an equivalent. Some state workers' compensation, insurance, or other reports may be acceptable substitutes. To be considered an equivalent form, any substitute must contain all the information asked for on this form.

According to Public Law 91-596 and 29 CFR 1904, OSHA's recordkeeping rule, you must keep this form on file for 5 years following the year to which it pertains.

If you need additional copies of this form, you may photocopy and use as many as you need.

Completed by ____________

Title ____________

Phone (____) ____-______ Date ___/___/___

Information about the employee

1) Full name ____________

2) Street ____________

City ________ State ____ ZIP ____

3) Date of birth ___/___/___

4) Date hired ___/___/___

5) ☐ Male
☐ Female

Information about the physician or other health care professional

6) Name of physician or other health care professional ____________

7) If treatment was given away from the worksite, where was it given?

Facility ____________

Street ____________

City ________ State ____ ZIP ____

8) Was employee treated in an emergency room?
☐ Yes
☐ No

9) Was employee hospitalized overnight as an in-patient?
☐ Yes
☐ No

Information about the case

10) Case number from the *Log* ________ *(Transfer the case number from the Log after you record the case.)*

11) Date of injury or illness ___/___/___

12) Time employee began work ________ AM / PM

13) Time of event ________ AM / PM ☐ Check if time cannot be determined

14) **What was the employee doing just before the incident occurred?** Describe the activity, as well as the tools, equipment, or material the employee was using. Be specific. *Examples:* "climbing a ladder while carrying roofing materials"; "spraying chlorine from hand sprayer"; "daily computer key-entry."

15) **What happened?** Tell us how the injury occurred. *Examples:* "When ladder slipped on wet floor, worker fell 20 feet"; "Worker was sprayed with chlorine when gasket broke during replacement"; "Worker developed soreness in wrist over time."

16) **What was the injury or illness?** Tell us the part of the body that was affected and how it was affected; be more specific than "hurt," "pain," or sore." *Examples:* "strained back"; "chemical burn, hand"; "carpal tunnel syndrome."

17) **What object or substance directly harmed the employee?** *Examples:* "concrete floor"; "chlorine"; "radial arm saw." *If this question does not apply to the incident, leave it blank.*

18) **If the employee died, when did death occur?** Date of death ___/___/___

Public reporting burden for this collection of information is estimated to average 22 minutes per response, including time for reviewing instructions, searching existing data sources, gathering and maintaining the data needed, and completing and reviewing the collection of information. Persons are not required to respond to the collection of information unless it displays a current valid OMB control number. If you have any comments about this estimate or any other aspects of this data collection, including suggestions for reducing this burden, contact: US Department of Labor, OSHA Office of Statistics, Room N-3644, 200 Constitution Avenue, NW, Washington, DC 20210. Do not send the completed forms to this office.

213F04.TIF

Figure 4 ◆ OSHA Form 301.

Instructor's Notes:

Summary

Accident investigations are an essential part of accident prevention. The information collected from accident investigations is not only used to determine the causes of accidents, it is used to provide corrective actions so the same accident won't happen again.

Accident investigations are usually done by first-line supervisors. Once the supervisor has completed an accident investigation form, he or she gives it to the safety technician to review and submit to the appropriate personnel. The safety technician also acts as a resource for the supervisor during the investigation. It is important for both first-line supervisors and safety technicians to know their role in accident investigations. It is also important for each to know what causes accidents, why investigations are conducted, how an accident investigation form is used, and how the information that is gathered will be used. Anyone who conducts interviews after and accident must be able to use effective interviewing techniques in order to gather information that will useful in the investigation. All of this knowledge helps to make work sites safer.

Review the material covered in the module.

Have the participants complete the Review Questions and go over the answers prior to administering the Module Examination.

Review Questions

1. The only difference between an accident and an incident is the _____.
   a. type of investigation conducted
   b. cause
   c. number of workers involved
   d. outcome

2. Incidents are events that result in property damage, personal injury, or death.
   a. True
   b. False

3. Which of the following is *not* a purpose of an accident investigation?
   a. Identifying underlying causes
   b. Reducing risk of a similar accident occurring
   c. Identifying the worker to blame
   d. Recommending corrective actions

4. Investigations are only necessary when a death has occurred.
   a. True
   b. False

5. The greatest disadvantage of a first-line supervisor conducting his or her own accident investigation is that he or she _____.
   a. is generally unfamiliar with the investigation process
   b. is usually too busy to conduct thorough investigations
   c. may find it difficult to remain unbiased
   d. cannot take immediate corrective actions

6. An investigator should begin investigating an accident _____.
   a. before injured workers have been removed
   b. immediately after the injured worker has been treated
   c. before the next shift of workers arrives
   d. within 24 hours

7. Second-level investigations should be considered when the nature of the accident is very complex.
   a. True
   b. False

8. Emergency responders should be interviewed to learn what they saw when they arrived at the scene.
   a. True
   b. False

9. Phrases such as "the worker did not plan the job properly" should be avoided when completing the accident investigation form because they assign blame instead of identifying root causes.
   a. True
   b. False

10. OSHA provides a mandatory format for accident investigation reports.
    a. True
    b. False

Administer the Module Examination. Record the results on Craft Training Report Form 200 and submit the results to the Training Program Sponsor.

Administer the Performance Test and fill out Performance Profile Sheets for each participant. If desired, participant proficiency noted during laboratory sessions may be used to complete the Performance Test. Record the results on Craft Training Report Form 200 and submit the results to the Training Program Sponsor.

ACCIDENT INVESTIGATION: POLICIES AND PROCEDURES 13.9

PROFILE IN SUCCESS

Richard S. Baldwin, BE&K Engineering and Construction

Corporate Safety Director

How did you choose a career in the Safety Industry?
I served two combat flying tours in Southeast Asia with the US Air Force. During my service, I was initially assigned as a Flight Safety Officer.

What types of training have you been through?
I have earned a BS Degree, Master's Degree, and have completed considerable graduate education courses in Occupational Safety and health. I have also taken part in several courses at the University of Southern California Safety Center, completed OSHA technical courses, and attended countless safety seminars. I learned the most as I prepared to speak at conferences such as ASSE, NSC, VPPPA, AIHA, and ABC.

What kinds of work have you done in your career?
I was an Air Force pilot for twenty years and while in a cockpit assignment, was a safety director or safety manager in some capacity. After the Air Force, I had three positions as a corporate safety manager or director; two were in chemical companies. My current position, where I have been for the last nine years, is with BE&K Engineering and Construction.

What does your present position involve?
I directly supervise 50 safety and health professionals who manage safety for over 10,000 construction, maintenance, and engineering personnel on worldwide paper mill, power plant, steel mill, and chemical plant construction and industrial maintenance projects. I administer all aspects of the safety and health process for the corporation. As well, I supervise four corporate safety and health staff members in addition to field personnel. I recruit, hire, and train new safety professionals, publish safety and health manuals and training materials, represent the company on three national construction safety committees (one as past chairperson) and several industry councils. I have authored safety articles in national trade magazines and journals. I have led the way for a reduction of the OSHA Recordable Incident Rate from 6.1 to 0.80, and the Lost Time Rate from 0.82 to 0.12. I've led the company to full membership in the OSHA Voluntary Protection Program (VPP), and was successful in achieving three OSHA Star construction sites. I am responsible for safety supervision at construction projects in several overseas locations. Additionally, I am a member of ASSE's Construction Safety Speaker's Bureau, and a recurring speaker at the National VPPPA Conference, ASSE Professional Development Conference, Texas Safety Association, Construction Project Managers' Academy, and numerous local safety council meetings and Regional VPPPA conferences.

What factors have contributed most to your success?
I have a strong interest in people and establishing the best safety systems to protect workers.

What advice would you give to those new to Safety industry?
Never cease to study technical safety requirements and the latest management techniques…especially the behavioral safety aspects of incident prevention.

Instructor's Notes:

GLOSSARY

Trade Terms Introduced in This Module

Accident: An event that results in property damage and/or personal injury or death.

Incident: An event that could have resulted in damage, injuries, or death, but did not. These serve as warnings of hazards that must be corrected.

Interviewee: Person being interviewed.

Reverse chronological order: Events told in order from last to first.

ACCIDENT INVESTIGATION: POLICIES AND PROCEDURES 13.11

Additional Resources

This module is intended to be a thorough resource for task training. The following reference works are suggested for further study. These are optional materials for continued education rather than for task training.

www.osha.gov

www.asse.org

The Psychology of Safety Handbook, 2001. E. Scott Geller, Ph.D. Boca Raton, FL: CRC/Lewis Publishers.

Root Cause Analysis Handbook: A Guide to Effective Incident Investigation, 1999. JBF Associates Division. Rockville, MD: Government Institutes.

Figure Credits

Brigid R. McKenna	213F01
Frank McDaniel	213F03

Instructor's Notes:

MODULE 75213-03 — ANSWERS TO REVIEW QUESTIONS

Answer		Section
1.	d	2.0.0
2.	b	2.0.0
3.	c	3.0.0
4.	b	3.1.0
5.	c	3.2.0
6.	b	3.4.0
7.	a	3.4.1
8.	a	3.5.0
9.	a	3.6.0
10.	b	3.6.2

CONTREN™ LEARNING SERIES — USER UPDATES

The NCCER makes every effort to keep these textbooks up-to-date and free of technical errors. We appreciate your help in this process. If you have an idea for improving this textbook, or if you find an error, a typographical mistake, or an inaccuracy in NCCER's Contren™ textbooks, please write us, using this form or a photocopy. Be sure to include the exact module number, page number, a detailed description, and the correction, if applicable. Your input will be brought to the attention of the Technical Review Committee. Thank you for your assistance.

Instructors – If you found that additional materials were necessary in order to teach this module effectively, please let us know so that we may include them in the Equipment/Materials list in the Instructor's Guide.

Write: Curriculum Revision and Development Department
National Center for Construction Education and Research
P.O. Box 141104, Gainesville, FL 32614-1104

Fax: 352-334-0932

E-mail: curriculum@nccer.org

Craft ______________________ Module Name ______________________

Copyright Date ____________ Module Number ____________ Page Number(s) ____________

Description

____

____

____

____

(Optional) Correction

____

____

____

(Optional) Your Name and Address

____

____

____

Module 75214-03

Accident Investigation: Data Analysis

NCCER STANDARDIZED CRAFT TRAINING PROGRAM

The National Center for Construction Education and Research (NCCER) provides a standardized national program of accredited craft training. Key features of the program include instructor certification, competency-based training, and performance testing. The program provides trainees, instructors, and companies with a standard form of recognition through a National Craft Training Registry. The program is described in full in the *Guidelines for Accreditation*, published by the NCCER. For more information on standardized craft training, contact the NCCER by writing us at P.O. Box 141104, Gainesville, FL 32614-1104; calling 352-334-0911; or e-mailing info@nccer.org. More information may be found at our Web site, www.nccer.org.

HOW TO USE THIS ANNOTATED INSTRUCTOR'S GUIDE

Each page presents two sections of information. The larger section displays each page exactly as it appears in the Participant Module. The narrow column ties suggested participant and instructor actions to each page and provides icons (detailed below) to call your attention to material, safety, audiovisual, or testing requirements. The bottom of each page includes space for your notes.

The **Audiovisual** icon indicates an appropriate time to show a transparency or other audiovisual aid.

The **Classroom** icon prompts you to define a term, stress a point, ask participants to explain a concept, or give examples.

The **Demonstration** icon directs you to show participants how to perform tasks.

The **Examination** icon tells you to administer the written module examination.

The **Homework** icon is placed where you may wish to assign reading for the next class, to assign a project, or to advise participants to prepare for an examination.

The **Laboratory** icon is used when participants are to practice performing tasks.

The **Materials** icon is a reminder for you to gather materials needed for classes, labs, and testing.

The **Performance Testing** icon tells you to administer a performance test or a portion thereof.

The **Safety** icon is used to emphasize safety issues. It is often keyed to *Caution* and *Warning* statements in the Participant Module.

The **Teaching Tip** icon indicates additional guidance is available, such as how to conduct an exercise, get the most educational value from a field trip, or encourage class participation. Teaching Tips may expand on a feature (*Think About It*, *Did You Know?*) or provide Quick Quizzes or similar exercises. You will be referred to the Teaching Tips section at the back of the module if there is additional material.

The **Combination** icon indicates that the laboratory listed corresponds with a performance task. If desired, you can note the proficiency of the participants during the laboratory and use it to satisfy performance testing requirements.

PREPARATION

Before teaching this module, you should review the Objectives, Performance Tasks, Materials and Equipment List, and the Module Outline. Be sure to allow ample time to prepare your own training or lesson plan and gather all required materials and equipment.

MODULE OVERVIEW

In this module, participants will learn the various methods for analyzing accident investigation data.

PREREQUISITES

Please refer to the Course Map in the Participant Module. Prior to training with this module, it is recommended that the participant shall have successfully completed the following:

Field Safety; Safety Technology, Modules 75201-03 through 75213-03

OBJECTIVES

Upon completion of this module, the participant will be able to:

1. Explain, in general, the methods commonly used for analyzing accident investigation information.
2. Explain at least three systematic approaches to accident investigation.

PERFORMANCE TASK

Under the supervision of the instructor, the participant should be able to:

1. Use one of the methods of analyzing accident data to determine the cause of an accident.

MATERIALS AND EQUIPMENT LIST

Overhead projector and screen
Transparencies
Whiteboard/chalkboard
Markers/chalk
Blank acetate sheets
Transparency pens
Pencils and scratch paper
Appropriate personal protective equipment
Module Examinations*
Performance Profile Sheets*

* Located in the Test Booklet.

SAFETY CONSIDERATIONS

Ensure that participants are equipped with appropriate personal protective equipment.

ADDITIONAL RESOURCES

This module is intended to present thorough resources for task training. The following reference works are suggested for both instructors and motivated participants interested in further study. These are optional materials for continued education rather than for task training.

Web site: www.osha.gov

Web site: www.asse.org

Root Cause Analysis Handbook: A Guide to Effective Incident Investigation, 1999. JBF Associates Division. Rockville, MD: Government Institutes.

TEACHING TIME FOR THIS MODULE

An outline for use in developing your lesson plan is presented below. Approximately 2½ hours are suggested to cover *Accident Investigation: Data Analysis*. This includes 10 minutes at the beginning of the session for administrative tasks and one 10-minute break. You may need to adjust the time required for hands-on activities and testing based on your class size and resources. Because laboratories often correspond to Performance Tasks, the proficiency of the participants may be noted during these exercises for Performance Testing purposes.

Topic	Planned Time
A. Introduction	____________
B. Accident Investigation Analysis	____________
C. Three Levels of Accident Causation Model	____________
1. Level I	____________
2. Level II	____________
3. Level III	____________
D. The Why Method	____________
E. Sequence of Events: Why Method	____________
F. OSHA's Problem-Solving Techniques	____________
1. Change Analysis	____________
2. Job Safety Analysis	____________
3. Organizing and Reporting the Data	____________
G. Trend Analysis	____________
H. Laboratory	____________
1. Provide each participant with an accident investigation scenario and have them determine the cause of the accident by using one of the accident investigation methods described in this module. This laboratory corresponds with Performance Task 1.	
I. Review	____________
J. Module Examination	____________
1. Participants must score 70% or higher to receive recognition from the NCCER.	
2. Record the testing results on Craft Training Report Form 200 and submit the results to the Training Program Sponsor.	
K. Performance Testing	____________
1. Participants must perform each task to the satisfaction of the instructor to receive recognition from the NCCER. If applicable, proficiency noted during laboratory exercises can be used to satisfy the Performance Testing requirements.	
2. Record the testing results on Craft Training Report Form 200 and submit the results to the Training Program Sponsor.	

Module 75214-03

Accident Investigation: Data Analysis

Instructor's Notes:

COURSE MAP

This course map shows all of the modules in Safety Technology. The suggested training order begins at the bottom and proceeds up. The local Training Program Sponsor may adjust the training order.

Assign reading of Module 75214-03.

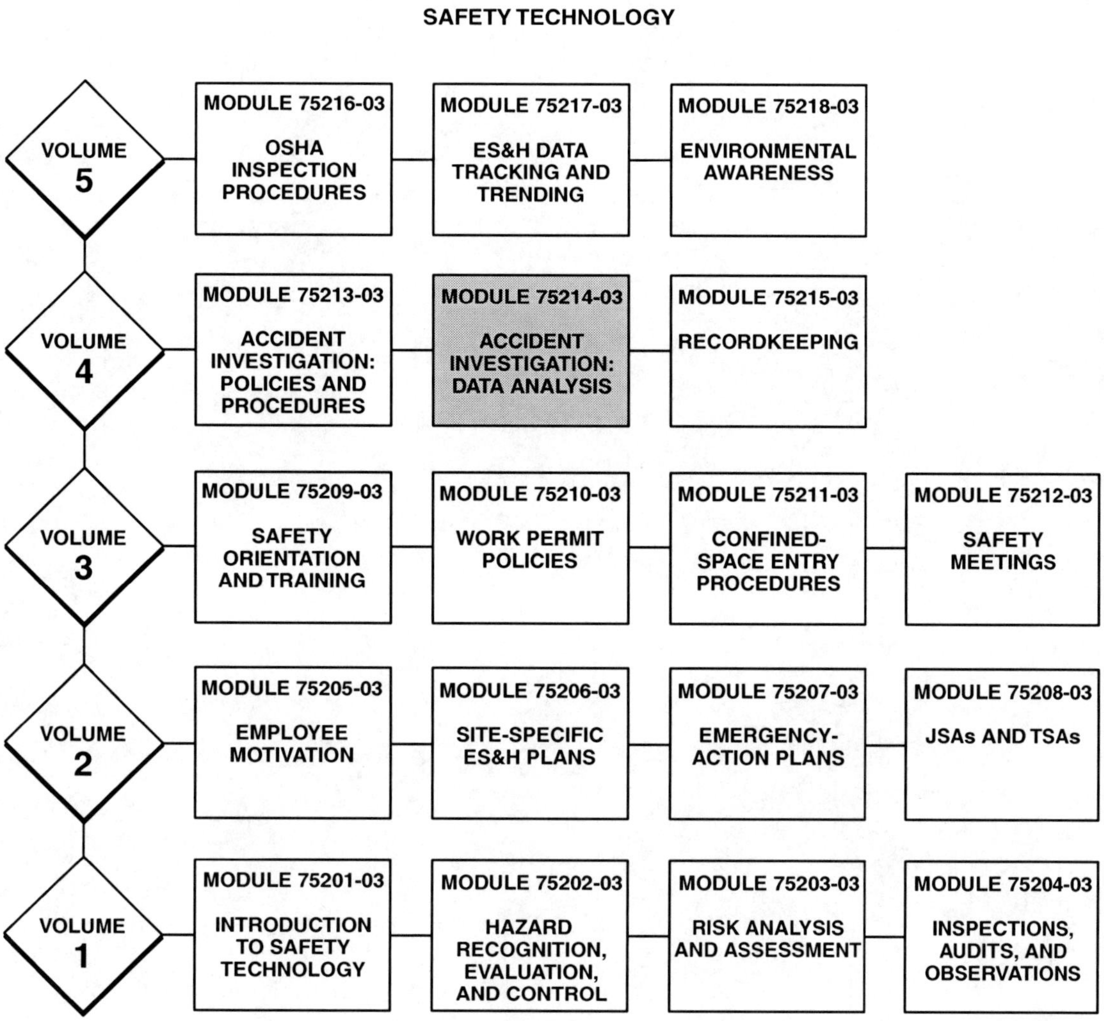

214CMAP.EPS

ACCIDENT INVESTIGATION: DATA ANALYSIS 14.iii

Instructor's Notes:

MODULE 75214-03 CONTENTS

Figures

ACCIDENT INVESTIGATION: DATA ANALYSIS 14.v

Instructor's Notes:

MODULE 75214-03

Accident Investigation: Data Analysis

Objectives

When you have completed this module, you will be able to do the following:

1. Explain, in general, the methods commonly used for analyzing accident investigation information.
2. Explain at least three systematic approaches to accident investigation.

Prerequisites

Before you begin this module, it is recommended that you successfully complete the following: Field Safety; Safety Technology Modules 75201-03 through 75213-03.

Required Materials

1. Pencil and paper
2. Appropriate personal protective equipment
3. Copy of *29 CFR 1926, OSHA Construction Industry Regulations*

1.0.0 ◆ INTRODUCTION

It is important to analyze accident investigation data in order to discover why accidents happen. Once you know how they happened, you can correct the problem. This helps prevent future accidents.

In the previous module, *Accident Investigation: Policies and Procedures*, you learned how to conduct an accident investigation. In this module, you will learn how to analyze accident investigation data. Data analysis is a special skill. As a safety technician, you will find that the ability to analyze data will be helpful in other areas of your work. This is because data analysis requires you to process information in a logical way. If you are able to do this, tasks such as conducting job safety analyses, planning safety meetings, and performing inspections and audits can be done quickly and effectively.

2.0.0 ◆ ACCIDENT INVESTIGATION ANALYSIS

Because every accident is different, there are different approaches to the challenge of analyzing accident investigation data. You may find that one approach is better or easier to use than another, or that a combination is best. The end objective of any analysis should be the same: finding out what caused the accident and what is necessary to fix the problem. The following are methods and models that can help you understand how and why accidents happen:

- The Three Levels of Accident Causation Model
- The Why Method
- Sequence of Events: Why Method
- OSHA's Problem-Solving Techniques

3.0.0 ◆ THREE LEVELS OF ACCIDENT CAUSATION MODEL

The causes of accidents can be classified by three different factors: direct causes, indirect causes, and root causes. These are also called the three levels of accident causation *(Figure 1)*. Each level of causation represents one of these factors. It's important to be able to clearly recognize and understand the components of these three levels for analyzing accident data. If you understand what causes accidents, you will be better able to identify the causes when reviewing accident investigation forms, witness interviews, or physical evidence.

Ensure that you have everything you need to teach the course. Check the Materials and Equipment list in the front of this module.

Explain that terms shown in bold are defined in the Glossary at the back of this module.

Review the goals of the course and explain what will be expected of the participant.

Show Transparency 1, Objectives.

Show Transparency 2, Performance Task.

Explain that data analysis is an important part of accident investigations.

Identify the four methods and models that can be used to analyze accident data.

Explain how the Three Levels of Accident Causation Model can be used to analyze accident data.

Show Transparency 3 (Figure 1).

Explain why identifying energy sources that can cause accidents is an important part of analyzing accident data.

Identify the common types of accidents that can occur as a result of uncontrolled releases of energy.

Explain that indirect causes are factors that contribute to an accident but aren't the main cause(s).

Explain the importance of identifying at-risk behavior and unsafe conditions. Provide examples of both.

Have the participants share work experiences in which they witnessed at-risk behavior and/or unsafe conditions. Ask them how the situation was corrected.

> **DID YOU KNOW?**
>
> ***Bureau of Labor Statistics***
>
> The Bureau of Labor Statistics (BLS) is the principal fact-finding agency for the federal government. The BLS operates an independent national statistical agency that collects, processes, analyzes, and disseminates essential statistical data to the American public, the U.S. Congress, other federal agencies, state and local governments, business, and labor. They are experts in the field of data analysis. *Appendix*, the BLS Data Integrity Guidelines, is an example of one procedure the BLS has in place to ensure the integrity of the data it collects, analyzes, and disperses.
>
> Source: U.S. Bureau of Labor Statistics

3.1.0 Level I

Level I represents direct causes of accidents. These are accidents involving the uncontrolled release of energy. These accidents may or may not cause injury or property damage, but they are still dangerous. When investigating an accident, be sure to look closely at energy sources that may have been released unintentionally. Examples of accidents caused by uncontrolled releases of energy include the following:

- Being struck by an object or equipment
- Being caught in between two objects
- Being struck by debris during the detonation of explosives
- Being cut or scraped by a bladed tool

These types of accidents are preventable if equipment is properly maintained and workers follow established safety guidelines.

3.2.0 Level II

Level II represents indirect causes of accidents. Indirect causes are factors that contribute to an accident but aren't the main cause. Indirect causes usually involve unsafe acts and conditions. In the past, investigators looked only for indirect causes of accidents, not the reason behind the unsafe act or condition. This proved to be ineffective because indirect causes are often just symptoms of a greater problem, but are not the actual cause.

3.2.1 Unsafe Acts

It's important to be able to recognize when a worker's behavior is at risk. The following is a list of the most common at-risk behaviors found on a job site:

- Failing to use personal protective equipment correctly
- Failing to warn co-workers of potentially hazardous conditions or unsafe behaviors
- Failing to follow instructions and procedures
- Using defective equipment
- Lifting improperly
- Taking an improper working position
- Making safety devices inoperable
- Operating equipment at improper speeds
- Operating equipment without authority
- Servicing equipment while it is in motion or energized
- Loading or placing equipment or supplies improperly or in a dangerous way

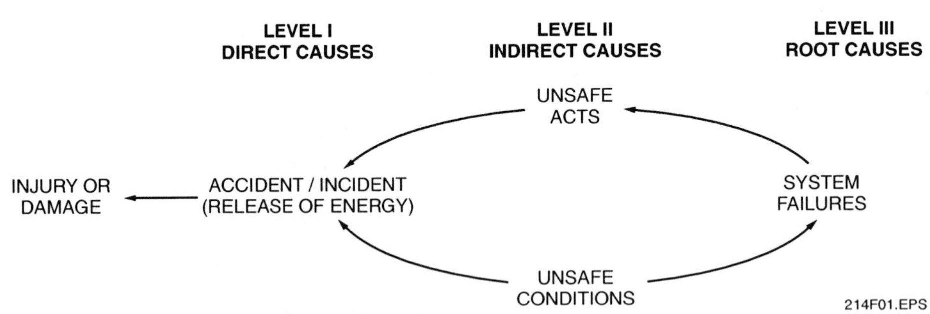

Figure 1 ◆ Three levels of causation.

Instructor's Notes:

- Using equipment improperly
- Working while impaired by alcohol or drugs
- Engaging in horseplay

3.2.2 Unsafe Conditions

Unsafe conditions are physical conditions that are different from acceptable, normal, or correct conditions. The following is a list of the most common unsafe conditions:

- Congested workplaces
- Defective tools, equipment, or supplies
- Excessive noise
- Fire and explosion hazards
- Hazardous atmospheric conditions
  - Gases
  - Dusts
  - Fumes
  - Vapors
- Inadequate supports or guards
- Inadequate warning systems
- Poor housekeeping
- Poor illumination
- Poor ventilation

Most accidents are caused by at-risk acts and conditions. It's important to be aware of both types of causes and use as many preventive measures as necessary, including proper training, appropriate equipment maintenance, and good work-site housekeeping.

3.3.0 Level III

Level III represents the root causes of accidents (*Figure 2*). The root or basic causes of accidents are the underlying reasons an accident happened. Root causes not only affect single accidents being investigated, they also affect other future accidents and work problems. Once root causes are fixed, the types of accidents that occurred because of them will not happen again. Root causes should be corrected as soon as it is practical to correct them. This often takes time because new procedures may have to be written and implemented, and new equipment and training may be required. Immediate temporary controls (ITCs) should be implemented until the root actions can be implemented. This will help prevent accidents and save lives.

4.0.0 ◆ THE WHY METHOD

The Why Method of accident investigation involves simply asking the question "Why did this happen?" and investigating until the question "Why?" can no longer be asked (*Figure 3*). The line of questioning may take several paths, each of which will lead to one or more basic or root causes. Once you discover the basic or root causes of accidents, you are better able to correct problems and create a safe work environment.

The Why Method can be enhanced by combining it with the Sequence of Events Method. This combined methodology is called the Sequence of Events: Why Method.

5.0.0 ◆ SEQUENCE OF EVENTS: WHY METHOD

The Sequence of Events: Why Method is a systematic method of objectively identifying the root cause(s) of unsafe acts, conditions, and incidents. It involves identifying the sequence of events leading to the unsafe act or condition. The following steps are used in the Sequence of Events: Why Method.

Step 1 Identify the actual or potential injury, damage, or near miss.

Step 2 Ask, "Why did this situation occur?"

Step 3 Work backwards in time asking "Why?" until you can no longer ask "Why?" or you reach a dead end.

Step 4 Identify the events leading up the incident. These preceding events can take one of two forms: they can be something that happened that should not have happened, or they can be something that did not happen but should have.

For example, a laborer on a construction site was told to go clean up the third floor of the building. He was specifically told to pick up all boards and scrap lumber. When the laborer got to the third floor, he found a pile of trash and scrap lumber covering a 4' × 8' piece of plywood. He picked up all of the loose pieces of wood and trash and put them down the trash chute. He then proceeded to pick up the plywood. He picked up the piece of plywood and held it in front of him, which obstructed his view. When he stepped forward, he fell through a 30" × 32" floor opening the plywood had been covering. He fell approximately 12' to the second floor and suffered serious hip and back injuries.

Explain how to identify the root causes of accidents.

Show Transparency 4 (Figure 2).

Describe the Why Method of accident investigation. Discuss the benefits of using this method.

Show Transparency 5 (Figure 3).

Explain the steps used in the Sequence of Events: Why Method.

Show Transparency 6, Sequence of Events: Why Method Steps.

ACCIDENT INVESTIGATION: DATA ANALYSIS 14.3

Provide an example of how the Sequence of Events: Why Method is used to analyze accident data. Show the participants the sequence of events that was determined using this method.

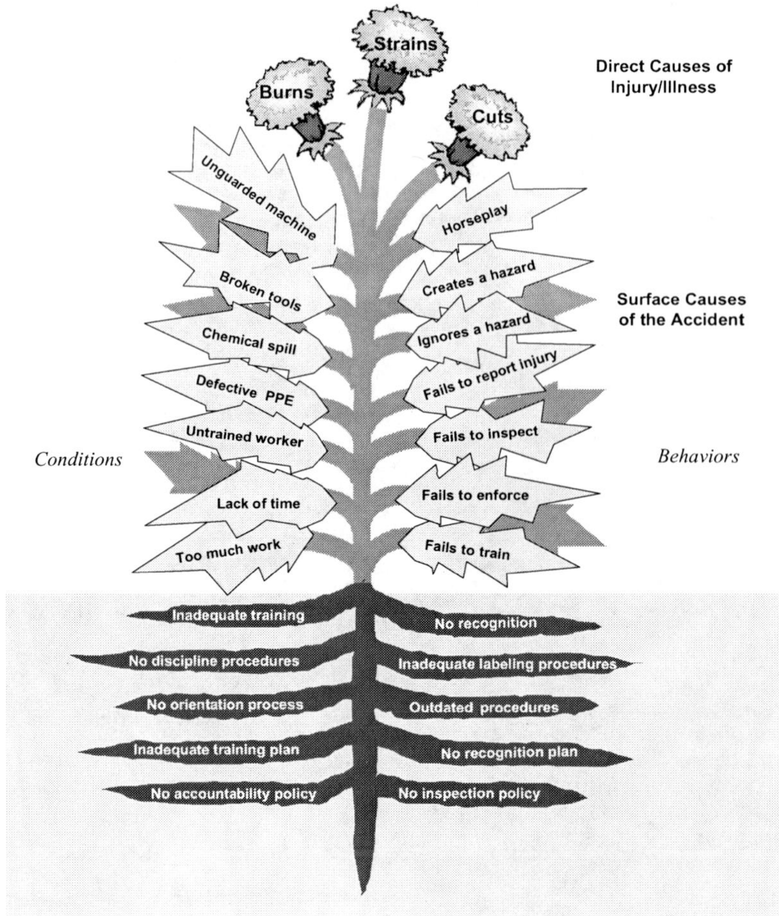

214F02.EPS

Figure 2 ◆ Root causes.

WHY DID BOB FALL?
ANSWER: THE LADDER BROKE.

WHY DID THE LADDER BREAK?
ANSWER: IT HAD A BAD RUNG.

WHY DID THE LADDER HAVE A BAD RUNG?
ANSWER: IT WASN'T STORED PROPERLY.

WHY WASN'T IT STORED PROPERLY?
ANSWER: THERE WAS NO PROCEDURE.

CONCLUSION: BOB FELL BECAUSE THERE WAS NO PROCEDURE FOR LADDER STORAGE

214F03.EPS

Figure 3 ◆ The Why Method.

The following sequence of events regarding this accident was determined by using the Sequence of Events: Why Method.

- *Accident event* – A laborer fell through an unlabeled floor opening.
- *First preceding event* – Another contractor's temporary employee who covered the hole failed to secure the plywood in place and label the hole with the word *cover* as required by the OSHA fall protection standard.
- *Second preceding event* – The temporary contractor employee had been on the job only two days and had no knowledge of the OSHA requirements for covering floor openings and received no specific instructions from his supervisor other than to cover up that hole with a sheet of plywood.

Instructor's Notes:

- *Root cause* – This accident was caused by the lack of training and insufficient instructions to the temporary worker.

This method is effective because it goes beyond simply asking why. It allows for a deeper level of analysis because it requires you to record the chain of events leading up to the accident. When you are able to see all of the pieces of the accident, you can more easily determine the root cause.

6.0.0 ◆ OSHA'S PROBLEM-SOLVING TECHNIQUES

OSHA documents thousands of accidents that occur daily throughout the United States. These result from a failure of people, equipment, supplies, or surroundings to behave as expected. Accidents represent problems that must be solved through investigations. A successful accident investigation determines not only what happened, but also finds how and why the accident occurred. Formal problem-solving techniques can be used to solve problems of any degree of complexity. This section discusses two of the most commonly used procedures suggested by OSHA to analyze accident data: change analysis and job safety analysis.

6.1.0 Change Analysis

As its name implies, the change analysis technique emphasizes change. In terms of accident analysis problem solving, you must look for deviations from the norm. In order to do this, all situations, especially accidents and incidents, that have resulted in an unanticipated change should be carefully identified and analyzed. Analyzing working conditions or behavior that were different from usual conditions or behavior can identify the cause of the accident or incident. Once you determine what was different or the source of the change, corrective actions should be taken.

The following steps are often used during the change analysis method:

Step 1 Define the problem. What happened?

Step 2 Establish the norm. What should have happened?

Step 3 Identify, locate, and describe the change.

Step 4 Specify what was and what was not affected.

Step 5 Identify the distinctive features of the change.

Step 6 List the possible causes.

Step 7 Select the most likely causes.

6.2.0 Job Safety Analysis

Job safety analysis (JSA) is part of many existing accident prevention programs. In general, a JSA breaks a job into basic steps and identifies the hazards associated with each step. A JSA also prescribes controls for each hazard. A JSA consists of a chart listing these steps, hazards, and controls. Review the JSA during the investigation if a JSA has been conducted for the job involved in an accident. Perform a JSA if one is not available. Perform a JSA as a part of the investigation to determine the events and conditions that led to the accident.

6.3.0 Organizing and Reporting the Data

An accident investigation is not complete until a report is prepared and submitted to proper authorities. These reports help provide thorough documentation of an accident as well as create a structure for organizing the accident data. The information collected from change analyses and job safety analyses should be used for accident investigation reports. The following outline can be used as a tool to organize information gathered by these methods:

1. *Background Information*
   a. Where and when the accident occurred
   b. Who and what were involved
   c. Operating personnel and other witnesses

2. *Account of the Accident (what happened?)*
   a. Sequence of events
   b. Extent of damage
   c. Accident type
   d. Agency or source, such as energy or hazardous material

3. *Discussion (analysis of the how and why the accident happened)*
   a. Direct causes, such as energy sources or hazardous materials
   b. Indirect causes, such as unsafe acts or conditions
   c. Basic causes, such as management policies, personal factors, or environmental factors

4. *Recommendations (to prevent a recurrence) for immediate and long-range action to remedy)*
   a. Basic causes
   b. Indirect causes
   c. Direct causes, such as reduced quantities, protective equipment, or protective structures

Describe OSHA's approach to analyzing accident data.

Discuss the steps used in OSHA's change analysis technique.

Show Transparency 7, Change Analysis Steps.

Describe how job safety analyses (JSAs) can be used to analyze accident data.

Discuss OSHA's method for organizing and reporting accident data. Explain the outline that is used with this method. Refer the participants to the Participant Module.

ACCIDENT INVESTIGATION: DATA ANALYSIS 14.5

Explain the types of trends that can be studied to help determine the causes of accidents.

Provide each participant with an accident investigation scenario and have him or her determine the cause of the accident by using one of the accident investigation methods described in this module. Note the proficiency of each participant. This laboratory corresponds with Performance Task 1.

Review the material covered in the module.

Have the participants complete the Review Questions and go over the answers prior to administering the Module Examination.

Administer the Module Examination. Record the results on Craft Training Report Form 200 and submit the results to the Training Program Sponsor.

7.0.0 ◆ TREND ANALYSIS

One of the many values of incident investigations is the trend analysis that can be studied after the investigation. By analyzing trends, one can get a much bigger and better picture of accident causation and take appropriate actions. By using completed accident and incident reports, one can analyze the data by such factors as:

- Time of day
- Length of employment
- Shift
- Length of time on the job
- Supervisor
- Type of supervision
- Job classification
- Task being performed
- Body part injured
- Type of injury
- Hours worked in the previous 24 hours

Keeping track of accident investigation data and then using it to find trends will help improve the overall safety process. Make sure all data is forwarded to the corporate office for an even broader look. The findings may surprise you.

Summary

Analyzing accident investigation data is a critical step in determining how and why accidents occur. If you are able to determine the causes of accidents, you will be able to prevent future accidents. This will save the lives of workers as well as prevent costly equipment damage. Certain methods and models of data analysis may be more effective than others based on the type of accident that occurred. No matter which method is used, the ultimate goal is finding the cause of the accident and fixing the problem.

Review Questions

For Questions 1 through 3, match the level of accident causation to the corresponding description.

Level of accident causation

1. Level I _____
2. Level II _____
3. Level III _____

Description

a. Represents the root causes of accidents.
b. Represents direct causes of accidents.
c. Represents indirect causes of accidents.

4. Level I accidents involve _____.
   a. unsafe acts
   b. unsafe conditions
   c. uncontrolled releases of energy
   d. the failure to correct known on-site hazards

5. The Why Method involves simply asking the question "Why did this happen?"
   a. True
   b. False

6. The Why Method can be enhanced by combining it with the Three Levels of Causation Model.
   a. True
   b. False

7. The length of time on the job should be considered when performing a trend analysis.
   a. True
   b. False

8. Data analysis is a simple task that be done by anyone on site.
   a. True
   b. False

9. The end objective of any accident data analysis is to _____.
   a. blame the person responsible for accident
   b. fine those responsible for the accident
   c. find out what caused the accident and fix it
   d. report the finding to OSHA

10. The Sequence of Events: Why Method is effective because it goes beyond simply asking why an accident happened.
    a. True
    b. False

Instructor's Notes:

GLOSSARY

Trade Terms Introduced in This Module

Trial and error: A method of reaching a correct solution or satisfactory result by trying out various means or theories until error is sufficiently reduced or eliminated.

Administer the Performance Test and fill out Performance Profile Sheets for each participant. If desired, participant proficiency noted during laboratory sessions may be used to complete the Performance Test. Record the results on Craft Training Report Form 200 and submit the results to the Training Program Sponsor.

Instructor's Notes:

APPENDIX

Bureau of Labor Statistics Data Integrity Guidelines

The following guidelines must be followed by all Bureau of Labor Statistics (BLS) program offices and BLS employees to ensure the integrity of information maintained and disseminated by the BLS.

Office of Management and Budget (OMB) information quality guidelines define *integrity* as the security of information—protection of the information from unauthorized access or revision to ensure that the information is not compromised through corruption or falsification.

Confidential Nature of BLS Records

Data collected or maintained by, or under the auspices of, the BLS under a pledge of confidentiality shall be treated in a manner that will ensure that individually identifiable data will be used only for statistical purposes and will be accessible only to authorized persons.

Pre-release economic series data prepared for release to the public will not be disclosed or used in an unauthorized manner before they have been cleared for release and will be accessible only to authorized persons.

Authorized persons include only those individuals who are responsible for collecting, processing, or using the data in furtherance of statistical purposes or for the other stated purposes for which the data were collected. Authorized persons are authorized access to only those data that are integral to the program on which they work and only to the extent required to perform their duties.

When non-BLS employees are granted access to confidential BLS data or Privacy Act data, they must be notified of their responsibility for taking specific actions to protect the data from unauthorized disclosure. The vehicle for providing this notification is the written contract or other agreement that authorizes them to receive the data. Accordingly, if a commercial contract, cooperative agreement, interagency agreement, letter of agreement, memorandum of understanding, or other agreement provides a non-BLS employee access to BLS confidential data or Privacy Act data, it must contain appropriate provisions to safeguard the data from unauthorized disclosure. The authorization document will state the purpose for which the data will be used and that all persons with access to the data will follow the BLS confidentiality policy, including signing the BLS non-disclosure affidavit. These provisions are required whether the data are accessed on or off BLS premises. They also are required when access to the data may be incidental to the work conducted under the contract or other agreement, such as in systems development projects, survey mail-out processing, etc.

Data Collection

The integrity of the BLS data collection process requires that all survey information be sound and complete. Data must be obtained from the appropriate company official or respondent, and the data entries must accurately report the data and responses they provided. The administrative aspects of the data collection process, such as work time reported and travel voucher entries, must be factually reported. Therefore, employees must not deliberately misrepresent the source of the data, the method of data collection, the data received from respondents, or entries on administrative reporting forms.

All BLS programs must follow the appropriate procedure for requesting authorization of processes for the electronic transmission of respondent-identifying data to or from respondents.

ACCIDENT INVESTIGATION: DATA ANALYSIS 14.9

Procedures for Safeguarding Confidential Information

Program office managers are responsible for implementing procedural and physical safeguards to protect confidential information from disclosure or misuse within their offices, including:

- Preparing written procedures for the identification, labeling, handling, and disposal of confidential data. Ensuring that all employees within their organizations are familiar with and understand these procedures.
- Ensuring that new employees are informed about the different types of confidential data maintained in their work areas and the special precautions that are to be taken with their use, storage, and disposal.
- Developing data collection instruments and collection methodology in conformance with OMB guidelines on confidentiality.
- Ensuring that commercial contracts, cooperative and inter-agency agreements, letters of agreement, and memoranda of understanding, which give non-BLS employees access to confidential data, contain the proper confidentiality- and security-related clauses.

All BLS employees are responsible for following the rules of conduct in the handling of personal information contained in the records covered under the Privacy Act of 1974, which are in the custody of the BLS.

Dissemination of News and Data Releases

Public information documents require advance bureau-level clearance through the Associate Commissioner for Publications, who is responsible for seeing that each publication meets BLS publication standards and also the standards set by the Department of Labor, the Congressional Joint Committee on Printing, and OMB. BLS offices also are required to consult the Associate Commissioner for Publications before instituting an automated process to disseminate news releases or other products to the public.

No advance release of embargoed data shall be made unless directed by the Commissioner of Labor Statistics under the discretion granted under OMB Statistical Directive Number 3. BLS organizations shall strictly follow the Commissioner's specifications in making an advance release.

Data Security

The BLS has established appropriate computer security measures to safeguard the BLS' data processing environment against destruction or corruption of data or systems, unauthorized disclosure of data, and loss of service. These security measures are part of an overall management control process that includes program management, financial management, physical and personnel security, statistical data security, and information technology (IT) security. Associate, Assistant and Regional Commissioners, and Directors are assigned overall responsibility for directing the application of such controls to the Automated Information Systems and/or application systems, which they manage. The BLS Data Security Steering Committee provides overall direction to BLS security efforts.

Instructor's Notes:

REFERENCES

Additional Resources

This module is intended to present thorough resources for task training. The following reference works are suggested for further study. These are optional materials for continued education rather than for task training.

www.osha.gov

www.asse.org

Root Cause Analysis Handbook: A Guide to Effective Incident Investigation, 1999. JBF Associates Division. Rockville, MD: Government Institutes.

ACCIDENT INVESTIGATION: DATA ANALYSIS 14.11

MODULE 75214-03 — ANSWERS TO REVIEW QUESTIONS

Answer	Section
1. b	3.1.0
2. c	3.2.0
3. a	3.3.0
4. c	3.1.0
5. a	4.0.0
6. b	4.0.0
7. a	7.0.0
8. b	1.0.0
9. c	2.0.0
10. a	5.0.0

CONTREN™ LEARNING SERIES — USER UPDATES

The NCCER makes every effort to keep these textbooks up-to-date and free of technical errors. We appreciate your help in this process. If you have an idea for improving this textbook, or if you find an error, a typographical mistake, or an inaccuracy in NCCER's Contren™ textbooks, please write us, using this form or a photocopy. Be sure to include the exact module number, page number, a detailed description, and the correction, if applicable. Your input will be brought to the attention of the Technical Review Committee. Thank you for your assistance.

Instructors – If you found that additional materials were necessary in order to teach this module effectively, please let us know so that we may include them in the Equipment/Materials list in the Instructor's Guide.

Write: Curriculum Revision and Development Department
National Center for Construction Education and Research
P.O. Box 141104, Gainesville, FL 32614-1104

Fax: 352-334-0932

E-mail: curriculum@nccer.org

Craft ______________________ Module Name ______________________

Copyright Date ____________ Module Number ____________ Page Number(s) ____________

Description ____

____

____

____

(Optional) Correction ____

____

____

(Optional) Your Name and Address ____

____

____

Module 75215-03

Recordkeeping

NCCER STANDARDIZED CRAFT TRAINING PROGRAM

The National Center for Construction Education and Research (NCCER) provides a standardized national program of accredited craft training. Key features of the program include instructor certification, competency-based training, and performance testing. The program provides trainees, instructors, and companies with a standard form of recognition through a National Craft Training Registry. The program is described in full in the *Guidelines for Accreditation*, published by the NCCER. For more information on standardized craft training, contact the NCCER by writing us at P.O. Box 141104, Gainesville, FL 32614-1104; calling 352-334-0911; or e-mailing info@nccer.org. More information may be found at our Web site, www.nccer.org.

HOW TO USE THIS ANNOTATED INSTRUCTOR'S GUIDE

Each page presents two sections of information. The larger section displays each page exactly as it appears in the Participant Module. The narrow column ties suggested participant and instructor actions to each page and provides icons (detailed below) to call your attention to material, safety, audiovisual, or testing requirements. The bottom of each page includes space for your notes.

The **Audiovisual** icon indicates an appropriate time to show a transparency or other audiovisual aid.

The **Classroom** icon prompts you to define a term, stress a point, ask participants to explain a concept, or give examples.

The **Demonstration** icon directs you to show participants how to perform tasks.

The **Examination** icon tells you to administer the written module examination.

The **Homework** icon is placed where you may wish to assign reading for the next class, to assign a project, or to advise participants to prepare for an examination.

The **Laboratory** icon is used when participants are to practice performing tasks.

The **Materials** icon is a reminder for you to gather materials needed for classes, labs, and testing.

The **Performance Testing** icon tells you to administer a performance test or a portion thereof.

The **Safety** icon is used to emphasize safety issues. It is often keyed to *Caution* and *Warning* statements in the Participant Module.

The **Teaching Tip** icon indicates additional guidance is available, such as how to conduct an exercise, get the most educational value from a field trip, or encourage class participation. Teaching Tips may expand on a feature (*Think About It*, *Did You Know?*) or provide Quick Quizzes or similar exercises. You will be referred to the Teaching Tips section at the back of the module if there is additional material.

The **Combination** icon indicates that the laboratory listed corresponds with a performance task. If desired, you can note the proficiency of the participants during the laboratory and use it to satisfy performance testing requirements.

PREPARATION

Before teaching this module, you should review the Objectives, Performance Tasks, Materials and Equipment List, and the Module Outline. Be sure to allow ample time to prepare your own training or lesson plan and gather all required materials and equipment.

MODULE OVERVIEW

In this module, participants will learn the OSHA recordkeeping requirements and how to manage on-site safety and health records.

PREREQUISITES

Please refer to the Course Map in the Participant Module. Prior to training with this module, it is recommended that the participant shall have successfully completed the following:

Field Safety; Safety Technology, Modules 75201-03 through 75214-03

OBJECTIVES

Upon completion of this module, the participant will be able to:

1. Identify and follow OSHA and company requirements for recordkeeping.
2. Properly document work-related illnesses and injuries using OSHA Forms 300, 300A, and 301.
3. Explain how to manage safety and health records for a job site.

PERFORMANCE TASKS

There are no performance tasks for this module.

MATERIALS AND EQUIPMENT LIST

Overhead projector and screen
Transparencies
Whiteboard/chalkboard
Markers/chalk
Blank acetate sheets
Transparency pens
Pencils and scratch paper
Appropriate personal protective equipment
Copy of *29 CFR 1926, OSHA Construction Industry Regulations*
OSHA Forms 300, 300A, and 301
Sample accident investigation report
Module Examinations*

* Located in the Test Booklet.

SAFETY CONSIDERATIONS

Ensure that participants are equipped with appropriate personal protective equipment.

TEACHING TIME FOR THIS MODULE

An outline for use in developing your lesson plan is presented below. Approximately 2½ hours are suggested to cover *Recordkeeping*. This includes 10 minutes at the beginning of the session for administrative tasks and one 10-minute break. You may need to adjust the time required for hands-on activities and testing based on your class size and resources.

Topic	Planned Time
A. Introduction	____________
B. OSHA Requirements	____________
1. Work-Related Illnesses or Injuries	____________
2. Recording Workplace Illness and Injury	____________
3. Cases Involving Restricted Work	____________
4. Determining Lost Work Days and Restricted Work Days	____________
5. Identifying Employees on OSHA Form 300	____________
C. Classifying Illnesses and Injuries	____________
1. Injuries	____________
2. Classifying Illnesses	____________
D. OSHA Forms	____________
1. OSHA Form 300	____________
2. OSHA Form 300A	____________
3. OSHA Form 301	____________
E. Other Needed Safety, Health, and Training Records	____________
F. Review	____________
G. Module Examination	____________

1. Participants must score 70% or higher to receive recognition from the NCCER.
2. Record the testing results on Craft Training Report Form 200 and submit the results to the Training Program Sponsor.

Module 75215-03

Recordkeeping

Instructor's Notes:

COURSE MAP

This course map shows all of the modules in Safety Technology. The suggested training order begins at the bottom and proceeds up. The local Training Program Sponsor may adjust the training order.

Assign reading of Module 75215-03.

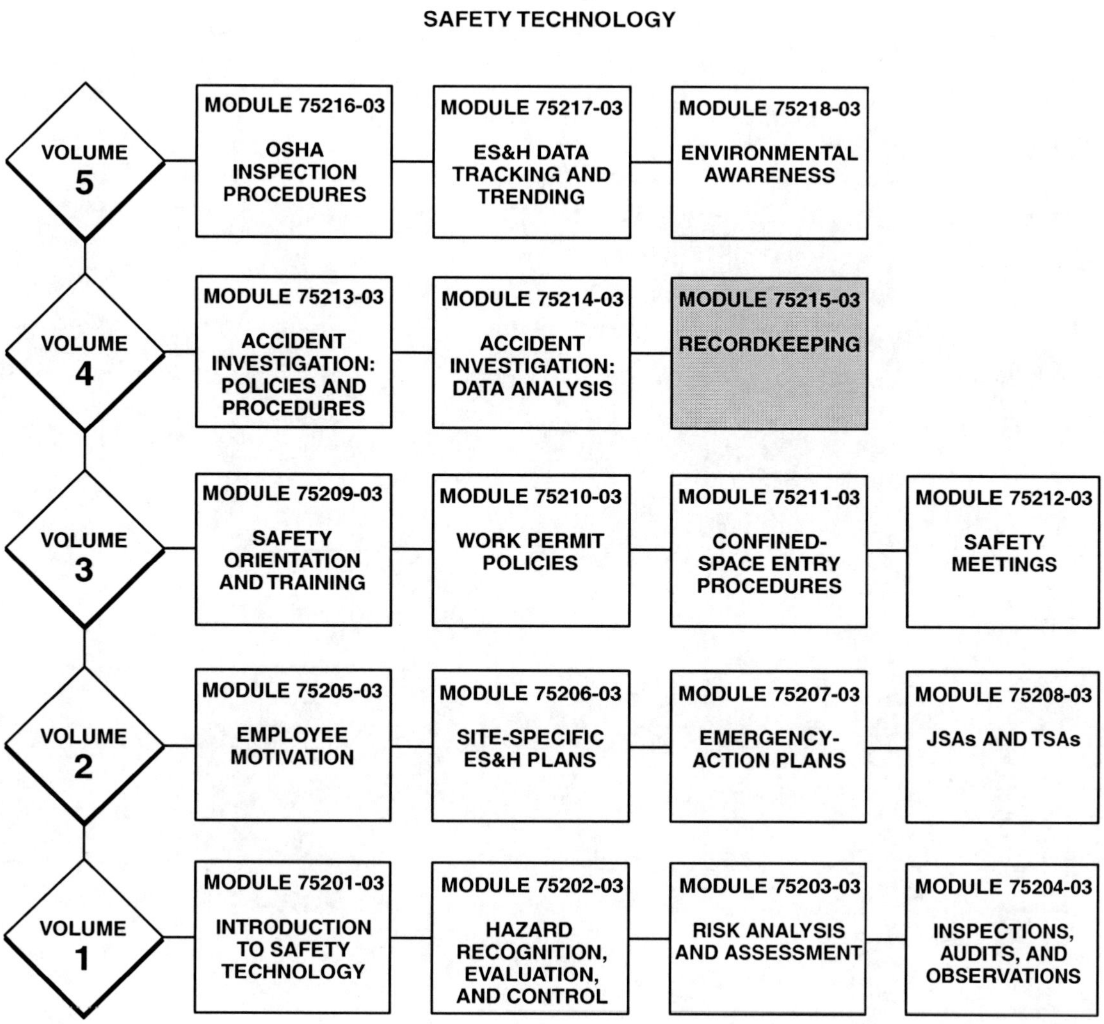

215CMAP.EPS

RECORDKEEPING 15.iii

Instructor's Notes:

MODULE 75215-03 CONTENTS

Figures

RECORDKEEPING 15.v

Instructor's Notes:

MODULE 75215-03

Recordkeeping

Ensure that you have everything you need to teach the course. Check the Materials and Equipment list in the front of this module.

Objectives

When you have completed this module, you will be able to do the following:

1. Identify and follow OSHA and company requirements for recordkeeping.
2. Properly document work-related illnesses and injuries using OSHA Forms 300, 300A, and 301.
3. Explain how to manage safety and health records for a job site.

Prerequisites

Before you begin this module, it is recommended that you successfully complete the following: Field Safety; Safety Technology, Modules 75201-03 through 75214-03.

Required Materials

1. Pencil and paper
2. Appropriate personal protective equipment
3. Copy of *29 CFR 1926, OSHA Construction Industry Regulations*

Explain that terms shown in bold are defined in the Glossary at the back of this module.

Review the goals of the course and explain what will be expected of the participant.

Show Transparency 1, Objectives.

1.0.0 ◆ INTRODUCTION

Recordkeeping is a critical part of a company's safety and health program. The data that is collected is used to help keep track of work-related injuries and illnesses. Once this data is gathered and analyzed, it can be used to help identify and correct problem areas to prevent future illnesses and injuries. Recordkeeping not only provides information about illness and injury to management, it also informs workers about incidents that happen in their work area. When workers are aware of injuries, illnesses, and hazards in the workplace, they are more likely to follow safe work practices and report workplace hazards.

Recordkeeping is required by OSHA. They use specific illness and injury information that is reported to them as part of the agency's site-specific inspection targeting program. The Bureau of Labor Statistics (BLS) also uses injury and illness records as the source data for the Annual Survey of Occupational Injuries and Illnesses. This report shows safety and health trends nationwide and industry wide.

As a safety technician, you are responsible for making sure recordkeeping on your site is done correctly. You must know all of the OSHA requirements for recording and classifying workplace illnesses and injuries. In addition, you may be responsible for coordinating and/or maintaining all job-site safety, health, and training records.

Explain why recordkeeping is a critical part of a company's safety and heath program.

Describe the safety technician's role in recordkeeping.

Explain how the 1970 enactment of OSHA affected the way work-site record-keeping is done.

2.0.0 ◆ OSHA REQUIREMENTS

All employers covered by the Occupational Safety and Health Act of 1970 (OSHA) are required to meet the recordkeeping regulations from *OSHA Part 1904*. If your company had more than ten employees at any time during the last **calendar year**, you must keep OSHA injury and illness records. To determine if your company is exempt because of size, figure out your company's highest number of employees working at any given time during the last calendar year. If the number is more than ten, you are required to keep OSHA illness and injury records.

If your company had ten or fewer employees at all times during the last calendar year, you do not need to keep OSHA injury and illness records unless OSHA or the BLS informs you in writing that you must do so. Even if your employer is not required to keep injury and illness records, they are still required to report to OSHA within in eight hours any workplace incident that results in a death or in the hospitalization of three or more employees.

RECORDKEEPING 15.1

Explain OSHA's recordkeeping requirements.

Define the term recordable.

Discuss the four factors that should be considered when determining if an illness or injury is recordable.

Show Transparency 2, Recordable Illness and Injury Factors.

Explain OSHA's requirements for determining if illnesses or injuries are work related.

Discuss the importance of conducting a thorough investigation before classifying an injury or illness as work related.

Show Transparency 3 (Figure 1).

Using Transparency 3, show the participants how to use a decision tree to determine if an illness or injury is recordable. Provide the participants with a scenario and have them use the decision tree to determine if the selected illness or injury is recordable.

> **NOTE**
> Check *OSHA Part 1904* if there is any question about whether your company is exempt from OSHA's recordkeeping rule.

OSHA also has specific requirements about the type of information that is recorded about job-related illnesses and injury, how it is collected, and how it is classified. OSHA Forms 300, 300A, and 301 are used to record this information. These forms are discussed in more detail later in this module.

Recordkeeping is a process that involves identifying and classifying specific information about an illness or injury to determine whether or not it is **recordable**. Consider these four factors when gathering and analyzing data during the recordkeeping process.

- Is the illness or injury work related?
- Is the illness or injury recordable?
- What types of medical treatment needs to be recorded?
- How many days were missed or involved restricted work after the incident?

2.1.0 Work-Related Illnesses or Injuries

According to OSHA, in order to be considered work related, an injury or illness must be related to conditions or events in the work environment that caused or contributed to the illness or injury. For example, when a worker strains his or her back as a result of picking up equipment, that injury is considered work related. When an illness or injury is considered work related and meets certain criteria, it must be properly recorded. Before recording an illness or injury, make sure a

> **NOTE**
> Illnesses and injuries that significantly aggravate a pre-existing condition are also considered work related if medical treatment is required or the injury results in lost workdays or restricted duty.

workers' compensation claims and will protect the company from legal liability.

2.2.0 Recording Workplace Illness and Injury

It's important to know when an illness or injury should be recorded. *Figure 1* shows a decision tree that can help you determine whether or not an ill-

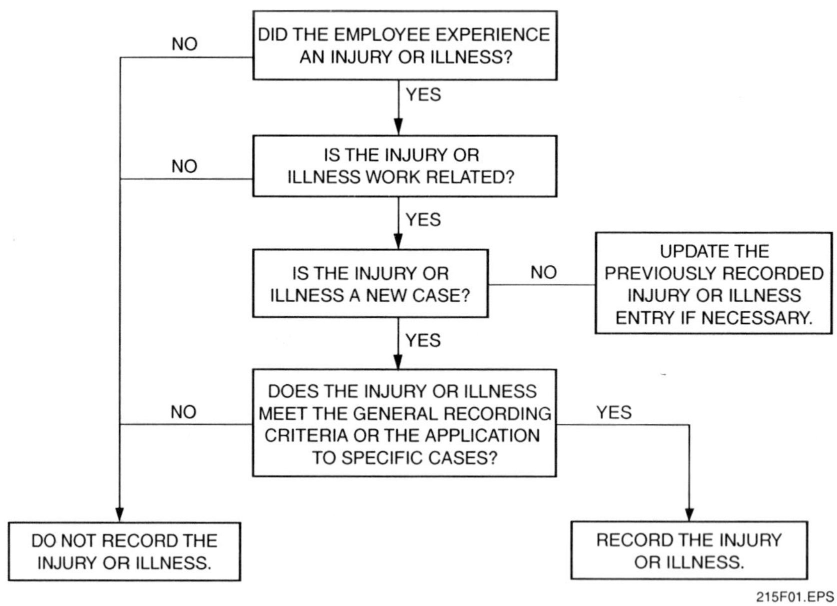

Figure 1 ◆ Recordable illness and injuries decision tree.

Instructor's Notes:

ness or injury is recordable. In addition to the decision tree, OSHA has some specific criteria for judging the types of illnesses and injuries that must be recorded. The following lists OSHA's criteria for illnesses and injuries that must be recorded.

- Work-related injuries and illnesses that result in death, loss of consciousness, days away from work, **restricted work activity**, job transfer, or medical treatment beyond first aid
- Work-related injuries or illness that have been diagnosed by a physician or other licensed health care professional
- Work-related cases involving cancer, **chronic irreversible disease**, a fractured or cracked bone, or a punctured eardrum

OSHA has established additional criteria for cases requiring recordkeeping. These cases must be work related in order to be reported. Record the following conditions when you have verified that they are work related:

- Any needle-stick injury or cut from a sharp object that is contaminated with another person's blood or other potentially **infectious material**
- Any case requiring an employee to be removed from the site for medical treatment
- Any case in which a positive skin test or diagnosis has been made by physician or other licensed health care professional after exposure to a known case of active tuberculosis

Additional criteria such as the types of medical treatment and first aid should also be considered when recording illness and injuries.

2.2.1 Medical Treatment

OSHA defines medical treatment as managing and caring for a patient for the purpose of fighting disease or illness. Certain types of medical treatment are not applicable to OSHA's recordkeeping requirement. In your role as safety technician, you must know what treatments or procedures are considered medical treatment in terms of the OSHA requirement. The following is a list of the types of medical treatments that are not considered medical treatments and are not recordable.

- Visits to a doctor or health care professional solely for observation or counseling
- Diagnostic procedures, including administering prescription medications that are used solely for medical testing
- Any procedure that can be labeled first aid

2.2.2 First Aid

First aid is emergency care or treatment given to an ill or injured person before regular medical aid can be given. Since first aid is not considered recordable, make sure you know the type of treatment that is given to an ill or injured worker.

The following types of treatment are considered first aid and do not need to be recorded.

- Using non-prescription medications at non-prescription strength
- Cleaning, flushing, or soaking wounds on skin surface
- Using wound coverings, such as bandages, gauze pads, or butterfly bandages
- Using hot or cold therapy
- Using any totally non-rigid means of support, such as elastic bandages, wraps, and non-rigid back belts
- Using temporary immobilization devices while transporting an accident victim such as splints, slings, neck collars, or back boards
- Drilling a fingernail or toenail to relieve pressure
- Draining fluids from blisters
- Using eye patches
- Using simple irrigation or a cotton swab to remove foreign bodies not embedded in the eye
- Using irrigation, tweezers, cotton swab, or other simple means to remove splinters or foreign material from areas other than the eye
- Using finger guards
- Using massages
- Administering fluids to relieve heat stress

2.3.0 Cases Involving Restricted Work

Workers can be placed on a restricted work schedule if they are ill or injured from a work-related incident and cannot perform the regular functions of their jobs. Keep in mind, however, that not all illness and injury cases result in restricted work. According to OSHA's recordkeeping requirement, restricted work activity should be assigned to only those workers who have experienced a work-related injury or illness that prevents them from doing the routine functions of their jobs or from working a full day. Only a health care professional or the worker's employer can recommend a restricted work schedule. Make sure workers provide proper documentation, usually in the form of a doctor's note, to support the request for restricted work.

Explain OSHA's criteria for illnesses and injuries that must be recorded.

Describe the types of conditions that OSHA requires employers to record. Refer the participants to the bulleted list in the Participant Module.

Explain OSHA's definition of medical treatment. Discuss the role of the safety technician in regard to the definition.

Identify common first aid treatments. Emphasize that these treatments are not considered recordable.

Show Transparency 4, First-Aid Treatments.

Explain OSHA requirements for cases involving restricted work.

RECORDKEEPING 15.3

Explain OSHA's requirements for determining lost work days and restricted work days.

Show the participants how to calculate lost work days and restricted work days. Have the participants calculate lost work days and restricted work days from a given scenario.

Explain the circumstances under which it is acceptable to omit an ill or injured worker's name on OSHA Form 300.

Explain that a separate, confidential list of case numbers and names must accompany OSHA Form 300 when names are omitted. Emphasize the use of good judgment with privacy matters.

Explain OSHA's definition of an injury. Provide examples.

Explain OSHA's definition of an illness. Provide examples.

Discuss the importance of classifying illnesses and injuries properly.

2.4.0 Determining Lost Work Days and Restricted Work Days

Determining lost workdays and restricted workdays is an important part of analyzing workplace illnesses and injuries. It is also an OSHA recordkeeping requirement.

The number of lost and restricted days can be calculated by counting calendar days from the day after the incident occurs. The day of the accident should not be included in this number. The totals should be separate if a single injury or illness involved both days away from work and days of restricted work activity. According to OSHA recordkeeping rule, the number of lost or restricted workdays should stop being counted once the total of either, or the combination of both, reaches 180 days.

2.5.0 Identifying Employees on OSHA Form 300

There are circumstances in which the employee's name should not be entered on OSHA Form 300 (Log of Work-Related Injuries and Illness). Because of privacy issues, it is acceptable to leave the worker's name off the form when the following types of injuries or illnesses have occurred:

- An injury or illness to an intimate body part or to the reproductive system
- An injury or illness resulting from a sexual assault, a mental illness, a case of HIV infection, hepatitis, or tuberculosis
- Needle-stick injury or cut from a sharp object that is contaminated with blood or other infectious materials
- Other illnesses, if the employee independently and voluntarily requests that his or her name not be entered on the log.

Enter the words "privacy case" in the space normally used for the employee's name when you cannot enter the worker's name on the OSHA 300 Log. When "privacy case" is entered on the form, a separate, confidential list of the case numbers and workers' names must be kept so that you can update the cases and provide information to OSHA if asked to do so.

Details of an intimate or private nature do not need to be included on these forms but enough information must be entered to identify the cause and general severity of the injury or illness. Use good judgment when describing the injury or illness on both the OSHA 300 and 301 forms when it is believed that information describing the case may identify the worker.

3.0.0 ◆ CLASSIFYING ILLNESSES AND INJURIES

OSHA has specific criteria for classifying illnesses and injuries. Each is based on what is considered recordable according to *OSHA Part 1904*. It's important to understand which types of injuries and illnesses are recordable so that they are recorded correctly.

3.1.0 Injuries

According to OSHA, an injury is any wound or damage to the body resulting from an event in the work environment. The following injuries are classified as recordable injuries:

- Cuts
- Punctures
- **Lacerations**
- **Abrasions**
- Fractures
- Bruises
- **Contusions**
- Chipped teeth
- Amputations
- Insect bites
- Electrocution
- Thermal, chemical, electrical, or radiation burns
- Sprains and strains resulting from a slip, trip, or fall

Any time any of these injuries happen on site, they must be recorded if they meet the appropriate criteria.

3.2.0 Classifying Illnesses

Illnesses can be defined as an unhealthy condition of body or mind. OSHA classifies recordable illness into these four types:

- Skin diseases or disorders
- Respiratory conditions
- Poisoning
- Occupational illnesses

Knowing how to properly classify illnesses can speed up the recording process, allowing more time for other safety concerns, such as training or auditing. Make sure all illnesses can be verified as work related before they are recorded.

3.2.1 Skin Diseases or Disorders

Skin diseases or disorders are illnesses involving a worker's skin that are caused by exposure to chemicals, plants, or other substances. The fol-

Instructor's Notes:

lowing are the types of skin diseases and disorders workers commonly get.

- **Contact dermatitis**
- **Eczema**
- Rashes caused by:
  - Irritants
  - Sensitizers
  - Poisonous plants
- Acne from oil
- Blisters
- **Ulcers**

3.2.2 Respiratory Conditions

Respiratory conditions are illnesses associated with breathing hazardous biological agents, chemicals, dust, gases, vapors, or fumes. Examples of the types of respiratory illnesses that workers commonly report include:

- *Silicosis* – A lung disease characterized by massive fibrosis of the lungs
- *Asbestosis* – A lung disease caused by inhaling asbestos particles
- *Pneumonitis* – An inflammation of the lungs
- *Pharyngitis* – An inflammation of the throat
- *Rhinitis* – An inflammation of the mucous membrane of the nose
- *Beryllium disease* – An inflammation of the lungs
- *Tuberculosis* – An infectious lung disease
- *Occupational asthma* – A breathing condition brought on by substances in the workplace
- *Chronic obstructive pulmonary disease (COPD)* – A chronic lung disease such as chronic bronchitis, emphysema, chronic asthma, and bronchiolitis
- *Hypersensitivity pneumonitis* – A medical syndrome caused by sensitization to and repeated inhalation of an inhaled organic **antigen** from dust
- *Chronic obstructive bronchitis* – A chronic cough lasting at least 3 months

3.2.3 Poisoning

Poisoning is caused when toxic or poisonous substances are ingested or absorbed into the body through skin or by breathing. The following are examples of the types of poisons that can cause recordable illnesses.

- Metals such as lead, mercury, cadmium, or arsenic
- Gases such as carbon monoxide or hydrogen sulfide
- Organic solvents such as benzene, benzol, or carbon tetrachloride
- Insecticide sprays such as parathion or lead arsenate
- Chemical preservatives such as formaldehyde

3.2.4 Occupational Illnesses

Other illnesses are directly related to workplace activities and conditions. For example, some work sites can be very hot or very cold. Others may expose workers to radiation. The illnesses that result from these conditions are called occupational illnesses. The following are the types of illnesses that are commonly classified as occupation illnesses.

- Heatstroke
- Sunstroke
- Heat exhaustion
- Heat stress
- **Hypothermia**
- Frostbite
- Radiation sickness
- Tumors

4.0.0 ◆ OSHA FORMS

OSHA requires the use of three forms to record injuries and accidents at the work site. They are forms 300, 300A, and 301.

- Form 300, the *Log of Work-Related Injuries and Illnesses*, is used to record specific details about work-related illnesses and injury.
- Form 300A, the *Summary of Work-Related Injuries and Illnesses*, shows the totals for the year in each category.
- Form 301, the *Injury and Illness Incident Report*, is used to record specific information about a work-related injury or illness.

Employees, former employees, and their representatives have the right to review the OSHA Forms 300 and 300A in their entirety. They also have limited access to the OSHA Form 301 or another accident investigation reports. See *29 CFR, Part 1904.35*, OSHA's recordkeeping rule, for further details on the access provisions for Form 301 and other accident investigation reports.

4.1.0 OSHA Form 300

Form 300, commonly called the log, is used to classify and record work-related injuries and illnesses and to note the extent and severity of each case. When an accident occurs, the log is used to

Explain the types of skin diseases or disorders that workers commonly contract.

Discuss the respiratory conditions that workers commonly contract.

Show Transparency 5, Respiratory Conditions.

Describe how poisons or toxins affect the human body. Identify common poisons.

Discuss the other occupational illnesses that affect workers. Discuss ways to prevent these conditions.

Identify the OSHA's required recordkeeping forms: 300, 300A, and 301A.

RECORDKEEPING 15.5

Explain the purpose, uses, and requirements of OSHA Form 300.

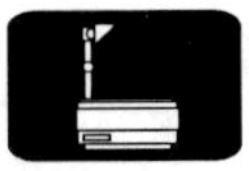

Show Transparency 6 (Figure 2).

Show the participants how to complete OSHA Form 300. Provide the participants with a scenario and have the participants make an entry on OSHA Form 300.

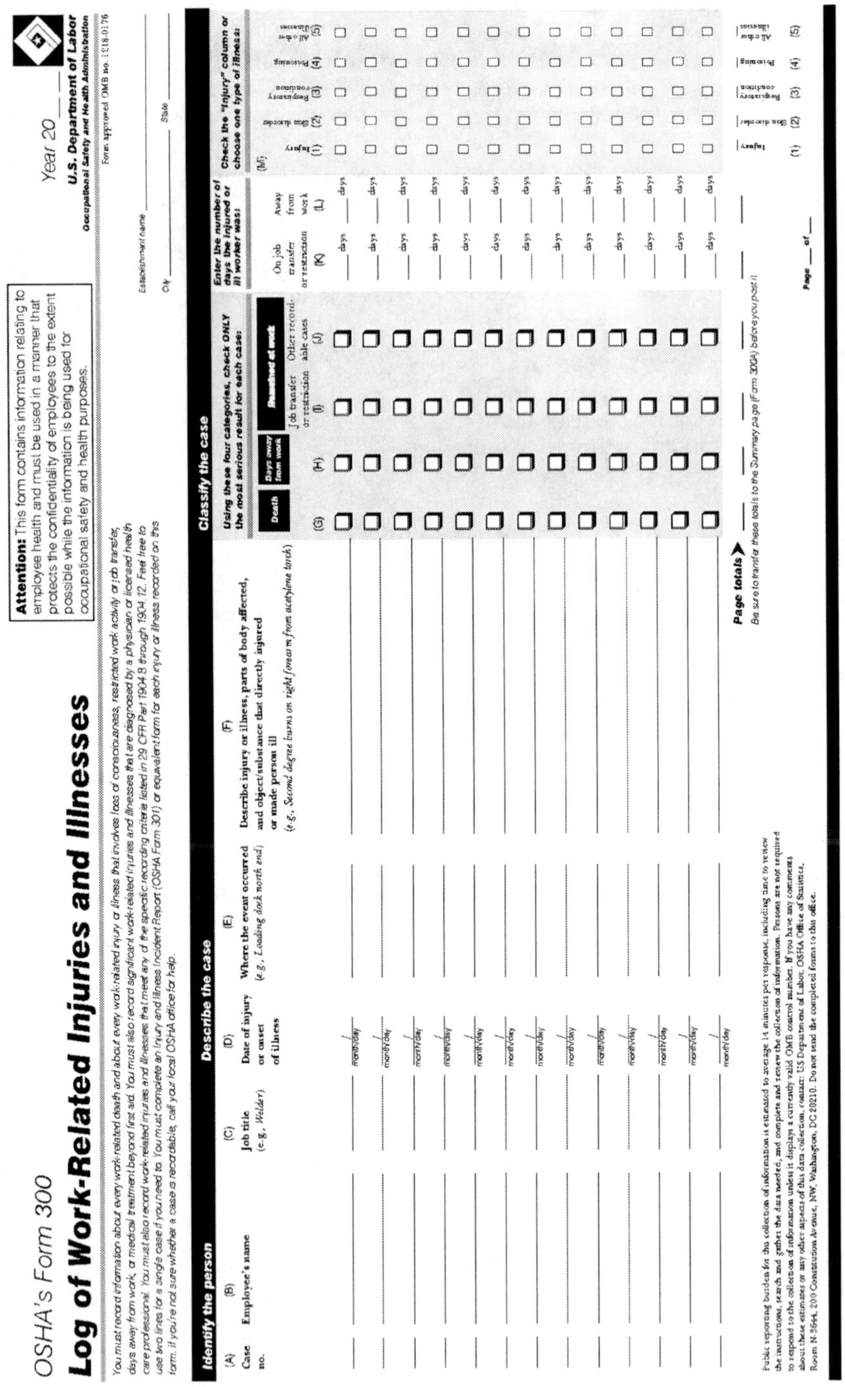
OSHA's Form 300

Log of Work-Related Injuries and Illnesses

Attention: This form contains information relating to employee health and must be used in a manner that protects the confidentiality of employees to the extent possible while the information is being used for occupational safety and health purposes.

Year 20 __ __

U.S. Department of Labor
Occupational Safety and Health Administration

Form approved OMB no. 1218-0176

You must record information about every work-related death and about every work-related injury or illness that involves loss of consciousness, restricted work activity or job transfer, days away from work, or medical treatment beyond first aid. You must also record significant work-related injuries and illnesses that are diagnosed by a physician or licensed health care professional. You must also record work-related injuries and illnesses that meet any of the specific recording criteria listed in 29 CFR Part 1904.8 through 1904.12. Feel free to use two lines for a single case if you need to. You must complete an Injury and Illness Incident Report (OSHA Form 301) or equivalent form for each injury or illness recorded on this form. If you're not sure whether a case is recordable, call your local OSHA office for help.

Establishment name ____________

City ____________ State ____________

Identify the person			Describe the case			Classify the case										
(A) Case no.	(B) Employee's name	(C) Job title (e.g., *Welder*)	(D) Date of injury or onset of illness	(E) Where the event occurred (*e.g., Loading dock north end*)	(F) Describe injury or illness, parts of body affected, and object/substance that directly injured or made person ill (*e.g., Second degree burns on right forearm from acetylene torch*)	Using these four categories, check ONLY the most serious result for each case: Death (G)	Days away from work (H)	Remained at work: Job transfer or restriction (I)	Remained at work: Other record-able cases (J)	Enter the number of days the injured or ill worker was: On job transfer or restriction (K)	Away from work (L)	Check the "Injury" column or choose one type of illness: (M) Injury (1)	Skin disorder (2)	Respiratory condition (3)	Poisoning (4)	All other illnesses (5)
			__/__ month/day			☐	☐	☐	☐	__ days	__ days	☐	☐	☐	☐	☐
			__/__ month/day			☐	☐	☐	☐	__ days	__ days	☐	☐	☐	☐	☐
			__/__ month/day			☐	☐	☐	☐	__ days	__ days	☐	☐	☐	☐	☐
			__/__ month/day			☐	☐	☐	☐	__ days	__ days	☐	☐	☐	☐	☐
			__/__ month/day			☐	☐	☐	☐	__ days	__ days	☐	☐	☐	☐	☐
			__/__ month/day			☐	☐	☐	☐	__ days	__ days	☐	☐	☐	☐	☐
			__/__ month/day			☐	☐	☐	☐	__ days	__ days	☐	☐	☐	☐	☐
			__/__ month/day			☐	☐	☐	☐	__ days	__ days	☐	☐	☐	☐	☐
			__/__ month/day			☐	☐	☐	☐	__ days	__ days	☐	☐	☐	☐	☐
			__/__ month/day			☐	☐	☐	☐	__ days	__ days	☐	☐	☐	☐	☐
			__/__ month/day			☐	☐	☐	☐	__ days	__ days	☐	☐	☐	☐	☐
			__/__ month/day			☐	☐	☐	☐	__ days	__ days	☐	☐	☐	☐	☐
			__/__ month/day			☐	☐	☐	☐	__ days	__ days	☐	☐	☐	☐	☐
					Page totals ➤	__	__	__	__	__	__	Injury (1)	Skin disorder (2)	Respiratory condition (3)	Poisoning (4)	All other illnesses (5)

Be sure to transfer these totals to the Summary page (Form 300A) before you post it.

Public reporting burden for this collection of information is estimated to average 14 minutes per response, including time to review the instructions, search and gather the data needed, and complete and review the collection of information. Persons are not required to respond to the collection of information unless it displays a currently valid OMB control number. If you have any comments about these estimates or any other aspects of this data collection, contact: US Department of Labor, OSHA Office of Statistics, Room N-3644, 200 Constitution Avenue, NW, Washington, DC 20210. Do not send the completed forms to this office.

Page __ of __

215F02.TIF

Figure 2 ◆ OSHA Form 300.

Instructor's Notes:

> **NOTE**
> If the outcome or extent of an injury or illness changes after you have recorded the case, simply draw a line through the original entry. Then write the new entry where it belongs. Remember, you need to record the most serious outcome for each case.

record specific details about what happened and how it happened. Employers, especially larger companies with more than one site, must provide a separate form for each site that is expected to operate for one year or longer. *Figure 2* shows an example of a log.

4.2.0 OSHA Form 300A

Form 300A, also called the summary, shows the work-related injury and illness totals for the year in each category. All establishments covered by *OSHA Part 1904* must complete this summary page, even if no work-related injuries or illnesses occurred during the year. *Figure 3* shows Form 300A. *Figure 4* shows the optional worksheet used to calculate the needed data for this form.

At the end of the year, the number of incidents in each category is counted, totaled, and transferred from the log (Form 300) to the summary (Form 300A). The summary is used to count individual entries for each category. The totals are then entered. If there were no cases on the log, a zero should be written in the total field.

After the summary has been completed, it is posted in a visible location so that workers can review and become aware of injuries and illnesses occurring in their workplace. Make sure all entries have been verified and are complete and accurate before posting the summary. The summary for the previous year must be posted from February first through the month of April. The accuracy of the summary must be certified by the senior management executive, who must then sign the form.

> **NOTE**
> Never post the log. Only post the summary at the end of the year.

4.3.0 OSHA Form 301

Form 301, the injury and illness incident report, is one of the first forms you must fill out when a recordable work-related injury or illness has occurred. It's important to note that the injury and illness information needed to complete Forms 300 and 300A can also be collected from accident investigation reports like the one shown in *Figure 5*. The form in *Figure 5* meets the recordkeeping requirements specified by OSHA Form 301 (*Figure 6*). You'll notice, however, that the accident investigation report includes more information than the OSHA form. This is because OSHA does not provide a mandatory format for accident investigation reports, but does require that specific information be provided.

Form 300, or another type of accident investigation report, must be completed within 7 calendar days after you receive information that a recordable work-related injury or illness has occurred. Workers' compensation, insurance, or other reports may be acceptable substitutes. To be considered an equivalent form, any substitute must contain all the information asked for on Form 300.

> **NOTE**
> According to *Public Law 91-596* and *29 CFR 1904*, OSHA's recordkeeping rule, you must keep this form on file for five years following the year to which it pertains.

Together with the log of work-related injuries and illnesses (Form 300) and the accompanying summary (Form 300A), these forms help the employer and OSHA develop an understanding of the extent and severity of work-related incidents.

5.0.0 ◆ OTHER NEEDED SAFETY, HEALTH, AND TRAINING RECORDS

In addition to injury and illness statistics, other records must be maintained. Some are required by OSHA or MSHA. Others are kept as a matter of good business practice. The safety technician is often required to maintain or coordinate the maintenance of these records. Records typically required by OSHA and/or MSHA include:

- Employee job-related medical records
- Employee job-related exposure records
- Job- or task-specific training records
- Crane and hoisting equipment inspection and maintenance records
- Respiratory protection inspection records
- Fire extinguisher inspection records

Explain the purpose, uses, and requirements of OSHA Form 300A.

Show Transparency 7 (Figure 3).

Explain that there is an optional worksheet that can be used to help complete OSHA Form 300A.

Show Transparency 8 (Figure 4).

Explain the purpose, uses, and requirements of OSHA Form 301. Explain that OSHA requires that specific information be provided.

Show Transparencies 9 and 10 (Figures 5 and 6).

Explain the other types of records that safety technicians are required to coordinate and maintain. Refer the participants to the Participant Module.

Show the participants how to complete OSHA Form 300A. Provide the participants with an example of year-end work-related illness and injury totals, and have the participants complete the OSHA Form 300A.

OSHA's Form 300A

Summary of Work-Related Injuries and Illnesses

Year 20__ __

U.S. Department of Labor
Occupational Safety and Health Administration

Form approved OMB no. 1218-0176

All establishments covered by Part 1904 must complete this Summary page, even if no work-related injuries or illnesses occurred during the year. Remember to review the Log to verify that the entries are complete and accurate before completing this summary.

Using the Log, count the individual entries you made for each category. Then write the totals below, making sure you've added the entries from every page of the Log. If you had no cases, write "0."

Employees, former employees, and their representatives have the right to review the OSHA Form 300 in its entirety. They also have limited access to the OSHA Form 301 or its equivalent. See 29 CFR Part 1904.35, in OSHA's recordkeeping rule, for further details on the access provisions for these forms.

Number of Cases

Total number of deaths	Total number of cases with days away from work	Total number of cases with job transfer or restriction	Total number of other recordable cases
______ (G)	______ (H)	______ (I)	______ (J)

Number of Days

Total number of days of job transfer or restriction	Total number of days away from work
______ (K)	______ (L)

Injury and Illness Types

Total number of . . .
(M)

(1) Injuries ______

(2) Skin disorders ______

(3) Respiratory conditions ______

(4) Poisonings ______

(5) All other illnesses ______

Post this Summary page from February 1 to April 30 of the year following the year covered by the form.

Public reporting burden for this collection of information is estimated to average 50 minutes per response, including time to review the instructions, search and gather the data needed, and complete and review the collection of information. Persons are not required to respond to the collection of information unless it displays a currently valid OMB control number. If you have any comments about these estimates or any other aspects of this data collection, contact: US Department of Labor, OSHA Office of Statistics, Room N-3644, 200 Constitution Avenue, NW, Washington, DC 20210. Do not send the completed forms to this office.

Establishment information

Your establishment name ______________________

Street ______________________

City ______________ State ____ ZIP ______

Industry description (*e.g., Manufacture of motor truck trailers*) ______________________

Standard Industrial Classification (SIC), if known (*e.g., SIC 3715*) ___ ___ ___ ___

Employment information (*If you don't have these figures, see the Worksheet on the back of this page to estimate.*)

Annual average number of employees ______

Total hours worked by all employees last year ______

Sign here

Knowingly falsifying this document may result in a fine.

I certify that I have examined this document and that to the best of my knowledge the entries are true, accurate, and complete.

______________________ Company executive ______________ Title

( ) - ______________ Phone ___/___/___ Date

215F03.TIF

Figure 3 ◆ OSHA Form 300A.

15.8 SAFETY TECHNOLOGY — PARTICIPANT MODULE 75215-03

Instructor's Notes:

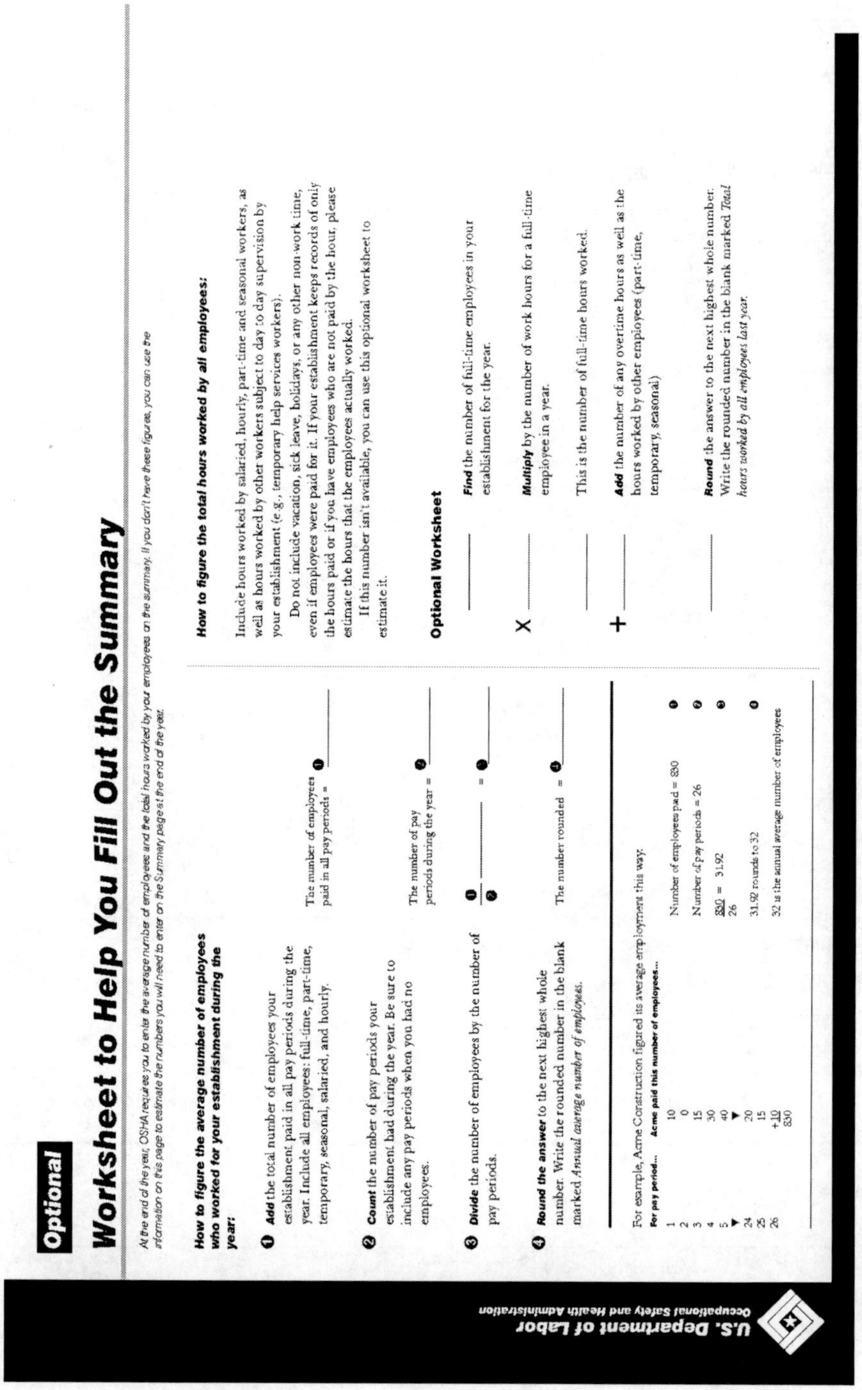

Optional

Worksheet to Help You Fill Out the Summary

At the end of the year, OSHA requires you to enter the average number of employees and the total hours worked by your employees on the summary. If you don't have these figures, you can use the information on this page to estimate the numbers you will need to enter on the Summary page at the end of the year.

How to figure the average number of employees who worked for your establishment during the year:

❶ **Add** the total number of employees your establishment paid in all pay periods during the year. Include all employees: full-time, part-time, temporary, seasonal, salaried, and hourly. — The number of employees paid in all pay periods = ❶ ______

❷ **Count** the number of pay periods your establishment had during the year. Be sure to include any pay periods when you had no employees. — The number of pay periods during the year = ❷ ______

❸ **Divide** the number of employees by the number of pay periods. — ❶ / ❷ ______ = ❸ ______

❹ **Round the answer** to the next highest whole number. Write the rounded number in the blank marked *Annual average number of employees*. — The number rounded = ❹ ______

For example, Acme Construction figured its average employment this way:

For pay period...	Acme paid this number of employees...
1	10
2	0
3	15
4	30
5	40
▼	▼
24	20
25	15
26	+10
	830

Number of employees paid = 830 ❶

Number of pay periods = 26 ❷

$\frac{830}{26}$ = 31.92 ❸

31.92 rounds to 32 ❹

32 is the annual average number of employees

How to figure the total hours worked by all employees:

Include hours worked by salaried, hourly, part-time and seasonal workers, as well as hours worked by other workers subject to day to day supervision by your establishment (e.g., temporary help services workers).

Do not include vacation, sick leave, holidays, or any other non-work time, even if employees were paid for it. If your establishment keeps records of only the hours paid or if you have employees who are not paid by the hour, please estimate the hours that the employees actually worked.

If this number isn't available, you can use this optional worksheet to estimate it.

Optional Worksheet

______ **Find** the number of full-time employees in your establishment for the year.

X ______ **Multiply** by the number of work hours for a full-time employee in a year.

______ This is the number of full-time hours worked.

\+ ______ **Add** the number of any overtime hours as well as the hours worked by other employees (part-time, temporary, seasonal)

______ **Round** the answer to the next highest whole number. Write the rounded number in the blank marked *Total hours worked by all employees last year*.

U.S. Department of Labor
Occupational Safety and Health Administration

215F04.TIF

Figure 4 ◆ 300A worksheet.

Show the participants how to complete the optional OSHA Form 300A Worksheet. Provide the participants with a scenario, including any necessary documentation, and have the participants complete the optional OSHA Form 300A Worksheet.

RECORDKEEPING 15.9

Download and show the archived webcast of OSHA's overview of the revised Record-keeping rule at: www.vodium.com/osha/live1.html.

Accident/Incident Investigation Form

▯ Rx/Treatment ▯ First Aid
▯ Property Damage Initial Report ▯ Follow-up

INJURED/INVOLVED EMPLOYEE INFORMATION

Employee #: __________ Name: ______________________________ S.S.# ______________

DOB: ____________ Date of Hire: __________ State Where Hired: __________ Occupation: ______________ Job #: ______________

Pay Rate: $______________ per hr./day/month/year Employees scheduled days off:______________

Workday begins: __________ am/pm Ends: __________ am/pm All involved Employees Drug tested? Yes No Paid for DOI? Yes No

INJURY/DAMAGE DESCRIPTION: Date of Incident: ______________ Time: __________ am/pm

Describe incident (include specific injury/damage detail)______________________________

Complete this section only if the incident resulted in injury.

MEDICAL INFORMATION: Physician Name: ______________________________ Phone: ______________

Physician address: ______________________________ Was Rx Prescribed? Y / N

Trmt. Provided: (Rx, X-ray, Sutures, Exam, Chiro, etc.) ______________________________

Date RTW: ______________ Regular? ❍ Restricted? ❍

RESTRICTED DUTY

LIFTING	❍ CANNOT	❍ CAN	IF "CAN" LIMITATION IN LBS ______________
STOOPING, BENDING, KNEELING	❍ CANNOT	❍ CAN	LIMITATION ______________
CLIMBING	❍ CANNOT	❍ CAN	LIMITATION ______________
REACHING	❍ CANNOT	❍ CAN	LIMITATION ______________
USE OF LEFT OR RIGHT UPPER EXTREMITY	❍ CANNOT	❍ CAN	LIMITATION ______________
USE OF LEFT OR RIGHT LOWER EXTREMITY	❍ CANNOT	❍ CAN	LIMITATION ______________

ANY OTHER RESTRICTIONS (INCLUDE RESTRICTIONS FROM PRESCRIBED MEDS)

ESTIMATED DURATION OF RESTRICTED DUTY

SUPERVISOR INFORMATION: Date reported: ____________ Time reported: __________ am/pm

Person reported to: ______________________________ Phone: ______________

Supervisor: ______________________________ Phone: ______________

Signed: ______________________________ Date: ______________

(Person completing report) (Report completed date)

Maintain the original report at the jobsite and transmit copy to the appropriate safety personnel.

DO NOT WRITE BELOW THIS LINE

Timed Received:		Forward to Carrier:	Yes / No	❍ This event resulted in Fatality
Number of incidents for this project:		❍ This event is medical recordable	❍ This event is lost work day	❍ This event is 1st Aid Only

SW-70 Revised 2/03

215F05.EPS

Figure 5 ◆ Accident investigation form.

Instructor's Notes:

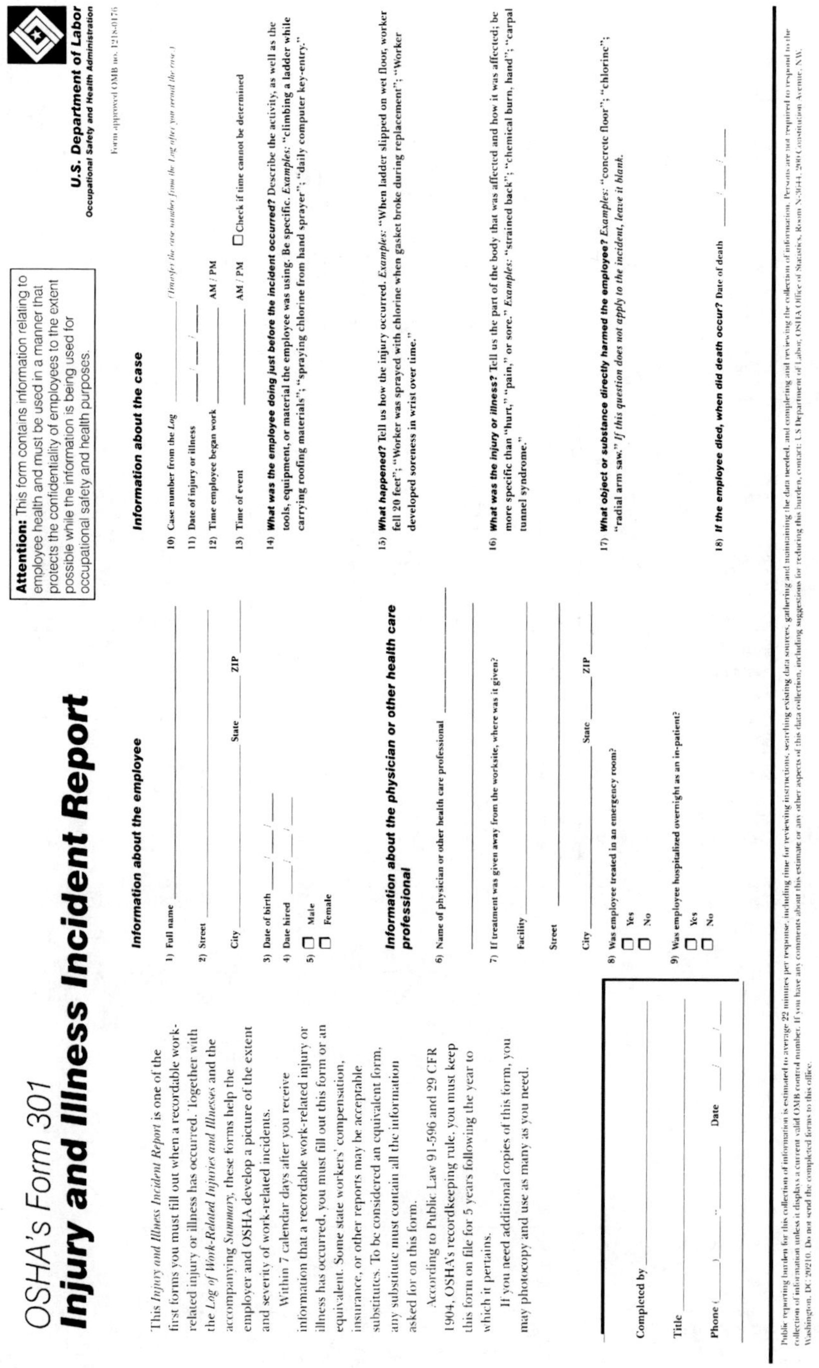

U.S. Department of Labor
Occupational Safety and Health Administration

Form approved OMB no. 1218-0176

Attention: This form contains information relating to employee health and must be used in a manner that protects the confidentiality of employees to the extent possible while the information is being used for occupational safety and health purposes.

OSHA's Form 301
Injury and Illness Incident Report

This *Injury and Illness Incident Report* is one of the first forms you must fill out when a recordable work-related injury or illness has occurred. Together with the *Log of Work-Related Injuries and Illnesses* and the accompanying *Summary*, these forms help the employer and OSHA develop a picture of the extent and severity of work-related incidents.

Within 7 calendar days after you receive information that a recordable work-related injury or illness has occurred, you must fill out this form or an equivalent. Some state workers' compensation, insurance, or other reports may be acceptable substitutes. To be considered an equivalent form, any substitute must contain all the information asked for on this form.

According to Public Law 91-596 and 29 CFR 1904, OSHA's recordkeeping rule, you must keep this form on file for 5 years following the year to which it pertains.

If you need additional copies of this form, you may photocopy and use as many as you need.

Completed by ______

Title ______

Phone (____) ____-______ Date ___/___/___

Information about the employee

1) Full name ______

2) Street ______

City ______ State ______ ZIP ______

3) Date of birth ___/___/___

4) Date hired ___/___/___

5) ☐ Male
☐ Female

Information about the physician or other health care professional

6) Name of physician or other health care professional ______

7) If treatment was given away from the worksite, where was it given?

Facility ______

Street ______

City ______ State ______ ZIP ______

8) Was employee treated in an emergency room?
☐ Yes
☐ No

9) Was employee hospitalized overnight as an in-patient?
☐ Yes
☐ No

Information about the case

10) Case number from the *Log* ______ (*Transfer the case number from the Log after you record the case.*)

11) Date of injury or illness ___/___/___

12) Time employee began work ______ AM / PM

13) Time of event ______ AM / PM ☐ Check if time cannot be determined

14) ***What was the employee doing just before the incident occurred?*** Describe the activity, as well as the tools, equipment, or material the employee was using. Be specific. *Examples:* "climbing a ladder while carrying roofing materials"; "spraying chlorine from hand sprayer"; "daily computer key-entry."

15) ***What happened?*** Tell us how the injury occurred. *Examples:* "When ladder slipped on wet floor, worker fell 20 feet"; "Worker was sprayed with chlorine when gasket broke during replacement"; "Worker developed soreness in wrist over time."

16) ***What was the injury or illness?*** Tell us the part of the body that was affected and how it was affected; be more specific than "hurt," "pain," or sore." *Examples:* "strained back"; "chemical burn, hand"; "carpal tunnel syndrome."

17) ***What object or substance directly harmed the employee?*** *Examples:* "concrete floor"; "chlorine"; "radial arm saw." *If this question does not apply to the incident, leave it blank.*

18) ***If the employee died, when did death occur?*** Date of death ___/___/___

Public reporting burden for this collection of information is estimated to average 22 minutes per response, including time for reviewing instructions, searching existing data sources, gathering and maintaining the data needed, and completing and reviewing the collection of information. Persons are not required to respond to the collection of information unless it displays a current valid OMB control number. If you have any comments about this estimate or any other aspects of this data collection, including suggestions for reducing this burden, contact: US Department of Labor, OSHA Office of Statistics, Room N-3644, 200 Constitution Avenue, NW, Washington, DC 20210. Do not send the completed forms to this office.

215F06.TIF

Figure 6 ◆ OSHA Form 301.

Explain that the illness and injury information can be collected from accident investigation reports as well as Forms 300 and 300A.

Have the participants use a completed accident investigation form to gather the information needed to complete OSHA Form 301.

Explain that even though a safety technician may not be responsible for maintaining records, he or she should know who is and where the records are kept.

Show Transparency 11 (Figure 7).

Explain how to use a records location list like the one shown in Transparency 11.

Explain OSHA's requirements regarding the maintenance and storage of records.

Other commonly kept safety, health, and training records include:

- Job applications
- Alcohol and substance abuse policy acknowledgement forms
- Job-site inspection, audits, and observation reports
- Disciplinary action reports

How and by whom these records are kept will vary depending on company policies and procedures. As a safety technician, you may have responsibility for some or all such records. Even if you are not responsible for keeping the records, you should know who does and where the records are kept. *Figure 7* is a sample records location list that can be used to document where records are kept and by whom.

OSHA has specific requirements for the maintenance of employee job-related medical and exposure records that are outlined in *29 CFR 1910.1020*. As a safety technician, you should be well versed

RECORDS LOCATION LIST

Type of Record	Responsible Party	Record Location
Employee Applications		
Employee Work-Related Medical Records		
Employee Work-Related Exposure Records		
Employee Training Records & Acknowledgement Forms		
OSHA 300 Logs		
Accident Report Forms & Investigating Report Documentation		
Inspection Report Forms & Inspection Documentation		
Safety Meeting Report Forms & Meeting Minutes		
Motor Vehicle and Mobile Equipment Maintenance & Inspection Report Forms		
Crane Repair & Annual Inspection Records		
Fire Extinguisher Inspection Records		
Job-Specific Work Permits (Confined Space Entry, LOTO, Hot Work)		

215F07.EPS

Figure 7 ◆ Records location list.

Instructor's Notes:

on these requirements, including employee access, which means the right to review and copy.

Employee job-related medical and exposure records should be kept confidential and under lock and key. Only those personnel authorized by senior management may review medical and exposure records. The purpose of such reviews should be limited to that which is necessary to manage the safety and health process and protect the health and well being of the employees. Critical, confidential employee medical and exposure records may only be released to members of management on a need-to-know basis. Such information should not include any physical conditions or medical problems. The information should only relate to the employee's fitness for duty. For example, the employee may not lift more than 15 pounds, or may only use a respirator for escape purposes only. Such information must be kept confidential.

Employees have a right to see and copy their personal medical and exposure records. Employee exposure records by job title or description should also be available to employees and their designated representatives. Employees should be informed of their right to see their job-related medical and exposure records and your procedures for granting access. This should be done at least annually.

Summary

Good recordkeeping is essential to an effective safety and health program. Without good records, it is difficult to analyze safety and health data and take the corrective actions to eliminate hazards. *OSHA Part 1904*, the recordkeeping rule, gives the requirements and criteria necessary for recording illness and injury information promptly and accurately. As a safety technician, you should have a good working knowledge of regulatory and company recordkeeping requirements.

Review Questions

1. Companies are required to keep OSHA injury and illness records if they have more than five employees at any time during the calendar year.
   a. True
   b. False

2. All of the following are recordable illnesses and injuries *except* _____.
   a. hay fever
   b. broken bones
   c. punctured eardrums
   d. cancer

3. Medical treatment is defined by OSHA as the managing and caring for a patient for the purpose of fighting disease or illness.
   a. True
   b. False

4. All of the following are considered non-recordable medical treatment *except* _____.
   a. first aid procedures
   b. diagnostic procedures
   c. visits to the doctor for counseling
   d. visits to the emergency room

5. Using hot or cold therapy on someone who is injured or ill is considered recordable first aid.
   a. True
   b. False

Explain the rights of employees in regard to accessing their own records.

Review the material covered in the module.

Have the participants complete the Review Questions and go over the answers prior to administering the Module Examination.

Administer the Module Examination. Record the results on Craft Training Report Form 200 and submit the results to the Training Program Sponsor.

RECORDKEEPING 15.13

Instructor's Notes:

GLOSSARY

Trade Terms Introduced in This Module

Abrasion: An irritation of the skin or mucous membrane that is caused by friction.

Antigen: A toxin or enzyme that is capable of stimulating an immune response, similar to an allergy.

Calendar year: The period of a year beginning and ending with January 1 and December 31 respectively.

Chronic irreversible disease: A long-lasting, permanent disease.

Contact dermatitis: An inflammation of the skin.

Contusion: An injury to tissue that causes bruising but usually not a wound.

Eczema: An inflammatory condition of the skin characterized by redness, itching, and oozing sores that become scaly, crusted, or hardened.

Hypothermia: A life-threatening medical condition that can happen in cold weather when someone's body temperature becomes too low.

Infectious material: Biological materials that carry contagious diseases and illnesses.

Laceration: A torn and ragged wound.

Recordable: An illness or injury is classified as *recordable* if it meets certain OSHA reporting guidelines.

Restricted work activity: Work activity assigned to those workers who have experienced a work-related injury or illness that prevents them from doing the routine functions of their jobs or from working a full workday.

Ulcer: A break in skin or mucous membrane that festers and erupts like an open sore.

RECORDKEEPING 15.15

Additional Resources

This module is intended to present thorough resources for task training. The following reference works are suggested for further study. These are optional materials for continued education rather than for task training.

www.osha.gov

www.asse.org

OSHA 2002 Recordkeeping Simplified, 2002. James E. Roughton, CRSP, CHMM, CSP. Burlington MA: Butterworth-Heinemann.

Figure Credits

Frank McDaniel	215F05
Professional Safety Associates, Inc.	215F07

Instructor's Notes:

MODULE 75215-03 — ANSWERS TO REVIEW QUESTIONS

Answer		Section
1.	b	2.0.0
2.	a	2.2.0
3.	a	2.2.1
4.	d	2.2.1
5.	b	2.2.2
6.	c	2.3.0
7.	b	2.4.0
8.	c	4.0.0
9.	a	4.0.0
10.	b	4.0.0

CONTREN™ LEARNING SERIES — USER UPDATES

The NCCER makes every effort to keep these textbooks up-to-date and free of technical errors. We appreciate your help in this process. If you have an idea for improving this textbook, or if you find an error, a typographical mistake, or an inaccuracy in NCCER's Contren™ textbooks, please write us, using this form or a photocopy. Be sure to include the exact module number, page number, a detailed description, and the correction, if applicable. Your input will be brought to the attention of the Technical Review Committee. Thank you for your assistance.

Instructors – If you found that additional materials were necessary in order to teach this module effectively, please let us know so that we may include them in the Equipment/Materials list in the Instructor's Guide.

Write: Curriculum Revision and Development Department
National Center for Construction Education and Research
P.O. Box 141104, Gainesville, FL 32614-1104

Fax: 352-334-0932

E-mail: curriculum@nccer.org

Craft ____________________ Module Name ____________________

Copyright Date ________ Module Number ________ Page Number(s) ________

Description ____________________

____

____

____

(Optional) Correction ____________________

____

____

(Optional) Your Name and Address ____________________

____

____

Module 75216-03

OSHA Inspection Procedures

NCCER STANDARDIZED CRAFT TRAINING PROGRAM

The National Center for Construction Education and Research (NCCER) provides a standardized national program of accredited craft training. Key features of the program include instructor certification, competency-based training, and performance testing. The program provides trainees, instructors, and companies with a standard form of recognition through a National Craft Training Registry. The program is described in full in the *Guidelines for Accreditation*, published by the NCCER. For more information on standardized craft training, contact the NCCER by writing us at P.O. Box 141104, Gainesville, FL 32614-1104; calling 352-334-0911; or e-mailing info@nccer.org. More information may be found at our Web site, www.nccer.org.

HOW TO USE THIS ANNOTATED INSTRUCTOR'S GUIDE

Each page presents two sections of information. The larger section displays each page exactly as it appears in the Participant Module. The narrow column ties suggested participant and instructor actions to each page and provides icons (detailed below) to call your attention to material, safety, audiovisual, or testing requirements. The bottom of each page includes space for your notes.

The **Audiovisual** icon indicates an appropriate time to show a transparency or other audiovisual aid.

The **Classroom** icon prompts you to define a term, stress a point, ask participants to explain a concept, or give examples.

The **Demonstration** icon directs you to show participants how to perform tasks.

The **Examination** icon tells you to administer the written module examination.

The **Homework** icon is placed where you may wish to assign reading for the next class, to assign a project, or to advise participants to prepare for an examination.

The **Laboratory** icon is used when participants are to practice performing tasks.

The **Materials** icon is a reminder for you to gather materials needed for classes, labs, and testing.

The **Performance Testing** icon tells you to administer a performance test or a portion thereof.

The **Safety** icon is used to emphasize safety issues. It is often keyed to *Caution* and *Warning* statements in the Participant Module.

The **Teaching Tip** icon indicates additional guidance is available, such as how to conduct an exercise, get the most educational value from a field trip, or encourage class participation. Teaching Tips may expand on a feature (*Think About It*, *Did You Know?*) or provide Quick Quizzes or similar exercises. You will be referred to the Teaching Tips section at the back of the module if there is additional material.

The **Combination** icon indicates that the laboratory listed corresponds with a performance task. If desired, you can note the proficiency of the participants during the laboratory and use it to satisfy performance testing requirements.

PREPARATION

Before teaching this module, you should review the Objectives, Performance Tasks, Materials and Equipment List, and the Module Outline. Be sure to allow ample time to prepare your own training or lesson plan and gather all required materials and equipment.

MODULE OVERVIEW

In this module, participants will gain a thorough understanding of the OSHA inspection process.

PREREQUISITES

Please refer to the Course Map in the Participant Module. Prior to training with this module, it is recommended that the participant shall have successfully completed the following:

Field Safety; Safety Technology, Modules 75201-03 through 75215-03

OBJECTIVES

Upon completion of this module, the participant will be able to:

1. Explain why OSHA inspects construction sites.
2. Describe the process for an on-site OSHA inspection.
3. Explain the role of the safety technician during an inspection.
4. Explain the difference between a focused inspection and a wall-to-wall inspection.
5. Explain suggested and required follow up resulting from an OSHA inspection.
6. Explain the consequences of OSHA citations, violations, and fines.
7. Explain the rights and responsibilities of employees and employers during an OSHA inspection.
8. Explain OSHA's multi-employer work site inspection and citation procedures.

PERFORMANCE TASKS

There are no performance tasks for this module.

MATERIALS AND EQUIPMENT LIST

Overhead projector and screen
Transparencies
Whiteboard/chalkboard
Markers/chalk
Blank acetate sheets
Transparency pens
Pencils and scratch paper
Appropriate personal protective equipment
Copy of *29 CFR 1926, OSHA Construction Industry Regulations*
TV (optional)
VCR (optional)
The Gulf Publishing Company video, *What Every Manager Should Know about OSHA Inspections*
Module Examinations*

* Located in the Test Booklet.

SAFETY CONSIDERATIONS

Ensure that participants are equipped with appropriate personal protective equipment.

ADDITIONAL RESOURCES

This module is intended to present thorough resources for task training. The following reference works are suggested for both instructors and motivated participants interested in further study. These are optional materials for continued education rather than for task training.

Web site: www.osha.gov

Web site: www.asse.org

OSHA Inspections: Preparation and Response, 1997. Rick Kaletsky. New York, NY: McGraw-Hill.

TEACHING TIME FOR THIS MODULE

An outline for use in developing your lesson plan is presented below. Approximately 2½ hours are suggested to cover *OSHA Inspection Procedures*. This includes 10 minutes at the beginning of the session for administrative tasks and one 10-minute break. You may need to adjust the time required for hands-on activities and testing based on your class size and resources.

Topic	Planned Time
A. Introduction	________
B. OSHA Inspections	________
1. Causes for OSHA Inspections	________
2. The Inspection Process	________
3. Violations, Citations, and Penalties	________
C. The Focused Inspection Program	________
1. Program Qualifications	________
2. Modified Inspection Procedure	________
D. Post Inspection Follow-Up	________
1. Reports to Management and Employees	________
2. Abatement and Abatement Certification	________
3. Corrective Actions	________
4. Informal Conference, Appeals, and Hearings	________
5. Affirmative Defenses	________
E. Voluntary Consultation Service	________
F. Review	________
G. Module Examination	________

1. Participants must score 70% or higher to receive recognition from the NCCER.
2. Record the testing results on Craft Training Report Form 200 and submit the results to the Training Program Sponsor.

Module 75216-03

OSHA Inspection Procedures

Instructor's Notes:

COURSE MAP

This course map shows all of the modules in Safety Technology. The suggested training order begins at the bottom and proceeds up. The local Training Program Sponsor may adjust the training order.

Assign reading of Module 75216-03.

SAFETY TECHNOLOGY

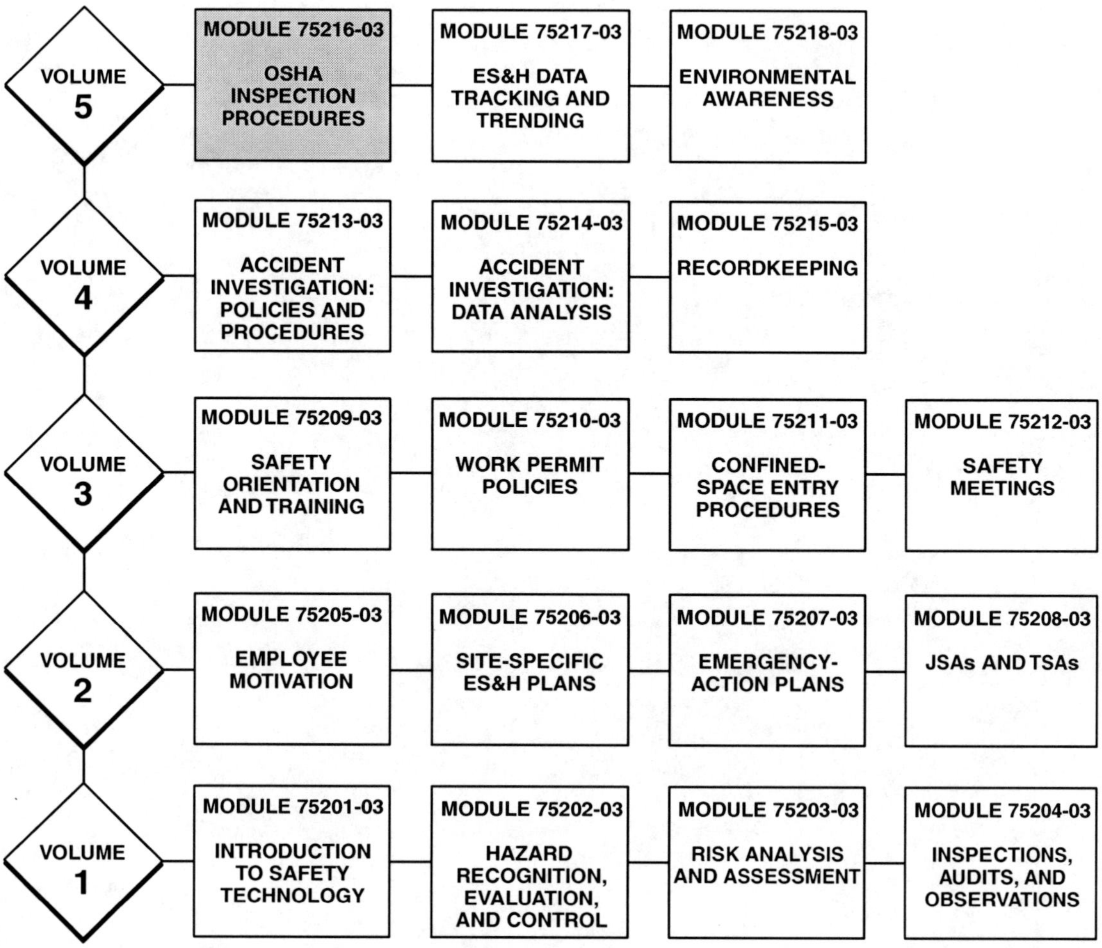

216CMAP.EPS

OSHA INSPECTION PROCEDURES 16.iii

Instructor's Notes:

MODULE 75216-03 CONTENTS

OSHA INSPECTION PROCEDURES 16.v

Figures

Tables

Instructor's Notes:

MODULE 75216-03

OSHA Inspection Procedures

Objectives

When you have completed this module, you will be able to do the following:

1. Explain why OSHA inspects construction sites.
2. Describe the process for an on-site OSHA inspection.
3. Explain the role of the safety technician during an inspection.
4. Explain the difference between a focused inspection and a wall-to-wall inspection.
5. Explain suggested and required follow-up resulting from an OSHA inspection.
6. Explain the consequences of OSHA citations, violations, and fines.
7. Explain the rights and responsibilities of employees and employers during an OSHA inspection.
8. Explain OSHA's multi-employer work site inspection and citation procedures.

Prerequisites

Before you begin this module, it is recommended that you successfully complete the following: Field Safety; Safety Technology, Modules 75201-03 through 75215-03.

Required Materials

1. Pencil and paper
2. Appropriate personal protective equipment
3. A copy of *29 CFR 1926 OSHA Construction Industry Regulations*

1.0.0 ◆ INTRODUCTION

OSHA periodically inspects work sites to make sure safety standards are met. They will also schedule an inspection after a serious accident. OSHA inspections should never be taken lightly because OSHA can levy large fines or shut down operations if **violations** are discovered.

A good safety management program is your best defense against fines. You must make sure all applicable OSHA rules are used and enforced on your job site. OSHA has developed a focused inspection program for construction sites. If you have a written program and a **competent person** on site, your company may qualify for a more limited inspection.

Twenty-six states currently have their own OSHA-approved health and safety programs. In order for states to have their own OSHA programs, their rules must be at least as tough as the federal standards. The state rules and procedures may differ from the federal standards explained in this module. You can find out which states have their own programs on OSHA's Web site at: http://www.osha.gov/fso/osp/index.html. It is your responsibility to be aware of the local regulations covering your job site.

2.0.0 ◆ OSHA INSPECTIONS

OSHA representatives, called **compliance safety and health officers (CSHOs)**, inspect job sites. They are health and safety professionals who help employers reduce job hazards and enforce health and safety regulations. Before an inspection, CSHOs review the history of the business, the

Ensure that you have everything you need to teach the course. Check the Materials and Equipment list in the front of this module.

Explain that terms shown in bold are defined in the Glossary at the back of this module.

Review the goals of the course and explain what will be expected of the participant.

Show Transparency 1, Objectives.

Explain the purpose of OSHA inspections.

Explain that a good safety management program is your best defense against citations and fines.

Explain the role of the OSHA compliance safety and health officer (CSHO).

OSHA INSPECTION PROCEDURES 16.1

operations, and appropriate standards. They may also bring testing equipment to check compliance. Always be polite and do not attempt to lie or mislead an official. If you do, any fines you receive will be multiplied.

Identify the causes for OSHA inspections.

Explain how OSHA prioritizes the scheduling of inspections.

2.1.0 Causes for OSHA Inspections

OSHA will schedule an inspection when they think it is necessary. There are six million workplaces covered by federal and state OSHA standards. This makes it difficult to inspect every site. As a result, they must prioritize inspections to deal with the most hazardous conditions first. The following lists how OSHA generally prioritizes the scheduling of inspections:

Show Transparency 2, OSHA Inspection Prioritization.

1. ***Imminent Danger*** – An imminent danger is a hazard that could cause death or serious physical harm immediately, or before the danger could be eliminated through normal enforcement procedures.
2. *Catastrophes and Fatal Accidents* – OSHA will investigate an accident if someone dies or three or more employees are hospitalized.
3. *Complaints* – These are verbal or written complaints received by OSHA concerning hazards that threaten serious physical harm to workers.
4. *Programmed Inspections* – OSHA develops a schedule for inspecting the most hazardous industries, including construction. This is based on statistical data, such as job injury/illness rates and worker compensation claims.
5. *Follow-Up Inspections* – OSHA may re-inspect firms to see if hazards identified during an initial inspection have been corrected.

In 2002 there were over 96,000 state and federal OSHA inspections. Over half of these were in the construction industry. Twenty-five percent of the inspections were related to complaints or accidents. Sixty percent were programmed inspections of high-hazard industries.

Inspections are usually done without giving any notice. Sometimes OSHA may give notice, but it will normally be less than 24 hours. In some cases, notice is required to ensure that key people will be at the site. An inspection can be delayed for good cause if it would produce a more effective inspection. In some cases, an inspection can be delayed until after business hours so that inspectors have access to all work areas.

> **NOTE**
> If OSHA sends a notice of inspection, it must be clearly posted in advance so that everyone on site is aware of the inspection.

Identify the five steps in the inspection process.

Show Transparency 3, Inspection Process.

2.2.0 The Inspection Process

OSHA's inspection process has five steps. The following steps reflect the order in which the inspection should take place.

Step 1 Presentation of the inspector's credentials

Step 2 Opening conference

Step 3 Selection of representatives

Step 4 Walk-around inspection

Step 5 Closing conference

Discuss the importance of asking for the OSHA inspector's creden-tials upon his or her arrival.

Explain how the Fourth Amendment can restrict OSHA from entering a site without a search warrant.

2.2.1 Presentation of the Inspector's Credentials

When an OSHA inspector arrives, ask to see his or her credentials. These credentials will show the inspector's photo and serial number. Write down his or her name, serial number, and his or her supervisor's name. Notify your office when the inspector arrives. The safety manager, company president, or other authorized person will determine if a search warrant is required.

Even though OSHA has the authority to enter and inspect a work site, the U.S. Supreme Court, in the 1978 Barlow Decision, said that employers could invoke the Fourth Amendment of the U.S. Constitution to prevent unwanted entry onto a site. When this is done, OSHA must get a search warrant before conducting an inspection. Few employers actually feel the need to exercise this right. Learning your company's policy on this issue will save valuable time for both your company and the inspector.

Explain what happens during the opening conference.

2.2.2 Opening Conference

After the inspector's credentials have been presented, the CSHO will explain the reason for the inspection and the applicable standards. Always make detailed notes during the opening conference and request a copy of an employee complaint, if there is one. The complaint may be edited to remove the employee's name, if he or she requests it. Be sure that your OSHA posters are up, assured grounding materials are available, and hazard communication and safety programs are available for review.

List the representatives who are allowed to go with the CSHO during an inspection.

2.2.3 Selection of Representatives

Certain people can go with the CSHO during an inspection. One person represents the employer. This can be the safety technician or manager. An employee representative also has the right to go along. The employee representative can be a member of the employees' union, trade association, or

Instructor's Notes:

employee safety committee. On large construction sites, the CSHO may request representatives from all sub-contractors in addition to the general contractor. The best practice is to choose representatives before you are inspected. This will limit confusion, save time, and help you make sure that all parties are properly represented.

2.2.4 Walk-Around Inspection

After the opening conference, the CSHO and the representatives go through the workplace, inspecting for workplace hazards. The CSHO may talk with several employees. It must be a reasonable number, and work interruptions should be limited. Workers are not required to discuss anything with the inspector. However, if they do, they cannot be punished because of anything they say or show the CSHO during the inspection.

The compliance officer will discuss any violations noted during the walk-around. Always make a note of every violation that the inspector points out. During this time, ask the CSHO for technical advice on how to correct the hazard. If possible, correct all violations on the spot. Be sure that the inspector notes your correction. You may still be cited for a violation even if it is corrected. However, immediate correction shows **good faith**, which will help minimize fines.

The following tips help to ensure that the inspection is done efficiently and effectively.

- Do not try to hide something you believe is a violation.
- Do not ask if something is or is not in compliance.
- Be polite.
- Do not hesitate to boast about the safety precautions taken on the project.
- Take photos of the same items photographed by the inspector.

2.2.5 Closing Conference

The closing conference will review the hazards found on the site and the CSHO will review any apparent violations with the employer. Check each problem against the standard and ask for an explanation if it is not absolutely clear. Be sure to ask the CSHO how to correct the problem. Should you disagree with the CSHO, be polite but explain your opinion. Discuss a possible time period to fix the problem and set realistic goals for doing so. Understand that if the problem is not fixed on time, the fines are increased.

The CSHO must explain that violations may result in **citations** and penalties. He or she must also answer all questions about the inspection. Do not discuss proposed fines with the CSHO. He or she will not tell you how much the fines will be because the OSHA Area Director sets the fines after the CSHO has submitted a report. The CSHO should, however, inform you of your rights and responsibilities. Your rights and responsibilities are listed in the official OSHA booklet, *Employer Rights and Responsibilities Following an OSHA Inspection, OSHA #300, 2002.*

2.3.0 Violations, Citations, and Penalties

A violation is the act of breaking or not following any applicable OSHA standard. This includes the general duty for employers to provide a safe workplace. A citation is a formal notice that a violation has taken place. A **penalty** is a fine to discourage employers from repeat violations.

> **NOTE**
> You can be cited for having an unsafe workplace even if the recommended safety measures are not expressly written in the rules. Employers have a general duty to provide a safe workplace and can be cited for violating that duty. Use best practices to make sure you comply with the general duty to provide a safe workplace.

2.3.1 Violations

You can be cited for violating any OSHA standard that applies to your job site. The violations are categorized depending on their gravity and the willingness of the employer to correct them. The penalties increase for each group. These groups include:

- *Other-Than-Serious Violations* – These are violations affecting safety and health, but they are unlikely to cause death or serious physical injury.
- *Serious Violations* – These violations are likely to cause death or serious physical injury. The employer should or does know about the hazards causing the violation.
- *Willful Violations* – These violations involve those situations in which the employer intentionally and knowingly commits a violation of the standards, or is aware of the hazard and makes no reasonable effort to fix it. Willful violations are divided into repeat, failure to abate, and egregious violations.
  - *Repeat Violation* – This is a violation of any standard, regulation, rule, or order where,

Describe what happens during the walk-around inspection.

Discuss some tips to ensure that the inspection is done efficiently and effectively.

Explain what is done in the closing conference.

Define the terms violation, citation, and penalty.

Explain that an employer can be sited for violating any OSHA standard that applies to the company.

Identify the groups of violations OSHA uses.

Show Transparency 4, Violation Groups.

OSHA INSPECTION PROCEDURES 16.3

Explain that a CSHO is required by law to issues citations for violations of safety and health standards.

Identify the information that is included on a citation.

Explain the requirements and procedures for dealing with citations.

Explain why OSHA fines companies.

Describe the penalties related to specific violations.

Show Transparency 5 (Table 1).

upon re-inspection, a similar violation is found. To be the basis of a repeat citation, the original citation must be final. A citation that is being contested may not serve as the basis for a subsequent repeat citation. Repeat violations mean the same rules are broken by one company. These problems could be on different job sites and at different times.

- *Failure-to-Abate Violations* – This is a violation for failure to correct a previous violation OSHA has specifically cited on a particular job site.
- *Egregious* – When the hazard is so great it seems intentional, it is considered egregious. When the egregious policy is applied, the fines will be multiplied by the number of exposed workers. This is the most serious category.

2.3.2 Citations

A CSHO is required by law to issue citations for violations of safety and health standards. They are not allowed to issue warnings. In most cases, the citations are prepared by the OSHA Area Director and mailed to the employer. Citations include:

- A description of the violation and the applicable OSHA standard allegedly violated
- The proposed penalty
- The date by which the hazard must be corrected

A copy of the citation must be posted at or near the place the violation occurred. The notice must be posted for at least three working days or until the problem is corrected, whichever is longer. You must correct the problems listed in the citation. Inspectors may come back to check that they have been corrected. If the problems are not corrected on time, stronger penalties apply.

An OSHA citation must be taken care of right away. The employer may request an informal conference with the OSHA Area Director. The purpose of this conference is to resolve any issues arising form the inspection. The employer may provide additional information at the conference. At the informal conference, the parties can enter into a settlement agreement that can revise the citations and penalties.

If the issues are not resolved at the informal conference, the employer may appeal the citation. Employers have 15 working days to appeal a citation. They must file an intention to contest the OSHA citation before the independent OSHA Review Commission. Employers should hire an attorney or other legal counsel if they decide to fight an OSHA citation.

> **NOTE**
>
> Always notify your main office of an OSHA inspection and any citations. The Safety Manager or Director in conjunction with senior management will make a decision about contesting citations. They will also notify other job sites about any citations. Similar conditions on other job sites can be considered repeat violations.

2.3.3 Penalties

OSHA fines companies to discourage them from breaking the rules. More than $140 million in OSHA fines are collected each year. A good safety program can help your company avoid these penalties.

OSHA regulations set specific penalties for violations, as shown in *Table 1*. These amounts may be adjusted by the Area Director. Some penalties can be lowered for good faith compliance. Good faith compliance means that the employer made a serious effort to follow the rules to the best of his or her ability. In short, you have made a good faith effort to comply.

To show good faith, you must have a written safety and health program. The program must be an active part of your daily operations. OSHA has voluntary guidelines that include required programs, such as Hazard Communication, Lockout/Tagout, and other programs for construction. Other trade associations and safety organizations have sample programs that you can use. You must adjust these programs to fit the safety needs of your company.

In an extreme case, you can go to jail. Willful violations can bring criminal charges. If an employee died and the employer is convicted of a willful violation, he or she faces a fine or imprisonment for up to six months, or both. Criminal convictions have stiff penalties. They can be up to $250,000 for an individual, or $500,000 for a corporation.

OSHA will refer a criminal case to the Criminal Division of the Department of Justice (DOJ). The DOJ will review the case and determine if prosecution is reasonable. The case will be referred to the appropriate U.S. Attorney, who will bring criminal charges. The case will be heard in federal court, not by a review board or an administrative law judge.

Twenty-four states are covered under the federal OSHA program. The violations and penalties issued by federal OSHA inspectors in 2001 are

Instructor's Notes:

Table 1 OSHA Violations and Penalties

VIOLATION	PENALTY	ADJUSTMENTS/COMMENTS
Other-Than-Serious Violation	A discretionary penalty of up to $7,000 for each violation.	May be adjusted downward by as much as 95 percent, depending on the employer's good faith, history of previous violations, and size of business. When the adjusted penalty is less than $50, no penalty is proposed.
Serious Violation	A mandatory penalty of up to $7,000 for each violation.	May be adjusted downward, based on the employer's good faith, history of previous violations, the gravity of the alleged violation, and size of business.
Willful Violation	Up to $70,000 for each willful violation, with a minimum penalty of $5,000 for each violation.	May be adjusted downward, depending on the size of the business and its history of previous violations. Usually, no credit is given for good faith.
Repeat Violation	Up to $70,000 for each such violation.	None.
Failure-to-Abate Prior Violation	Up to $7,000 for *each day* the violation continues beyond the abatement date.	None.
Egregious	Fines will be multiplied by the number of exposed workers.	None.
Failure to properly maintain OSHA logs	$1,000	None.
Failure to promptly and properly report a fatality/catastrophe	$5,000	None.
Falsifying reports, records, or applications	$10,000 or up to six months in jail, or both.	Issued upon conviction.
Violations of posting requirements	Civil penalty of up to $7,000.	None.
Assaulting a CSHO, or otherwise resisting, opposing, intimidating, or interfering with a CSHO in the performance of his or her duties	A criminal offense, subject to a fine of not more than $5,000 and imprisonment for not more than three years.	Issued upon conviction.

216T01.EPS

Summarize the federal and state violations and the penalties issued by federal OSHA inspectors in 2001.

Show Transparencies 6 and 7 (Tables 2 and 3).

Identify the average penalty for a willful violation.

Explain the difference in the penalties between willful and serious violations.

Explain when and why OSHA started the Focused Inspections Initiative.

listed in *Table 2*. Twenty-six states have their own OSHA programs. The violations and penalties assessed by state OSHA programs in 2001 are listed in *Table 3*.

For both federal and state OSHA programs, the average serious violation had a penalty of $900. The average penalty for a willful violation was $28,000. That is 30 times more than the penalty for serious violations. The difference between a willful and a serious violation is the employer's willingness to make a reasonable effort to fix hazardous conditions. A good safety program can prevent expensive willful violation citations.

3.0.0 ◆ THE FOCUSED INSPECTION PROGRAM

On October 1, 1994 OSHA started the Focused Inspections Initiative to recognize responsible contractors who have effective safety and health programs. The program encourages contractors to develop effect health and safety programs. It is very different from a regular OSHA inspection. The new policy aims for an overall improvement in construction job-site safety.

Previously, construction inspections were all-inclusive. The CSHO would try to find all the

OSHA INSPECTION PROCEDURES 16.5

Describe the criteria that allow a company to qualify for the Focused Inspections Initiative.

Explain that the CSHO may use a written guideline to determine if a project qualifies for the program.

Show Transparency 8 (Figure 1).

Explain that one of the requirements for participation in the Focused Inspections Initiative program is ensuring that the OSHA requirements for effective safety and health plans are met, including the requirement for a written safety and health program.

Explain the required provisions in the plan. Refer the trainees to the bulleted list in the Participant Module.

Explain that another requirement for participation in the Focused Inspections Initiative program is having a competent person implement and monitor the safety and health plan.

Discuss OSHA's definition of a competent person.

Table 2 Federal OSHA Violations and Penalties for 2001

VIOLATIONS	TYPE	PENALTIES
416	Willful	$11,799,539
54,842	Serious	$48,312,043
1,969	Repeat	$7,710,736
231	Failure-to-abate	$597,301
20,749	Other (discretionary)	$2,145,151
226	Unclassified	$2,268,508
78,433	Total	$72,827,278

216T02.EPS

Table 3 State OSHA Violations and Penalties for 2001

VIOLATIONS	TYPE	PENALTIES
219	Willful	$6,444,925
58,476	Serious	$55,480,047
2,490	Repeat	$5,477,34
646	Failure-to-abate	$3,185,193
81,219	Other (discretionary)	$4,690,873
23	Unclassified	$593,500
144,075	Total	$75,871,882

216T03.EPS

violations on a construction site. As a result, CSHOs spent too much time and effort on a few projects. The contractor was likely to be cited for many violations, but these violations were not the cause of construction fatalities. The Focused Inspections Initiative directs CSHOs to spend more time inspecting workplaces with inadequate safety programs, looking for hazards that are most likely to cause fatalities and serious injuries to workers. OSHA has determined that this approach is effective in protecting overall worker safety.

3.1.0 Program Qualifications

A CSHO will decide if a project qualifies for a focused inspection by reviewing the project's safety and health plan. In order to qualify, the project must have a written safety and health program or plan, and a person responsible for and capable of implementing the program or plan. The CSHO must note in each case file why a focused inspection was or was not conducted.

The CSHO may use a written guideline to determine if a project qualifies for the program (*Figure 1*). You should ask for a copy of the completed form at the time of the inspection. All contractors and employee representatives must be informed why a focused or a comprehensive inspection is being conducted. This may be accomplished either by personal contact or by posting the guidelines. A request for a search warrant should not affect the determination as to whether a project will receive a focused inspection.

If the OSHA inspector is investigating a complaint or accident, the CSHO may conduct a focused inspection after the complaint or fatality has been addressed. If there is no coordination by the general contractor, prime contractor, or other such entity to ensure that all employers provide adequate protection for their employees, a comprehensive inspection will be conducted.

3.1.1 Effective Written Safety Program

An effective safety and health plan must meet the requirements of *29 CFR 1926 Subpart C, General Safety and Health Provisions*. Subpart C contains the basic requirements of a safety and health program on construction sites. It also contains the definitions for key terms used in the construction standards, such as competent person, qualified person, approved, and suitable. The plan must include provisions for:

- Project safety analysis
- Evaluation of subcontractors to conform to the plan
- Supervision and training
- Control of hazardous operations
- Documentation
- Employee involvement
- Emergency response

The plan must also be assessed based on the size and complexity of the project.

In addition, the employer must have a written and implemented safety and health program such as OSHA's voluntary "Safety and Health Management Guidelines." This includes programs required under the OSHA standards, such as Hazard Communication, Lockout/Tagout, and other programs for construction.

3.1.2 Competent Person

The program also requires that a competent person be responsible for implementation and monitoring of the plan. A competent person is a company-designated person who is capable of identifying existing and predictable hazards, and who has authorization to take prompt corrective measures to eliminate them. The on-site safety technician may fulfill this role with the support of additional safety managers as needed.

Instructor's Notes:

CONSTRUCTION FOCUSED INSPECTION GUIDELINE

This guideline is to assist the professional judgment of the compliance officer to determine if there is an effective project plan, to qualify for a Focused Inspection.

YES/NO

PROJECT SAFETY AND HEALTH COORDINATION; are there procedures in place by the general contractor or other such entity to ensure that all employers provide adequate protection for their employees? ________

Is there a DESIGNATED COMPETENT PERSON responsible for the implementation and monitoring of the project safety and health plan who is capable of identifying existing and predictable hazards and has authority to take prompt corrective measures? ________

PROJECT SAFETY AND HEALTH PROGRAM/PLAN* that complies with 1926 Subpart C and addresses, based upon the size and complexity of the project, the following: ________

________ Project Safety Analysis at initiation and at critical stages that describes the sequence, procedures, and responsible individuals for safe construction.

________ Identification of work/activities requiring planning, design, inspection or supervision by an engineer, competent person or other professional.

________ Evaluation/monitoring of subcontractors to determine conformance with the Project Plan.

________ (The Project Plan may include, or be utilized by subcontractors.)

________ Supervisor and employee training according to the Project Plan including recognition, reporting and avoidance of hazards, and applicable standards.

________ Procedures for controlling hazardous operations, such as: cranes, scaffolding, trenches, confined spaces, hot work, explosives, hazardous materials, leading edges, etc.

________ Documentation of: training, permits, hazard reports, inspections, uncorrected hazards, incidents and near misses.

________ Employee involvement in hazard: analysis, prevention, avoidance, correction and reporting.

________ Project emergency response plan.

216F01A.TIF

Figure 1 ◆ The Construction Focused Inspection Guideline (1 of 2).

Gulf Publishing Company's video, ***What Every Manager Should Know about OSHA Inspections***, can be used to enhance the topics in this module.

OSHA INSPECTION PROCEDURES 16.7

Identify the four leading hazards in the construction industry that the CSHO examines as part of the focused inspection.

Explain that your company's safety and health plan should cover how to handle these hazards.

Identify the steps in the modified inspection procedure.

* For examples, see owner and contractor association model programs, ANSI A10.33, A10.38, etc.

The walkaround and interviews confirmed that the Plan has been implemented, including:

_______ The four leading hazards are addressed: falls, struck by, caught in/between, electrical.

_______ Hazards are identified and corrected with preventative measures instituted in a timely manner.

_______ Employees and supervisors are knowledgeable of the project safety and health plan, avoidance of hazards, applicable standards, and their rights and responsibilities.

THE PROJECT QUALIFIED FOR A FOCUSED INSPECTION.

216T01B.TIF

Figure 1 ◆ The Construction Focused Inspection Guideline (2 of 2).

3.1.3 Four Leading Hazards

The inspection will focus on the four leading hazards that cause 90% of deaths on construction job sites. The four causes of death are:

- Falling from elevations
- Being struck by falling objects or vehicles
- Being caught in/between machinery, equipment, or soil
- Receiving an electrical shock

Show Transparency 9, Modified Inspection Procedure.

The inspector will focus attention on hazardous conditions that cause these types of accidents. For example, the CSHO will check possible sources of electrical shock, overhead power lines, power tools and cords, outlets, and temporary wiring.

Your safety program should cover all of these hazards or determine that they do not apply to your work site. You must be able to show the CSHO that you have training programs in place to deal with these hazards. The workers must be following the job safety programs.

Explain the differences between the two types of inspections.

Describe how the opening conference is conducted in a modified inspection.

Explain how the walk-around step is different in a modified inspection.

3.2.0 Modified Inspection Procedure

If your company qualifies for the Focused Inspection Program, the inspection process will be modified. The inspection will include the same five steps that are used in a regular inspection such as:

Step 1 Presentation of the inspector's credentials

Step 2 Opening conference

Step 3 Selection of representatives

Step 4 Walk-around inspection

Step 5 Closing conference

The difference between the regular inspection and the modified inspection will be that the opening conference will include a broader review of safety documentation and the walk-around inspection will be limited.

3.2.1 Opening Conference

Review of the site safety and health plan and supporting documentation will be a critical step in the inspection process. The inspector will review all appropriate documentation, including the log of injuries and illness, which most employers with more than 10 employees are required to keep. A good safety program will help you avoid violations and citations.

3.2.2 Walk-Around

The walk-around inspection will be limited. The CSHO will check that the safety and health plan is in force through interviews and observations. They will inspect for the four leading hazards and other serious hazards. Workers must be following standards or best practices listed in the safety plan. Inspectors may interview workers to see if they know the safety rules. The CSHO is not required to inspect the entire project.

If the CSHO sees that workers are not following the safety program, he or she will stop the focused inspection. The company then must face the usual wall-to-wall inspection. A serious violation does not automatically trigger a wall-to-wall inspection. The CSHO will use his or her professional judgment to decide.

Instructor's Notes:

Although the walk-around inspection will focus on the four leading hazards, citations will be issued for any serious violations. Citations will also be issued for other-than-serious violations that are not fixed immediately. Other-than-serious violations that are corrected will not usually be cited or documented.

After the walk-around, there will be a post-inspection conference. This is similar to the regular inspection process just described. The CSHO will tell you of any violations. Ask for the OSHA booklet that explains your rights and responsibilities.

NOTE

If there are violations, ask the CSHO to fax you the citation when it is available. That gives you some extra time to correct the problem. The official copy of the citation must be sent by certified mail.

4.0.0 ◆ POST INSPECTION FOLLOW-UP

If a violation is found by the CSHO, you need to record the problem and correct it. If the problem is not fixed right away, the CSHO will likely schedule another inspection. Normal follow-up to an inspection includes:

- A report to management
- A plan to correct any problems
- A date for re-inspection

The best way to prevent repeat violations is to do a self-inspection before OSHA returns.

NOTE

All violations must be corrected by the date set in the citation. If they are not, the fines will be much more expensive.

4.1.0 Reports to Management and Employees

A written report must be prepared as soon as possible so that nothing is overlooked. The report should give management a record of what the inspection covered. It should mirror what the CSHO has on file, including:

- Documents reviewed by the CSHO
- Photographs
- A summary of employee statements
- Noted violations
- Violations corrected on the spot
- Any outstanding violations noted by the CSHO

Explain the conditions that require a post inspection follow-up.

Describe the report that must be completed after an inspection.

Explain the requirements for handling violations found during a focused inspection.

THINK ABOUT IT

OSHA's Top 20 List

These are the top 20 standards that were cited most frequently by OSHA from October 2001 to September 2002. The total number of citations is in parentheses. How many of these standards apply at your work site?

1926.0451 — General Requirements for All Types of Scaffolding (8,423)
1910.1200 — Hazard Communication (6,951)
1926.0501 — Fall Protection Scope/ Applications/ Definitions (5,461)
1910.0134 — Respiratory Protection (4,250)
1910.0147 — The Control of Hazardous Energy, Lockout/Tagout (3,973)
1910.0305 — Electrical, Wiring Methods, Components and Equipment (3,202)
1910.0212 — Machines, General Requirements (2,878)
1910.0178 — Powered Industrial Trucks (2,574)
1910.0303 — Electrical Systems Design, General Requirements (2,291)
1910.0219 — Mechanical Power-Transmission Apparatus (2,088)
1926.0651 — Excavations, General Requirements (2,062)
1910.1030 — Bloodborne Pathogens (2,005)
1910.0132 — Personal Protective Equipment, General Requirements (1,847)
1926.1053 — Ladders (1,755)
1926.0100 — Head Protection (1,614)
1910.0023 — Guarding Floor and Wall Openings and Holes (1,535)
1926.0652 — Excavations, Requirements for Protective Systems (1,433)
1910.0157 — Portable Fire Extinguishers (1,429)
1926.0404 — Electrical, Wiring Design and Protection (1,406)
1926.0405 — Electrical Wiring Methods, Components and Equipment, General Use (1,390)

OSHA INSPECTION PROCEDURES 16.9

Explain abatement and the abatement certification process.

Explain how corrective actions are used to fix problems found during an inspection.

Explain the employees' right to know about corrective actions and abatement plans.

Discuss the process for appealing citations.

Define the term affirmative defense. Explain how affirmative defenses can be used in an appeals hearing.

Just as with an all-inclusive inspection, violations stemming from focused inspections must be corrected or appealed within 15 working days from receipt of the citation. If there are items that were not fixed, a plan must be made to correct them. Safety training or more equipment may be needed.

When a citation is received, it must be posted until the violation is corrected, or for three working days, whichever is longer. If the employer chooses to appeal, the appeal document must be posted at the same location. Any settlement agreements must also be posted.

4.2.0 Abatement and Abatement Certification

Abatement is the correction of the safety hazards or violations that led to an OSHA citation. **Abatement certification** is a written record that states that the problem noted in the citation has been fixed. If the hazard is corrected during an OSHA inspection, an abatement certification does not need to be filed. Abatement certification can be a simple one-page letter or a more complex plan. In either case, each of the violations must be listed, along with the corrective action.

Employers must submit an abatement certification to OSHA within 10 working days after the abatement date. The abatement date is set in the citation. This date is the deadline to correct the violation.

> **NOTE**
> The date on the citation can only be changed by OSHA or the court.

4.3.0 Corrective Actions

Corrective actions are steps taken to fix the problem. Corrective actions can include training or retraining workers on safety procedures, or establishing toolbox talks. They may also require additional safety equipment, signs, or personal protective equipment. Corrective actions must be recorded. Keeping records of these actions helps provide proof that the needed changes have been made. This makes the OSHA's re-inspection process faster and more efficient.

Employees have a right to know of any corrective action or abatement plans. The documents must be posted with the other notices. Involve employees in correcting safety problems. This will help ensure that every worker becomes responsible for job-site safety.

> **NOTE**
> Failure to correct a violation has a penalty of up to $7,000 for every day the violation continues beyond the abatement date.

4.4.0 Informal Conference, Appeals, and Hearings

If the employer believes that the citation is unreasonable, he or she may ask for an informal conference. This must happen within 15 working days of receipt of the citation. The informal conference is held to work out any disagreements. OSHA and the employer may make a settlement agreement to resolve the dispute. The primary goal is to protect workers and eliminate hazards. The settlement agreement must be posted.

If the employer is not satisfied after the informal hearing, he or she may request a formal hearing or appeal. This must be done by filing a Notice of Contest within 15 working days from receipt of the citation. The matter is then assigned to an administrative law judge who will formally review the case. At the appeal, the employer must be able to show why the citation was unreasonable.

4.5.0 Affirmative Defenses

At the appeal hearing, the employer can present an affirmative defense. An affirmative defense is an explanation or reason that the situation cited as a violation exists. It is a reason that will excuse the employer from a violation which has otherwise been proven by the CHSO. The employer must prove all aspects of the affirmative defense at the appeal hearing. Common affirmative defenses include:

- Unpreventible employee misconduct
- Impossibility
- Greater hazard
- Multi-employer work site

To show unpreventable employee misconduct, the employer must show that the hazard was unknown to the employer and in violation of a work rule that was effectively communicated and uniformly enforced.

Impossibility means that compliance with the regulation is impossible. The employer must show that compliance with the standard is functionally impossible or would prevent performance of required work. The employer must also prove that there is no alternative means of employee protection.

Instructor's Notes:

Another affirmative defense is that following the standard would be a greater hazard to employees than noncompliance. The employer must prove that compliance would be more dangerous and that there are no alternative means of employee protection. The employer must also prove that the application of a variance from the standard is inappropriate.

4.5.1 Multi-Employer Work Sites

A multi-employer work site is a job site where several employers are working. This is common in large construction projects. During an inspection, several employers can be cited for violations. These include:

- The employer whose employees are exposed to the hazard
- The employer who actually created the hazard
- The employer who is responsible, by contract or actual practice, for safety and health conditions at the work site
- The employer who is responsible for correcting the hazard

To prove an affirmative defense of a multi-employer work site, the employer must prove all five of the elements. These elements are as follows:

- The employer did not create the hazard.
- The employer did not have the responsibility or authority to remove the hazard.
- The employer did not have the ability to correct or remove the hazard.
- The employer can demonstrate that those responsible for controlling or correcting the hazard have been specifically notified of the hazard to which his or her employees are/were exposed.
- The employer has instructed his or her employees to recognize the hazard and where necessary, informed them how to avoid the dangers associated with it.
  - Where feasible, an exposing employer must have taken appropriate alternative means of protecting employees from the hazard.
  - When extreme circumstances justify it, the exposing employer shall have removed his or her employees from the job to avoid citation.

5.0.0 ◆ VOLUNTARY CONSULTATION SERVICE

OSHA offers a voluntary compliance assistance program to small- and medium-sized businesses. Any firm with 250 or fewer employees may receive free assistance from OSHA to identify workplace hazards. A full-service consultation is available to help employers establish effective workplace health and safety programs. It can cover the entire work site or be limited to a few specific problems.

The compliance assistance program is similar to an inspection in that it can include an opening conference, a walk-around, and a closing conference. The benefit is that no citations are issued. It will help you to recognize the hazards in your workplace. The CHSO will suggest options for improving workplace safety. The CHSO will also suggest possible sources for further assistance or training. Under certain circumstance, your firm can be granted a one year exemption from general enforcement inspections.

Summary

An OSHA inspection is a formal process. Recording each step of the inspection process is important. Both management and employees must be notified at certain points in the process, and an employer representative must be present. Any violations found in an inspection should be corrected immediately to reduce potential fines.

An OSHA inspection can result in stiff fines. Your best defense against hazards, citations, and fines is a good safety program. Your safety program should be your company's best effort to keep workers safe from hazards. If you have a written safety program and a safety person on site, you could qualify for a focused inspection. This program allows CSHOs to focus on the most hazardous conditions in the construction industry rather than smaller, less important issues. This allows companies to fix problems quickly and maintains a safe working environment for everyone.

OSHA offers many resources online and in written booklets to help you comply with safety regulations. They also offer a free consultation service that can help you avoid expensive fines and increase the safety of your job site.

Explain why several employers can be cited during a single inspection.

Explain the elements that must be present to prove an affirmative defense of a multi-employer work site.

Describe OSHA's voluntary consultation service.

Review the material covered in the module.

Have the participants complete the Review Questions and go over the answers prior to administering the Module Examination.

Administer the Module Examination. Record the results on Craft Training Report Form 200 and submit the results to the Training Program Sponsor.

OSHA INSPECTION PROCEDURES 16.11

Review Questions

1. Only federal OSHA rules apply to construction sites.
   a. True
   b. False

2. A Compliance Safety and Health Officer is an OSHA official who _____.
   a. sets fines for violations
   b. inspects workplaces
   c. corrects violations
   d. writes rules and standards

3. Any of the following can trigger an OSHA inspection *except* a(n) _____.
   a. accident in which someone dies
   b. accident in which only one person is hospitalized
   c. complaint by an employee
   d. previous OSHA inspection

4. The following people are usually part of an inspection *except* the _____.
   a. OSHA representative
   b. employer representative
   c. employee representative
   d. injured worker

5. You should hide violations from the OSHA inspector, if possible.
   a. True
   b. False

6. A(n) _____ is a formal notice from OSHA telling an employer that he or she has allegedly violated an OSHA standard.
   a. penalty
   b. violation
   c. citation
   d. abatement

7. A citation will include all of the following *except* _____.
   a. the proposed penalty
   b. the date by which the hazard must be corrected
   c. how to correct the violation
   d. a description of the violation

8. If you know about a law and break it anyway, the fine can be ten times greater.
   a. True
   b. False

9. To qualify for a focused inspection, a construction site must have a _____ on site.
   a. certified industrial inspector
   b. safety plan and personal protection equipment
   c. written plan and a designated, competent person
   d. qualified person and testing equipment

10. A citation must be posted for three working days or until the violation is corrected, whichever is longer.
   a. True
   b. False

Instructor's Notes:

PROFILE IN SUCCESS

Chad Wilson, CSST, Teton Industrial Group, Inc.
Safety Manager

How did you choose a career in the Safety Industry?
It was a natural progression from the craft area into safety. As an apprentice I was very interested in safety and when I was promoted to a crew leader, I was more than ready to take on the challenge of protecting my crew. From there I was presented the opportunity to manage a craft training program on a large project. After a couple of years in the training field I was offered a safety position with an industrial contractor.

What types of training have you been through?
In addition to the Electrical apprenticeship, I have also taken various OSHA training programs and have opportunity to attend NCCER construction management academies at Clemson University. My training and education in the safety field has been based on the NCCER's Safety programs where I have received the foundation to build a career in safety. I am a certified Construction Site Safety Technician and currently working toward the Construction Site Safety Master Instructor recognition.

What kinds of work have you done in your career?
From laborer to apprentice to training manager to safety manager, I have had a wide varied of experiences in the construction industry. Just when I think that I have done everything a new opportunity pops up. I started in the residential segment of the industry and moved to commercial and now into heavy industrial.

Tell us about your present job.
I am the safety manager of a large power project in the eastern part of the country. The project is called a turn-around due to the short term of the work, in this case about 12 weeks. We move in fast do our work, get the project back up and move on to other projects in the country. The pay is great and it never gets boring.

What factors have contributed most to your success?
A true commitment to ensuring that our employees are protected and go home every night in the same condition that they come to work in. Being able to communicate with the crews, project managers, and the corporate office is key to achieving success. Technical writing capabilities. Understanding that safety is part of the business and that doing the right thing is critical of both the company's and individual's long term success. Last but not least in the continuing self improvement that is required to stay at the top of the profession. To stop learning is to stop growing in the safety field.

What advice would you give to those new to safety industry?
Learn - Learn – Learn! Take every training opportunity you can find. Work with seasoned professional to understand and deal with the nuances of the industry. Work with and not against all parties involved. Everyone on a project can offer insight into dealing with problems. Developing relationships with employees, management, and other contractors, can produce results that everyone can live with.

OSHA INSPECTION PROCEDURES 16.13

Instructor's Notes:

GLOSSARY

Trade Terms Introduced in This Module

Abatement: The correction of the safety hazards or violations resulting from an OSHA citation.

Abatement certification: A written record that shows that the safety hazard or standard violation noted in the citation has been corrected.

Citation: A formal notice from OSHA that the company has allegedly violated a federal safety regulation and may be fined.

Corrective action: Actions the company must take to correct safety violations.

Competent person: A person who is capable of identifying existing and predictable hazards, and who has authority to take prompt corrective measures to eliminate them.

Compliance safety and health officers (CSHOs): OSHA officials who conduct work-site inspections to enforce safety regulations.

Good faith: Good faith means that the employer makes a serious effort to comply with relevant OSHA standards to the best of his or her ability.

Imminent danger: A hazard that could cause death or serious physical harm immediately, or before the danger could be eliminated through normal enforcement procedures.

Penalties: Fines imposed on a company for violating a federal or state safety regulation. They can also include prison time for criminal actions.

Violation: The act of violating federal or state safety regulations by failing to obey a safety standard or failing to keep the workplace free from hazards.

REFERENCES

Additional Resources

This module is intended to present thorough resources for task training. The following reference works are suggested for further study. These are optional materials for continued education rather than for task training.

www.osha.gov

www.asse.org

OSHA Inspections: Preparation and Response, 1997. Rick Kaletsky. New York, NY: McGraw-Hill.

Instructor's Notes:

MODULE 75216-03 — ANSWERS TO REVIEW QUESTIONS

Answer		Section
1.	b	1.0.0
2.	b	2.0.0
3.	b	2.1.0
4.	d	2.2.3
5.	b	2.2.4
6.	c	2.3.0
7.	c	2.3.2
8.	a	2.3.3
9.	c	3.1.0
10.	a	4.1.0

CONTREN™ LEARNING SERIES — USER UPDATES

The NCCER makes every effort to keep these textbooks up-to-date and free of technical errors. We appreciate your help in this process. If you have an idea for improving this textbook, or if you find an error, a typographical mistake, or an inaccuracy in NCCER's Contren™ textbooks, please write us, using this form or a photocopy. Be sure to include the exact module number, page number, a detailed description, and the correction, if applicable. Your input will be brought to the attention of the Technical Review Committee. Thank you for your assistance.

Instructors – If you found that additional materials were necessary in order to teach this module effectively, please let us know so that we may include them in the Equipment/Materials list in the Instructor's Guide.

Write: Curriculum Revision and Development Department
National Center for Construction Education and Research
P.O. Box 141104, Gainesville, FL 32614-1104

Fax: 352-334-0932

E-mail: curriculum@nccer.org

Craft __________ Module Name __________

Copyright Date __________ Module Number __________ Page Number(s) __________

Description __________

(Optional) Correction __________

(Optional) Your Name and Address __________

Module 75217-03

ES&H Data Tracking and Trending

NCCER STANDARDIZED CRAFT TRAINING PROGRAM

The National Center for Construction Education and Research (NCCER) provides a standardized national program of accredited craft training. Key features of the program include instructor certification, competency-based training, and performance testing. The program provides trainees, instructors, and companies with a standard form of recognition through a National Craft Training Registry. The program is described in full in the *Guidelines for Accreditation*, published by the NCCER. For more information on standardized craft training, contact the NCCER by writing us at P.O. Box 141104, Gainesville, FL 32614-1104; calling 352-334-0911; or e-mailing info@nccer.org. More information may be found at our Web site, www.nccer.org.

HOW TO USE THIS ANNOTATED INSTRUCTOR'S GUIDE

Each page presents two sections of information. The larger section displays each page exactly as it appears in the Participant Module. The narrow column ties suggested participant and instructor actions to each page and provides icons (detailed below) to call your attention to material, safety, audiovisual, or testing requirements. The bottom of each page includes space for your notes.

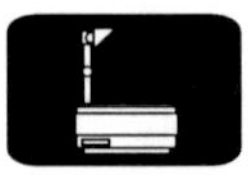

The **Audiovisual** icon indicates an appropriate time to show a transparency or other audiovisual aid.

The **Classroom** icon prompts you to define a term, stress a point, ask participants to explain a concept, or give examples.

The **Demonstration** icon directs you to show participants how to perform tasks.

The **Examination** icon tells you to administer the written module examination.

The **Homework** icon is placed where you may wish to assign reading for the next class, to assign a project, or to advise participants to prepare for an examination.

The **Laboratory** icon is used when participants are to practice performing tasks.

The **Materials** icon is a reminder for you to gather materials needed for classes, labs, and testing.

The **Performance Testing** icon tells you to administer a performance test or a portion thereof.

The **Safety** icon is used to emphasize safety issues. It is often keyed to *Caution* and *Warning* statements in the Participant Module.

The **Teaching Tip** icon indicates additional guidance is available, such as how to conduct an exercise, get the most educational value from a field trip, or encourage class participation. Teaching Tips may expand on a feature (*Think About It, Did You Know?*) or provide Quick Quizzes or similar exercises. You will be referred to the Teaching Tips section at the back of the module if there is additional material.

The **Combination** icon indicates that the laboratory listed corresponds with a performance task. If desired, you can note the proficiency of the participants during the laboratory and use it to satisfy performance testing requirements.

PREPARATION

Before teaching this module, you should review the Objectives, Performance Tasks, Materials and Equipment List, and the Module Outline. Be sure to allow ample time to prepare your own training or lesson plan and gather all required materials and equipment.

MODULE OVERVIEW

In this module, participants will learn how to track and trend ES&H data for the purpose of making workplaces safer.

PREREQUISITES

Please refer to the Course Map in the Participant Module. Prior to training with this module, it is recommended that the participant shall have successfully completed the following:

Field Safety; Safety Technology, Modules 75201-03 through 75216-03

OBJECTIVES

Upon completion of this module, the participant will be able to:

1. List or describe traditional methods of measuring safety performance.
2. List or describe proactive methods of measuring safety performance.
3. Use benchmarks established by the participant's firm, or corporate and industry best practices, analyze the data, and report the program strengths and areas needing improvement to site management.

PERFORMANCE TASKS

There are no performance tasks for this module.

MATERIALS AND EQUIPMENT LIST

Overhead projector and screen

Transparencies

Whiteboard/chalkboard

Markers/chalk

Blank acetate sheets

Transparency pens

Pencils and scratch paper

Appropriate personal protective equipment

Copy of *29 CFR 1926, OSHA Construction Industry Regulations*

Module Examinations*

* Located in the Test Booklet.

SAFETY CONSIDERATIONS

Ensure that participants are equipped with appropriate personal protective equipment.

ADDITIONAL RESOURCES

This module is intended to present thorough resources for task training. The following reference works are suggested for both instructors and motivated participants interested in further study. These are optional materials for continued education rather than for task training.

Web site: www.osha.gov

Web site: www.asse.org

Fundamentals of Industrial Hygiene, 2002. Edited by Barbara A. Plog, MPH, CIH, CSP, and Patricia J. Quinlan, MPH, CIH. Itasca, IL: National Safety Council.

Encyclopedia of Occupational Health and Safety, 1998. International Labor Office. Albany, NY: Boyd Printing.

TEACHING TIME FOR THIS MODULE

An outline for use in developing your lesson plan is presented below. Approximately 2½ hours are suggested to cover *ES&H Data: Tracking and Trending*. This includes 10 minutes at the beginning of the session for administrative tasks and one 10-minute break. You may need to adjust the time required for hands-on activities and testing based on your class size and resources.

Topic	Planned Time
A. Introduction	____________
B. Traditional Methods of Measuring Safety Performance	____________
1. OSHA Recordable Incidence Rates	____________
2. Analyzing Injuries	____________
3. Workers' Compensation Experience Modification Rate (EMR)	____________
C. Proactive Methods for Measuring Safety Performance	____________
1. Behavior Observations	____________
2. Safety Audits and Inspections	____________
3. Root Cause Analysis	____________
D. Analysis of Safety Data	____________
1. Accident Analysis	____________
2. Job Safety Analysis	____________
3. Safety Management	____________
E. Review	____________
F. Module Examination	____________

1. Participants must score 70% or higher to receive recognition from the NCCER.
2. Record the testing results on Craft Training Report Form 200 and submit the results to the Training Program Sponsor.

Module 75217-03

ES&H Data Tracking and Trending

Instructor's Notes:

COURSE MAP

This course map shows all of the modules in Safety Technology. The suggested training order begins at the bottom and proceeds up. The local Training Program Sponsor may adjust the training order.

Assign reading of Module 75217-03.

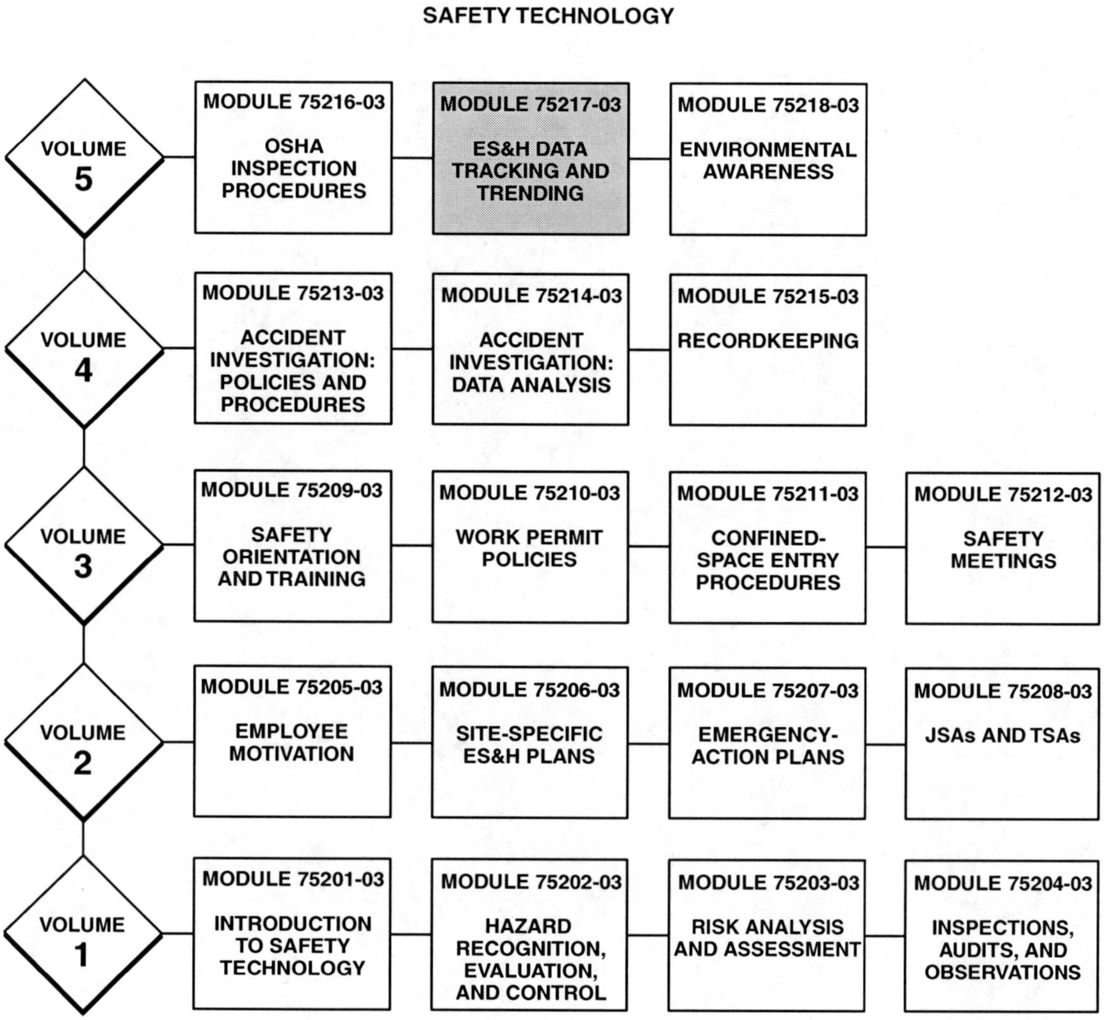

217CMAP.EPS

ES&H DATA TRACKING AND TRENDING 17.iii

Instructor's Notes:

MODULE 75217-03 CONTENTS

Figures

ES&H DATA TRACKING AND TRENDING 17.v

Instructor's Notes:

MODULE 75217-03

ES&H Data Tracking and Trending

Objectives

When you have completed this module, you will be able to do the following:

1. List or describe traditional methods of measuring safety performance.
2. List or describe proactive methods of measuring safety performance.
3. Use benchmarks established by the participant's firm, or corporate and industry best practices, analyze the data, and report the program strengths and areas needing improvement to site management.

Prerequisites

Before you begin this module, it is recommended that you successfully complete the following: Field Safety; Safety Technology, Modules 75201-03 through 75216-03.

Required Materials

1. Pencil and paper
2. Appropriate personal protective equipment
3. A copy of *29 CFR 1926 OSHA Construction Industry Regulations*

1.0.0 ◆ INTRODUCTION

Over the past few decades, health professionals have realized that chemicals and other issues that affect the natural environment will ultimately affect people's health. Safety professionals are dedicated to protecting individuals' health. It is therefore a natural development for traditional safety programs to expand into environmental safety and health programs. Many firms now use the term environmental safety and health (ES&H) to reflect these broader concerns.

Growing problems with safety are reflected in skyrocketing workers' compensation premiums. Poor safety performance can also lead to lost contracts. In order to make the workplace safer, problem areas must be studied. Traditionally, historical studies are used. Tracking accident trends provides clues for reducing future accidents. It's also one way insurance premiums are calculated.

The modern approach focuses on current conditions and behaviors. For example, hazardous work areas are inspected, first-line workers are observed and interviewed to provide current data, and procedural changes and behavior modification are based on analysis of this data. These changes are designed to prevent accidents from ever occurring.

2.0.0 ◆ TRADITIONAL METHODS OF MEASURING SAFETY PERFORMANCE

The effectiveness of a safety program is difficult to measure directly because when a safety program is effective there aren't any accidents. There are indicators, however, that are used to show performance trends. Traditional safety indicators include:

- **Incidence rates**
- Insurance company loss runs and loss ratios
- Workers' compensation **experience modification rates (EMRs)**
- Workers' compensation costs in dollars and cents per man-hour

These indicators provide a good way to study past performance. In addition, they provide an indication of future safety issues but are not always reliable or definitive.

Ensure that you have everything you need to teach the course. Check the Materials and Equipment list in the front of this module.

Explain that terms shown in bold are defined in the Glossary at the back of this module.

Review the goals of the course and explain what will be expected of the participant.

Show Transparency 1, Objectives.

Explain why the term environmental safety and health (ES&H) program replaced the term safety program.

Discuss how the modern approach to ES&H differs from previous approaches.

Identify the traditional methods of measuring safety performance.

Define the term lagging indicators.

Identify the types of recordable incidence rates that OSHA requires companies to collect and record.

Explain that OSHA offers a free computer program to aid in calculating incidence rates.

Explain the Bureau of Labor Statistics (BLS) method for calculating incidence rates. Show the participants how to use the BLS calculation. Use the equation in the Participant Module or create a new scenario.

Describe how incidence rate statistics are used.

DID YOU KNOW?

Deaths and Injuries in the Workplace

In 2001:

- There were 5,300 workplace fatalities due to unintentional injuries.
- There were 3.9 deaths per 100,000 workers.
- On the job, 3.9 million American workers suffered disabling injuries.
- Motor vehicle crashes accounted for 2,200 of the 5,300 workplace fatalities.
- The agriculture industry accounted for 700 deaths and 130,000 disabling injuries.
- Work injuries cost Americans $132.1 billion. That amounts to $970 per worker.

Source: Bureau of Labor Statistics

These methods are all known as **lagging indicators**. The data is analyzed after the mishap has occurred. Much can be gained from studying past outcomes, but there may not always be a direct relationship to future safety performance. Decisions based on past performance may lag well behind current and future problems.

2.1.0 OSHA Recordable Incidence Rates

OSHA requires the collection and recording of certain data. You can evaluate an organization's injury and illness record using this data. Types of recordable incidence rates include:

- Total injury and illness cases
- Lost workday injury and illness cases
- Lost workday injury-only cases
- Lost workday illness-only cases
- Injury and illness cases without lost workdays
- Injury-only cases without lost workdays
- Illness-only cases without lost workdays

These rates can show safety trends within a single organization. You can compare them to standards in your industry or across different industries. This will highlight problem areas. You can use this data to prevent work-related injuries and illnesses.

OSHA offers a free computer program to aid you in calculating incidence rates. The Safety Pays program can be downloaded from the osha.gov Web site. The program calculates incidence rates and other statistical data. You can use this information as one aspect of analyzing your company's safety performance.

2.1.1 Calculating Incidence Rates

Incidence rates are computed using a Bureau of Labor Statistics (BLS) method. The BLS uses a basis of 200,000 man-hours. This is based on 100 employees working fifty 40-hour weeks per year. This standardizes the rate for all industries, regardless of size, so they can be compared directly. Incidence is considered in terms of a rate per 100 workers.

The calculation is as follows:

Step 1 Count the total number of OSHA recordable injuries and illnesses for one year. Do not include fatalities. Include statistics both with and without lost workdays. This information should be available from OSHA logs.

Step 2 Calculate the number of hours actually worked in that year. Use payroll records. Do not include any paid time off.

Step 3 Incidence Rate = (number of injuries and illnesses × 200,000) ÷ (hours worked × number of employees)

For example, XYZ Construction has 77 employees. They each logged 2,000 hours in one year. There were a total of five recordable injuries and illnesses that year. Thus the formula is:

(5 incidents × 200,000) ÷ (2,000 hours × 77 employees)

The incidence rate of total cases for XYZ Construction is 6.5.

The same formula is used to calculate different incidence rates. For example, you might need to know the incidence rate of lost workday injury-only cases. In that instance, you would count only those specific cases in Step 1.

2.1.2 Evaluating Incidence Rate Statistics

Incidence rate data is compared to the previous year's rates. To some personnel, this provides a quick measure of the success of a company's ES&H program. If the incidence rate shows a decline, then ES&H practices can be deemed successful. If there

Instructor's Notes:

is no change, or an increase in rates, then ES&H practices require a closer look.

In spite of the fact that OSHA incidence rates are one of the most widely used measures of safety performance, most safety professionals agree that they are really poor indicators. There is a great deal of pressure on companies to creatively classify injuries to avoid recordability. OSHA has rigid recordkeeping requirements, but they are not always followed to the letter of the law. Incentive programs sometimes unintentionally discourage workers from reporting job-related injuries. Supervisors are reluctant to report injuries to the office for fear of discipline or a poor performance appraisal.

Incidence rates will also be compared across several firms. BLS provides rates for a wide variety of workplace environments (*Figure 1*). Statistics are broken down, both by type of industry and number of employees.

BLS data also shows how rates are distributed through an industry. The BLS table includes data titled first quartile, median, and third quartile. These tell you what proportion of comparable firms have incidence rates lower than yours.

Refer to the example in Section 2.1.1 and the excerpt from the BLS table (*Figure 1*). The BLS reports data by Standard Industrial Code (SIC). The SIC for construction is 154. XYZ Construction had a total incidence rate of 6.5. For a firm with 77 employees, the average incidence rate (from the table) is 8.9. Reading straight across the table, XYZ is not in the first quartile, which would require a rate below 2.4. It is in the median range, below 7.5. Therefore, XYZ's incidence rate is lower than at least half of all similar firms.

2.2.0 Analyzing Injuries

Incidence statistics help to determine historical and cross-industry trends. More specific information is also often available. Reports dealing with the nature of injuries may prove useful. Often these reports deal with the type of injury or part of the body injured. These forms may be available through BLS or state safety boards. You can create them for your firm using a chart or graph (*Figure 2*).

These reports provide indicators of the types of safety measures needed. In the chart, slips, trips, and falls are common injury types. Strains are not as common. That means the safety program would need to address slip, trip, and fall hazards. These could be lessened by using different footwear or employing better housekeeping practices. Analyzing the nature of an injury will often result in identifying personal protective equipment that can help prevent similar injuries.

Nonfatal occupational injury and illness incidence rates of total recordable cases by quarts distribution and employment size group, private industry, 2001

Industry, SIC Code and employment size group	Average incidence rates for all establishments: (Mean)	¼ of the establishments had a rate lower than or equal to: (1st Quartile)	½ of the establishments had a rate lower than or equal to: (Median)	¾ of the establishments had a rate lower than or equal to: (3rd Quartile)
Residential building construction (SIC 152)				
50 – 249	8.5	1.3	4.4	13.5
1,000+	11.5	–	–	–
Operative builders (SIC 153)				
Total all sizes	3.5	.0	.0	.0
1 – 10	2.9	.0	.0	.0
11 – 49	5.5	.0	3.6	9.7
50 – 249	2.8	.0	.0	4.9
250 – 999	3.9	–	–	–
Nonresidential building construction (SIC 154)				
Total all sizes	7.6	.0	.0	8.8
11 – 49	8.0	.0	5.8	14.0
50 – 249	8.9	2.4	7.5	15.3
250 – 999	6.3	1.9	5.0	9.4
1,000+	3.1	–	–	–

217F01.EPS

Figure 1 ◆ Excerpt from BLS incidence rate statistics.

Explain why most safety professionals agree that OSHA incidence rates are not good indicators of safety performance.

Explain how the BLS compares incidence rates across several firms.

Show Transparency 2 (Figure 1).

Explain the different ways the BLS categorizes incidence rates.

Explain why analyzing injuries is a useful technique for determining historical and cross-country trends.

Show Transparency 3 (Figure 2).

ES&H DATA TRACKING AND TRENDING 17.3

Explain why determining historical and cross-country trends can help prevent future accidents and incidents.

Explain how the experience modification rate (EMR) is used by the insurance industry to calculate the premiums firms pay for workers' compensation insurance.

Explain how an EMR is calculated.

Discuss the pros and cons of using EMRs to establish insurance premiums.

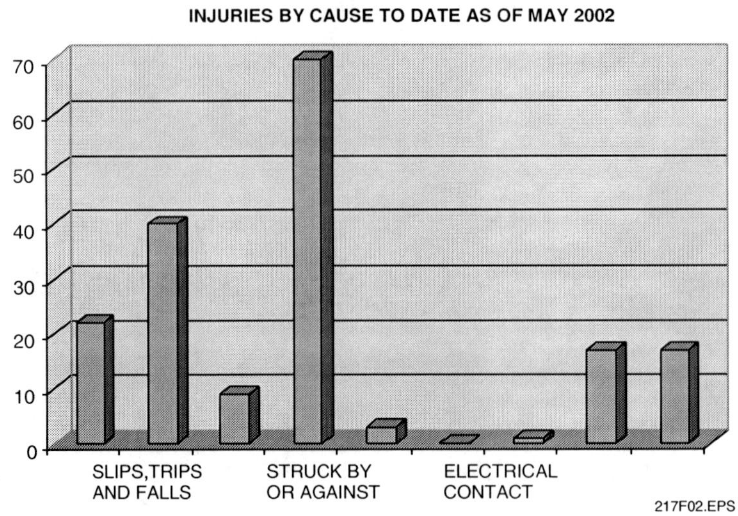

Figure 2 ◆ Excerpt from injury report by cause.

The sample chart was compiled with data from a particular company. Similar tables are also available with other breakdowns. For example, injuries may be examined as a function of a particular craft or of worker age. These tables can span several industries.

Internal safety histories must also include a wide variety of statistics. Incidence reports are not sufficient. Each incident reported also requires as much data as possible. Include age, type, and nature of injury, work accomplished, and environmental conditions. Careful study of past indicators can provide invaluable assistance in preventing future mishaps.

2.3.0 Workers' Compensation Experience Modification Rate (EMR)

A firm's safety record can be analyzed using the EMR. The EMR is an insurance industry indicator. It's used to calculate the premiums firms pay for workers' compensation insurance. The National Council on Compensation Insurance (NCCI) typically provides it.

The EMR compares actual losses, adjusted plus ballast in insurance claims against expected losses plus ballast. A three-year period, not including the current year, is used for calculations. Injuries that lead to insurance claims will increase the EMR. A clean record over a period of time will decrease it.

When safety is compared across similar industries, the average EMR is always 1.0. An EMR higher than 1.0 means a firm has greater than average losses in claims. A better than average safety record results in an EMR less than 1.0.

The EMR is used to calculate workers' compensation premiums. The industry standard rate is multiplied by the firm's EMR. An EMR of 0.7 means a 30% discount on the premium. At 1.25, a firm is paying a 25% surcharge.

In addition to insurance premiums, the EMR is used as a general safety indicator. A decreasing EMR over a period of years points to a successful safety program. The opposite is also true. As the EMR rises, improving safety must become a priority.

EMR provides a measure of safety program success across industries. It is an industry average, which tells you how you compare to similar firms. Many contracts now require that a winning bidder have an EMR below 0.99.

The use of EMR as an indicator of industry safety has become common. However, it is not without problems. The EMR uses data from past years, so it is immediately dated. It does not take fraudulent claims into account. It also does not consider accidents that could not have been prevented. Some companies attempt to keep the EMR low by paying for the smaller medical claims. If the insurance company does not pay a claim, your EMR is not affected. Also, the EMR is skewed in favor of larger contractors. By the nature of the formula, a small contractor with no claims could still have an EMR of 0.85. The EMR is simply one indicator to check. It is not only or primary safety performance measure.

Instructor's Notes:

DID YOU KNOW?

The Most Dangerous Industry

The death rates (per 100,000 workers) for 2001 are:

- Mining 31.8
- Agricultural 21.3
- Construction 13.3
- Utilities 11.4
- Manufacturing 3.3
- Government......................... 2.4
- Retail and wholesale trade .. 1.7
- Services............................. 1.4

Source: National Safety Council, *Injury Facts*

3.0.0 ◆ PROACTIVE METHODS FOR MEASURING SAFETY PERFORMANCE

Modern methods for measuring safety take a predictive approach. Some analysis may be based on past performance, but the data is used differently. Past observations may well lead to future changes. **Proactive indicators** attempt to prevent mishaps before they occur. These include:

- Behavior observations
- Audits
- Root cause analyses

Proactive indicators look at potential causes of safety mishaps. **Hazardous conditions** identify those things or objects that lead to injury. **Unsafe behaviors**, also called at-risk behaviors, reveal those actions taken or not taken that lead to injury. Reducing hazardous conditions and unsafe actions can prevent injury and illness.

3.1.0 Behavior Observations

Behavioral based safety (BBS) is a method of applying psychology to workplace safety. Unsafe behavior causes at least 80% of all workplace accidents. Some estimates go as high as 96%. Improving safety performance can best be achieved by focusing on unsafe behavior.

BBS is best suited for actions directly controlled by an individual. It includes actions such as ducking under a ladder or engaging in horseplay. It also includes failure to act, such as not properly stowing equipment or not wearing a hard hat.

3.1.1 Participation by Workforce and Management

BBS programs require the involvement of everyone in a company. Every level of workforce and management plays a part. The majority of accidents occur at the workforce level. These people are most likely to be hurt. They have the strongest interest in being involved in the elimination of unsafe behavior.

BBS is not implemented as a top-down strategy. Instead, both workforce and managers are seen as team members. Employee feedback is welcomed and utilized, not ignored. All workers are involved in the process.

Management must also be committed to the program. BBS requires that observers watch workers to improve safety performance. Observers must be trained and given sufficient time and materials to complete their tasks. Feedback and positive reinforcement must be provided. BBS is an ongoing program; it cannot work with a half-hearted commitment.

3.1.2 Targeting Unsafe Behaviors

In most industries, there are a few unsafe behaviors that result in many accidents. These behaviors are observable. Studying safety history, such as incidence reports, can help identify these behaviors. Determining the causes of unsafe behavior requires more analysis. For example, unsafe behaviors can be compiled into a checklist (*Figure 3*). The checklist should constantly be revised. Any worker might enter new, updated unsafe behaviors. Conversely, old checklist items can be removed as behavioral hazards are eliminated.

SAFETY PRACTICE	Y	N	COMMENTS (NO NAMES!)
EYES ON PATH OR WORK			
PROPER BODY MECHANICS			
CLEAR LINE OF SIGHT			
PINCH POINTS			
WORK PACE			
GETTING HELP			

217F03.EPS

Figure 3 ◆ Sample safety behavior checklist.

Note that the comments column in the sample includes a note stating that names are not to be used. This is because BBS works when behaviors are observed, not individuals. Punishment is not part of BBS. In fact, it tends to have a negative effect. Instead of encouraging safe behavior, it discourages honest observation. It also builds resentment. Workforce support disappears, and the process becomes meaningless.

Define the term proactive indicator. Identify typical proactive indicators.

Explain how proactive indicators, identified from previous data analysis, are used to prevent accidents before they occur.

Describe how the behavioral based safety (BBS) method is used to apply psychology to workplace safety.

Point out the level of participation that is required by management to make BBS successful.

Explain the BBS techniques used for observing and targeting unsafe behavior.

Explain how to use a safety behavior checklist.

Show Transparency 4 (Figure 3).

ES&H DATA TRACKING AND TRENDING 17.5

Describe the role of the observer in monitoring worker behavior. Explain how observation information is collected and analyzed.

3.1.3 Data Collection and Analysis

The observer's task is to monitor a worker's safety behavior. This must be done on a regular basis. Frequent observation provides more accurate data. It also affects behavior. Workers are more likely to engage in safe behavior when aware that they are being observed. This creates safer habits, even when no observation is taking place.

Decisions can be based on behavioral observation. This requires turning checklist results into measurable qualities. The percentage of safely performed observed behaviors is a common possibility (*Figure 4*). Analyzing trends leads to safety process improvement. The reasons behind unsafe practices often become evident through analysis.

Show Transparency 5 (Figure 4).

Explain how the steps of the BBS pull together to provide a thorough analysis of accidents as well as a process to make corrective actions.

3.1.4 Making the Process Work

Improvement begins to occur after BBS has been put into action. There's an introductory period of about four weeks near the beginning. People adjust to their new roles. Observers are trained. Unsafe behaviors are identified. Checklists are developed. Baseline behavior is established.

After the initial period, goals are set. These are not executive-level goals, but realistic targets set by the workforce. Observation continues, and goals are reexamined regularly. Improvement becomes a continuous process.

Everyone needs to know what's happening for BBS to work. Observers should talk to workers on the spot about their behavior. Larger scale findings can be displayed on posters or discussed in regular meetings. Once at-risk behaviors are identified, input is requested from employees as to possible causes and solutions. When recommendations are implemented, follow-up observations are made to see if there has been a positive change in the percentage of safe observations. As one behavior is corrected, the team can focus on others. This makes it a continuous process.

BBS is not a magic bullet. It is a process that must be managed and adjusted on a regular basis.

Describe the acceptance of the BBS in the construction industry.

3.1.5 Current Use of the BBS in the Construction Industry

Many supporters of the BBS process believe the process works well in an industrial setting. Still, they do not think it will work in the construction industry because of high labor turnover and jobs of short duration. However, one recent research projects found that the use of BBS process in the construction industry is on the rise.

Explain how safety audits are used to reveal safety issues.

3.2.0 Safety Audits and Inspections

Safety issues, especially recurring problems, can be revealed through **safety audits** and inspections. Frequent in-house inspections, coupled with independent audits, provide an excellent means of detecting hazardous trends.

OVERVIEW REPORT

Dates From: 01 Jan 00 To: 01 Apr 00	Total People Observed 575		Total Observations 575	
	SAFES	**CONCERNS**	**% SAFES**	**SAMPLE**
BODY / WORK POSITION				
EYES ON PATH OR WORK	471	28	94%	495
PROPER BODY MECHANICS WHEN LIFTING REACH	203	35	85%	238
CLEAR LINE OF SIGHT	244	32	88%	275
PINCH POINTS	124	14	90%	137
WORK PACE	406	7	98%	413
GETTING HELP	106	5	95%	111
Category Totals:	**1554**	**121**	**92%**	
TOOLS AND EQUIPMENT				
SELECTION AND USE OF EQUIPMENT	120	3	98%	123
FORKLIFT OPERATION	63	16	80%	76

217F04.EPS

Figure 4 ◆ Overview report sample.

Instructor's Notes:

OSHA Fatality Report

A laborer was killed when a gasoline storage tank he was cutting with a portable power saw exploded. The worker's company was involved in installing, removing, and junking gasoline pumps and underground tanks.

Although he had experience working with the saw and scrap materials, the worker did not adequately purge the tank and test for vapors before beginning to cut. The 18' × 6', 3,000 gallon tank had been used recently for underground storage at a service station. At the time of the explosion, the mechanic was cutting on the tank with a gasoline-powered portable saw equipped with an abrasive epoxy disk for cutting metal. The explosion propelled the worker 10' to 15' from the tank into another tank.

The Bottom Line: Adequate training to identify unsafe conditions and adequate oversight to avoid unsafe behaviors might have saved this worker's life.

Source: Occupational Safety and Health Administration (OSHA)

Explain how audits are used to examine workplace behavior.

Explain the safety technician's role as an auditor.

Use the Case History to open a discussion about unsafe acts and conditions the participants have witnessed.

3.2.1 Safety Audits

A safety audit is a systematic examination of workplace behavior and safety practices (*Figure 5*). The auditor checks that policies and procedures have been implemented and are adequate. Auditors should be independent. Here independent can mean a separate agency within the company or an outside firm. Government agencies or insurance carriers can also order independent audits.

An audit will compare current practices with other standards. It will also compare current practices to the past practices within the same company, using historical audit data. The audit will also compare the company to other companies. BLS data can identify standard risks associated with a task across a range of industries.

An audit should include every component of the safety process. This can be done collectively or one component at a time. At a minimum, management, training, and operating procedures should be covered. The short-term goal of an audit is to reveal strengths and weaknesses of the whole program. The long-term goal is to reduce losses through accident prevention activities.

An audit needs to consider historical safety data first. It must verify that problems uncovered by previous audits and inspections have been resolved. Once the audit begins, don't allow the checklist to run the audit. Well-trained auditors keep their eyes open for much more than what is written on the clipboard.

A successful audit examines unsafe behavior and hazardous conditions as well as the policies and procedures. It's not enough to only note physical hazards. Unsafe acts in the workplace must also be considered. An audit must involve more than a cursory examination. Frequently, direct questioning of workers is the best tool for gathering data.

ELEVATED WORK AUDIT UNGUARDED 8' OR ABOVE

PLANT LOCATION ______________

AUDITOR ____________ DATE___/___/___ SHIFT 1 2 3

OBSERVATION	1	2	3	4	5
*** EMPLOYEE USING FALL PROTECT SYSTEM**					
FULL BODY HARNESS USED					
LIFELINE/LANYARD NOT TOO LONG					
LIFELINE/LANYARD HOOKED NEAR EMPLOYEE					
LIFELINE/LANYARD PROPERLY TIED OFF					
LIFELINE/LANYARD SECURED TO ANCHOR POINT					
ANCHOR POINT WILL SUPPORT WEIGHT					
ANCHOR POINT DOES NOT HAVE SHARP EDGES					
RETRACTABLE DEVICE WITHIN CERTIFICATION DATE					
WAS PPP USED TO MOVE TO ELEVATED WORK					
MARK: YES= Y NO= N NOT APPLY= N/A (*) CIRCLE SIGNIFICANT VIOLATIONS= 0	COMMENTS:				

CONFORMANCE=OBSERVATIONS (-) VIOLATIONS (X) 100 / OBSERVATIONS

217F05.TIF

Figure 5 ◆ Sample audit checklist.

ES&H DATA TRACKING AND TRENDING 17.7

Explain how safety inspections are used to identify unsafe conditions and physical hazards. Discuss the use of checklists during safety inspections.

Explain why root cause analysis is a critical factor in measuring safety performance.

Define the terms direct cause, indirect cause, and root cause.

Explain the relationship between the three types of causes. Focus on why discovering and correcting root causes is critical to making the workplace safer.

Show Transparency 6 (Figure 6).

Use the scenario given in the Think About It to lead a discussion on accident causation. Have participants identify the direct, indirect, and root causes of this fatality. Offer accident scenarios from your own experience and discuss the causes and solutions.

The safety technician is an auditor. First-line supervisors should implement safety polices and procedures in their areas. The safety technician should audit the implementation and results. For example, the safety technician may audit compliance with lockout/tagout, confined-space entry, elevated work, or fall-protection procedures.

3.2.2 Safety Inspections

Work supervisors normally carry out **safety inspections**. A safety inspection tends to follow a set checklist format. The checklist may include OSHA violations. It will also likely include unsafe conditions and physical hazards.

The inspection process works hand-in-hand with less frequent audits. An audit may uncover an item requiring improvement. Inspections can then be used to monitor progress toward that improvement.

First-line supervisors are key players in the inspection process. They are familiar with the physical work area and the immediate dangers to workers. An auditor may work closely with the workforce for one part of an audit. The supervisor is there on a daily basis, and is best prepared to recognize old and new hazards.

Middle and upper management also need to play a role. Inspections work only when everyone is behind them. Line managers must work to implement changes suggested through the inspection process. Executive-level decisions need to take safety audit and inspection findings into account.

3.3.0 Root Cause Analysis

This subject is covered in depth in *Hazard Recognition, Evaluation, and Control*, but bears further discussion here because it is a critical factor in measuring safety performance. Initial analysis of a safety incident will show the **direct cause** of the injury. This is the actual action that caused an injury.

Behind every direct cause is an **indirect cause**. Indirect causes include hazardous conditions and unsafe behaviors. Steps can usually be taken immediately to prevent both types of indirect causes.

There is often a deeper cause behind an indirect cause. A **root cause** is an inadequacy in the safety system or process. Identifying a root cause is more difficult. Removing one, however, can improve a safety record faster than removing many indirect causes.

Consider the following example. A worker accidentally splashes caustic soda on his face. The direct cause of the injury is the chemical's reaction with his skin. The indirect cause may be a hazardous condition, like an improperly closed container. It might also be behavioral. Perhaps this worker does not practice safe handling techniques. The root cause might then be a company-wide lack of training on chemical handling.

The relationship between the various causes of an accident can be seen in a weed analogy (*Figure 6*). The direct causes are at the flowering top, impossible not to notice. Indirect causes form the stem and leaves, quickly observable to the trained eye. Root causes run deeper, and require digging to uncover and correct. Correcting an indirect cause without touching a root cause is like mowing a dandelion. The untouched roots will simply spring forth a new problem.

Root causes always precede indirect causes. A poorly designed safety system leads to unsafe conditions and acts. Some include obvious safety-related factors such as training programs, hazard identification, and accountability. They also include procedures, budgets, and processes not typically associated with safety. You need to examine all possibilities to find root causes.

Root causes intertwine throughout an organization. Eliminating a root cause that affects one hazard may reduce other hazards in unexpected ways. The flip side is that there may be more than one root cause tied to an indirect cause. Several roots may have to be eliminated before the unsafe condition is reduced.

THINK ABOUT IT

OSHA Fatality Report

Two remodeling construction employees were building a wall. One of the workers was killed when he was struck by a nail fired from a powder-actuated tool. The tool operator, while attempting to anchor plywood to a 2 × 4 stud, fired the tool. The nail penetrated the stud and the plywood partition prior to striking the victim.

What was the direct cause of this fatal accident?

What are the potential indirect causes?

What are the potential root causes?

Instructor's Notes:

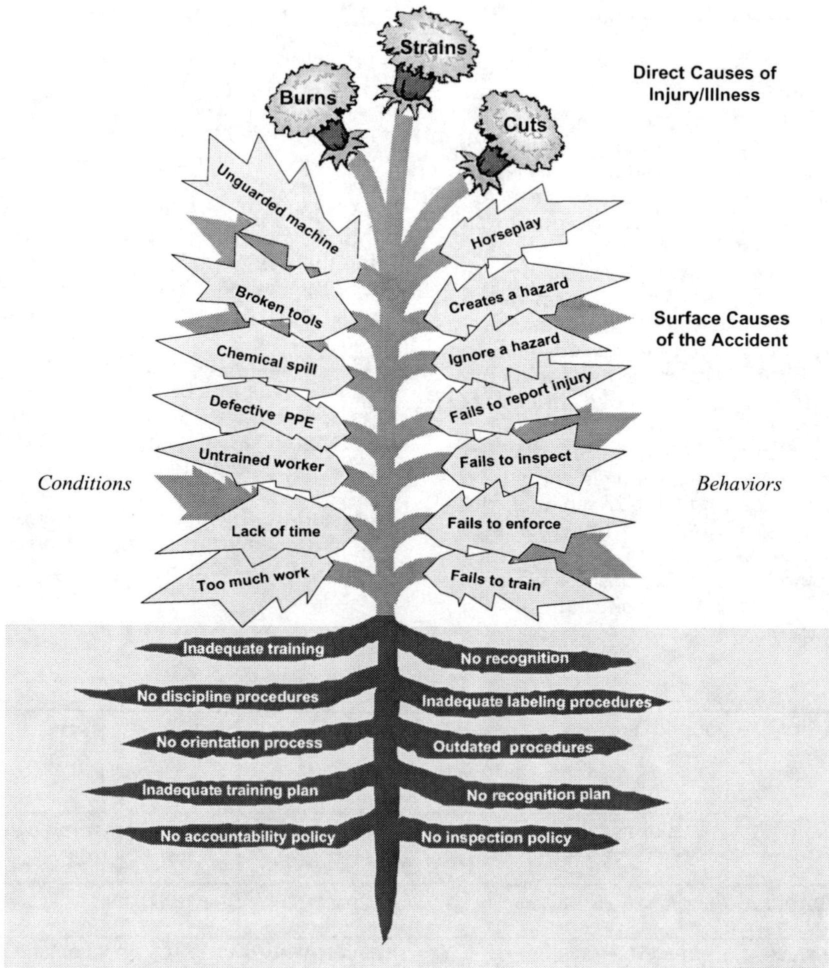

217F06.EPS

Figure 6 ◆ Root and indirect causes of incidents.

All levels of the workforce must participate in root cause analysis. Craft workers are often the best source for information. Interviews may reveal, for example, that unsafe practices are followed because of a perception of company policy.

Because root causes can be policy issues, upper management ultimately must be responsible for implementing change. Feedback and continuous analysis policies must be followed. A change in procedure to get rid of the root cause for one hazard may well create another. Every policy change must be examined in practice.

4.0.0 ◆ ANALYSIS OF SAFETY DATA

There are many ways of gathering and examining safety data. Both historical (lagging) and predictive (proactive) indicators can be used. Data analysis must be done before making procedural changes. Changes are proven effective when injury and illness rates drop.

4.1.0 Accident Analysis

Accident analysis is a reactive examination of the factors that led to an injury or illness. An accident can be caused by a variety of factors, requiring separate but interconnected study. Injury, event, and system analysis are all necessary.

Explain why craft workers may be the best sources of information when trying to determine the root causes of accidents.

Explain why data analysis is an important part of making procedural changes.

ES&H DATA TRACKING AND TRENDING 17.9

Explain why accident analysis is a key factor in overall safety analysis.

Describe the purpose of a job safety analysis. Explain why it is an important tool in overall safety analysis.

Show Transparency 7 (Figure 7).

Injury analysis focuses on the injury itself. The concentration is on how the injury was sustained, not on what caused the accident. These are the direct causes of the injury.

Event analysis looks into the indirect, or surface, causes of an accident. Hazardous conditions and unsafe behaviors are uncovered at this stage. Those conditions and behaviors may be corrected based on this analysis, but it may also point to a deeper problem.

This is where system analysis takes over. It uses the facts and observations gathered from injury and event analyses. Basic, or root, causes are uncovered. Indirect causes are traced back to problems in the system. This is the most challenging stage of analysis, because it looks critically at company policies.

An accident generally occurs as a result of many factors coming together. Studies have shown, on the average, anywhere from 10 to 27 separate contributing factors. Any analysis of workplace injuries must recognize that level of complexity. Only by understanding the relationships between contributing factors can analysis contribute toward better safety practices.

4.2.0 Job Safety Analysis

A **job safety analysis** (JSA), or job hazard analysis (JHA), is a proactive method for studying a job prior to commencement. Specifically, hazards and potential accidents are identified prior to breaking ground. A completed JSA describes a number of ways to avoid accidents (*Figure 7*).

Completing a JSA involves first gathering data about the job. Historical data is analyzed to determine the risks associated with the type of work. All of the key people involved in the job need to be identified. They will each have a say in JSA development.

Hazards are identified based on everyone's feedback. Safe practices in response to those hazards come next. Clear communication is essential here. Do not say, "Use hand tool," when "Use pliers," is more accurate.

The JSA should be open to continuous alteration and improvement. If, for example, the environment changes, then the agreed-upon content may also have to change.

JOB SAFETY ANALYSIS	JOB TITLE: Compactor Operator Page 1 of 1	JSA No.	DATE: 8/23/03	NEW ❑ REVISED ❑
	TITLE OF PERSON WHO DOES JOB: Grounds Crew	SUPERVISOR:	ANALYSIS PERFORMED BY: Troy Tepp, Sentry Insurance	
ORGANIZATION: State Fair Park	LOCATION: Refuse Collection	DEPARTMENT: Grounds	REVIEWED BY:	
SEQUENCE OF BASIC JOB STEPS	POTENTIAL HAZARDS	RECOMMENDED ACTION OR PROCEDURE		
1. Pull/push full dumpster into hydraulic lift gate.	• Lower back, arm, and shoulder strain. • Pinched hand between gate and rolling dumpster. • Hitting head on horizontal gate member.	• Repair/improve condition of lot pavement. • Install larger wheels or properly inflate existing • Push dumpster to move, pull to steer/guide. • Wear protective canvas or leather gloves. • Install plugs to drain water from dumpster. • Use a tugger or other motorized cart to move		
2. Raise hydraulic lift gate to "UP" position, dumping contents.	• None observed.	• None observed.		
3. Spray remaining trash residue from raised dumpster with water.	• Trash particles and water spray into eyes.	• ANSI approved eye or face protection when spraying/cleaning the dumpster.		

217F07.TIF

Figure 7 ◆ Sample JSA excerpt.

Instructor's Notes:

> **NOTE**
> Job safety analysis is covered in more detail in Volume 2, Module 75208-03, *JSAs and TSAs.*

4.3.0 Safety Management

Proactive safety measures require taking action before accidents occur. All methods described require continuous support. A successful safety management program will include:

- Scheduling and performing inspections, audits, and observations
- Tracking inspections, audits, and observations (ensuring that they are performed as scheduled)
- Tracking the status of open action items resulting from accident investigations and safety inspections
- Tracking and trending near-miss and first-aid reports
- Tracking the promptness of accident reporting and investigations
- Verifying that the required JSAs, Task Hazard Analyses, or Pre-Job Safety checklists are being used as intended
- Reporting on the completeness and accuracy of the incident investigation reports and JSAs
- Making full use of employee perception surveys

Summary

Workplace safety continues to grow in importance. There is a corresponding growth in interest regarding accurate tracking and trending of safety- and health-related issues.

The goal of any safety program is to reduce the number of injuries and illnesses. Every method is measured in order to determine the best ways of reducing accident occurrence.

Both historical (lagging) and predictive (proactive) data can be used in a successful safety management program. Only careful analysis of all available data, coupled with continuous dynamic support for safety measures, will reduce workplace injury statistics.

Describe management's role in safety. Explain the elements of an effective safety management program.

Review the material covered in the module.

Have the participants complete the Review Questions and go over the answers prior to administering the Module Examination.

Administer the Module Examination. Record the results on Craft Training Report Form 200 and submit the results to the Training Program Sponsor.

ES&H DATA TRACKING AND TRENDING 17.11

Review Questions

1. Insurance premiums are traditionally calculated using _____.
   a. historical studies of claims
   b. interviews of front-line workers
   c. observations and inspections
   d. safety audits

2. Traditional safety methods that study past performance are known as _____.
   a. hazardous conditions
   b. audits
   c. root cause analysis
   d. lagging indicators

3. XYZ Corp. has 36 employees. Last year, XYZ had a total of four OSHA recordable injury and illness cases. Their incidence rate for the year would be calculated using the following formula:
   a. $(36 \times 2{,}000) \div (200{,}000 \times 4)$
   b. $(4 \times 2{,}000) \div (200{,}000 \times 36)$
   c. $(4 \times 200{,}000) \div (2{,}000 \times 36)$
   d. $(36 \times 200{,}000) \div (2{,}000 \times 4)$

4. When analyzing accident data, comparison by _____ would be of *little* benefit.
   a. age
   b. race
   c. nature of injury
   d. length of employment

5. XYZ Corp. has an EMR of 1.25. That means XYZ has _____.
   a. a surcharge on its insurance premium
   b. a better than average safety record
   c. an OSHA-certified trainer on staff
   d. lower than average insurance claims

6. Proactive indicators attempt to prevent mishaps _____.
   a. through training
   b. after they occur
   c. before they occur
   d. only in the construction industry

7. Behavioral based safety programs only work if punishment is meted out fairly.
   a. True
   b. False

8. Which of the following is a potential root cause of an accident?
   a. The equipment was not locked out.
   b. The employee did not verify the lockout.
   c. The employee used a tag and not a lock.
   d. The company has no lockout/tagout procedure.

9. A JSA includes all of the following fields *except* _____.
   a. organization name
   b. potential hazards
   c. sequence of job steps
   d. injury report

10. A safety management program would include all of the following activities *except* _____.
    a. tracking of non-job related medical claims
    b. tracking the status of open action items resulting from accident investigations
    c. tracking the methods used to report accidents
    d. tracking promptness of accident reporting

Instructor's Notes:

GLOSSARY

Trade Terms Introduced in This Module

Behavioral based safety (BBS): A proactive method of safety management based on psychology. It requires systematic workplace observation and analysis of unsafe behaviors, resolution of problems, and is coupled with training and incentives for behavior modification.

Direct cause: The immediate cause of an injury or illness, not accounting for any underlying unsafe behaviors or conditions.

Experience modification rate (EMR): A lagging indicator of illness and injury rates based on insurance claims and predicted claims over a three-year period. It is most often applied by the insurance industry.

Hazardous conditions: Circumstances or objects that cause injury or illness. Most hazardous conditions arise as a result of unsafe (at-risk) behaviors.

Incidence rate: A lagging indicator of illness and injury rates based on a Bureau of Labor Statistics (BLS) formula. It is measured in annual incidents per 100 workers.

Indirect cause: The underlying cause of an injury or illness. Categories include hazardous conditions and unsafe (at-risk) behaviors.

Job safety analysis (also Job Hazard Analysis): A method for studying a job to identify hazards and potential accidents associated with each step and developing solutions that will eliminate, minimize, and prevent hazards and accidents.

Lagging indicator: A measure of performance based only on historical reporting.

Proactive indicator: A measure of performance based on predictive observations and analysis.

Root cause: The deepest level, system-related cause of an injury or illness. Requires in-depth analysis to discover. Also referred to as a basic cause.

Safety audit: A proactive method of safety management requiring observation and reporting by an unbiased individual.

Safety inspection: A proactive method of safety management requiring observation and reporting by a workforce supervisor.

Unsafe behavior (also At-risk behavior): Action taken or not taken that increases risk of injury or illness.

ES&H DATA TRACKING AND TRENDING 17.13

Additional Resources

This module is intended to present thorough resources for task training. The following reference works are suggested for further study. These are optional materials for continued education rather than for task training.

www.osha.gov

www.asse.org

Fundamentals of Industrial Hygiene, 2002. Edited by Barbara A. Plog, MPH, CIH, CSP, and Patricia J. Quinlan, MPH, CIH. Itasca, IL: National Safety Council.

Encyclopedia of Occupational Health and Safety, 1998. International Labor Office. Albany, NY: Boyd Printing.

Figure Credits

Nudatum Products 217F04

Instructor's Notes:

MODULE 75217-03 — ANSWERS TO REVIEW QUESTIONS

Answer	Section
1. a	1.0.0
2. d	2.0.0
3. c	2.1.1
4. b	2.2.0
5. a	2.3.0
6. c	3.0.0
7. b	3.1.2
8. d	3.3.0
9. d	4.2.0
10. c	4.3.0

CONTREN™ LEARNING SERIES — USER UPDATES

The NCCER makes every effort to keep these textbooks up-to-date and free of technical errors. We appreciate your help in this process. If you have an idea for improving this textbook, or if you find an error, a typographical mistake, or an inaccuracy in NCCER's Contren™ textbooks, please write us, using this form or a photocopy. Be sure to include the exact module number, page number, a detailed description, and the correction, if applicable. Your input will be brought to the attention of the Technical Review Committee. Thank you for your assistance.

Instructors – If you found that additional materials were necessary in order to teach this module effectively, please let us know so that we may include them in the Equipment/Materials list in the Instructor's Guide.

Write: Curriculum Revision and Development Department
National Center for Construction Education and Research
P.O. Box 141104, Gainesville, FL 32614-1104

Fax: 352-334-0932

E-mail: curriculum@nccer.org

Craft ______________________ Module Name ______________________

Copyright Date __________ Module Number __________ Page Number(s) __________

Description

____

____

____

____

(Optional) Correction

____

____

____

(Optional) Your Name and Address

____

____

____

Module 75218-03

Environmental Awareness

NCCER STANDARDIZED CRAFT TRAINING PROGRAM

The National Center for Construction Education and Research (NCCER) provides a standardized national program of accredited craft training. Key features of the program include instructor certification, competency-based training, and performance testing. The program provides trainees, instructors, and companies with a standard form of recognition through a National Craft Training Registry. The program is described in full in the *Guidelines for Accreditation*, published by the NCCER. For more information on standardized craft training, contact the NCCER by writing us at P.O. Box 141104, Gainesville, FL 32614-1104; calling 352-334-0911; or e-mailing info@nccer.org. More information may be found at our Web site, www.nccer.org.

HOW TO USE THIS ANNOTATED INSTRUCTOR'S GUIDE

Each page presents two sections of information. The larger section displays each page exactly as it appears in the Participant Module. The narrow column ties suggested participant and instructor actions to each page and provides icons (detailed below) to call your attention to material, safety, audiovisual, or testing requirements. The bottom of each page includes space for your notes.

The **Audiovisual** icon indicates an appropriate time to show a transparency or other audiovisual aid.

The **Classroom** icon prompts you to define a term, stress a point, ask participants to explain a concept, or give examples.

The **Demonstration** icon directs you to show participants how to perform tasks.

The **Examination** icon tells you to administer the written module examination.

The **Homework** icon is placed where you may wish to assign reading for the next class, to assign a project, or to advise participants to prepare for an examination.

The **Laboratory** icon is used when participants are to practice performing tasks.

The **Materials** icon is a reminder for you to gather materials needed for classes, labs, and testing.

The **Performance Testing** icon tells you to administer a performance test or a portion thereof.

The **Safety** icon is used to emphasize safety issues. It is often keyed to *Caution* and *Warning* statements in the Participant Module.

The **Teaching Tip** icon indicates additional guidance is available, such as how to conduct an exercise, get the most educational value from a field trip, or encourage class participation. Teaching Tips may expand on a feature (*Think About It*, *Did You Know?*) or provide Quick Quizzes or similar exercises. You will be referred to the Teaching Tips section at the back of the module if there is additional material.

The **Combination** icon indicates that the laboratory listed corresponds with a performance task. If desired, you can note the proficiency of the participants during the laboratory and use it to satisfy performance testing requirements.

PREPARATION

Before teaching this module, you should review the Objectives, Performance Tasks, Materials and Equipment List, and the Module Outline. Be sure to allow ample time to prepare your own training or lesson plan and gather all required materials and equipment.

MODULE OVERVIEW

In this module, participants will learn to identify, evaluate, and correct environmental problems that might arise on a typical construction site.

PREREQUISITES

Please refer to the Course Map in the Participant Module. Prior to training with this module, it is recommended that the participant shall have successfully completed the following:

Field Safety; Safety Technology, Modules 75201-03 through 75217-03

OBJECTIVES

Upon completion of this module, the participant will be able to:

1. List or describe at least five types of environmental problems or issues that might arise on a typical construction site.
2. List or describe methods to prevent soil and water contamination when handling fuels and chemicals commonly found or used on construction sites.
3. List or describe ways to minimize the production of hazardous wastes on construction sites.
4. In general terms, explain hazardous waste shipping and manifest requirements.
5. List or describe the training and medical surveillance requirements for personnel who work with lead, asbestos, silica, or hazardous wastes.

PERFORMANCE TASKS

There are no performance tasks for this module.

MATERIALS AND EQUIPMENT LIST

Overhead projector and screen
Transparencies
Whiteboard/chalkboard
Markers/chalk
Blank acetate sheets
Transparency pens
Pencils and scratch paper
Appropriate personal protective equipment
Copy of *29 CFR 1926, OSHA Construction Industry Regulations*
Uniform hazardous waste form
Module Examinations*

* Located in the Test Booklet.

SAFETY CONSIDERATIONS

Ensure that participants are equipped with appropriate personal protective equipment.

ADDITIONAL RESOURCES

This module is intended to present thorough resources for task training. The following reference works are suggested for both instructors and motivated participants interested in further study. These are optional materials for continued education rather than for task training.

Web site: www.osha.gov

Web site: www.epa.gov

Web site: www.asse.org

Hazardous Materials Behavior and Emergency Response Operations, 2000. Denis E. Zeimet, Ph.D., CIH and David N. Ballard. Des Plaines, IL: The American Society of Safety Engineers (ASSE).

Hazardous Materials Management Desk Reference, 2000. Edited by Doye B. Cox, P.E., CHMM. New York, NY: McGraw-Hill.

TEACHING TIME FOR THIS MODULE

An outline for use in developing your lesson plan is presented below. Approximately 2½ hours are suggested to cover *Environmental Awareness*. This includes 10 minutes at the beginning of the session for administrative tasks and one 10-minute break. You may need to adjust the time required for hands-on activities and testing based on your class size and resources.

Topic	Planned Time
A. Introduction	______
B. Chemical-Specific Programs	______
1. Asbestos	______
2. Lead	______
3. PCB	______
C. General Environmental Laws	______
1. Clean Water Act	______
2. Federal Insecticide, Fungicide, and Rodenticide Act	______
3. Comprehensive Environmental Response, Compensation, and Liability Act	______
D. Chemical Management	______
1. Chemical Handling	______
2. Pollution Prevention	______
E. Resource Conservation and Recovery Act	______
1. Identifying Hazardous Wastes	______
2. Shipping Hazardous Waste	______
F. Medical Monitoring Programs	______
1. Medical Testing	______
2. Other Requirements	______
G. Training Requirements	______
H. Review	______
I. Module Examination	______
1. Participants must score 70% or higher to receive recognition from the NCCER.	
2. Record the testing results on Craft Training Report Form 200 and submit the results to the Training Program Sponsor.	

Module 75218-03

Environmental Awareness

Instructor's Notes:

COURSE MAP

This course map shows all of the modules in Safety Technology. The suggested training order begins at the bottom and proceeds up. The local Training Program Sponsor may adjust the training order.

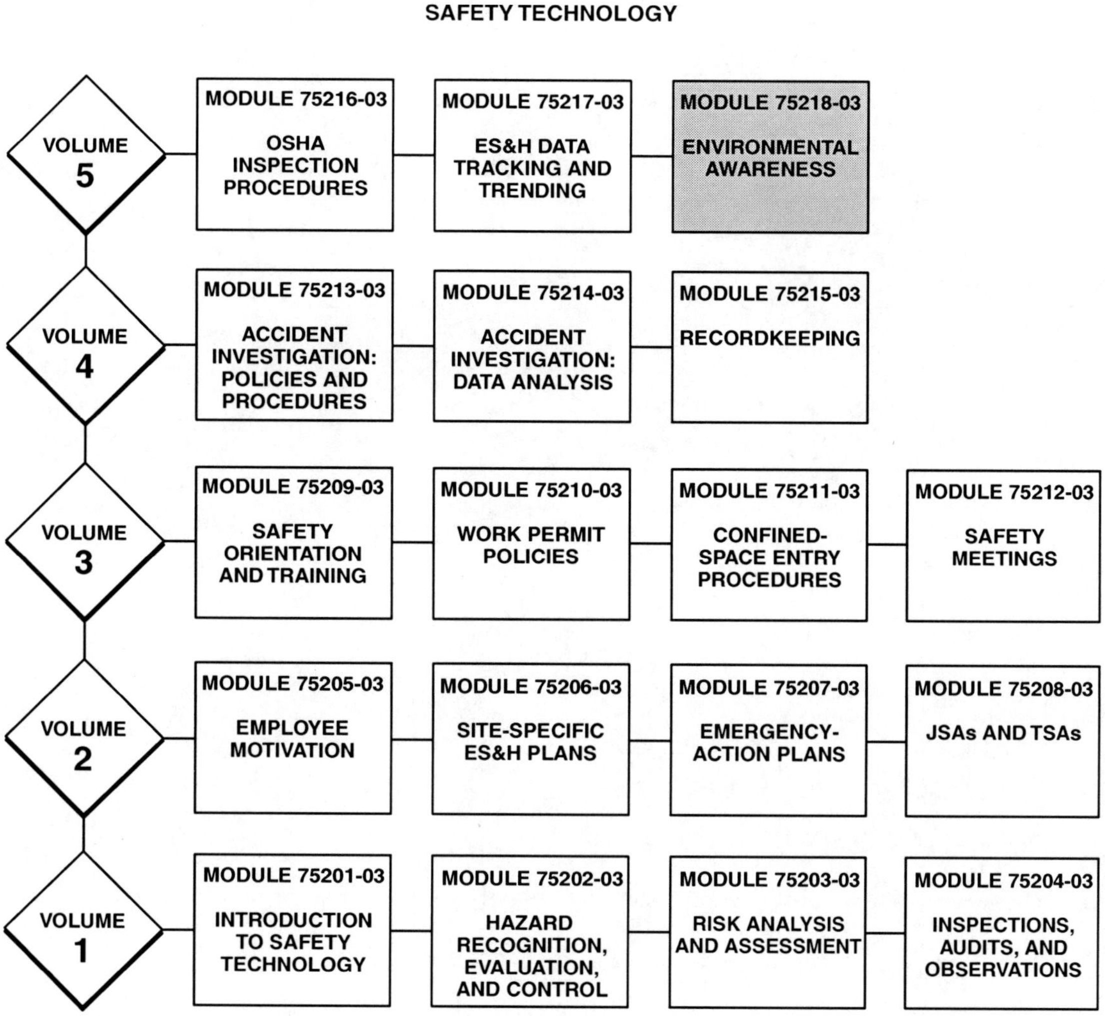

218CMAP.EPS

Assign reading of Module 75218-03.

Instructor's Notes:

MODULE 75218-03 CONTENTS

ENVIRONMENTAL AWARENESS 18.v

Figures

Tables

Instructor's Notes:

MODULE 75218-03

Environmental Awareness

Objectives

When you have completed this module, you will be able to do the following:

1. List or describe at least five types of environmental problems or issues that might arise on a typical construction site.
2. List or describe methods to prevent soil and water contamination when handling fuels and chemicals commonly found or used on construction sites.
3. List or describe ways to minimize the production of hazardous wastes on construction sites.
4. In general terms, explain hazardous waste shipping and manifest requirements.
5. List or describe the training and medical surveillance requirements for personnel who work with lead, asbestos, silica, or hazardous wastes.

Prerequisites

Before you begin this module, it is recommended that you successfully complete the following: Field Safety; Safety Technology, Modules 75201-03 through 75217-03.

Required Materials

1. Pencil and paper
2. Appropriate personal protective equipment
3. A copy of *29 CFR 1926 OSHA Construction Industry Regulations*

1.0.0 ◆ INTRODUCTION

Since the first environmental laws were passed in 1970, many programs have been created to protect the environment. Communities and governments are paying more attention to pollution in our air, soils, and water. You must be aware of how these programs can affect your job site. Some of the major environmental issues that affect construction trades include:

- **Asbestos** abatement
- Lead abatement
- **Polychlorinated biphenyls (PCB)** removal
- Chemical spills
- Water pollution
- **Hazardous waste** disposal

Poor handling of these materials can bring legal and financial problems. Soil and water pollution from chemical spills will result in fines, legal actions, and project delays. Waste containing asbestos, PCB, or other chemicals cannot be combined with other debris. These wastes must be properly marked and taken to an approved facility.

Many general construction firms prefer not to deal with these matters. They hire subcontractors who handle all of the details. For example, if an underground chemical storage tank was found while excavating, many contractors would hire an environmental firm to remove it, test soils, and file the proper reports with the government.

Construction company estimators and safety personnel must be familiar with these environmental issues. Pre-bid planning will help identify possible pitfalls. A good plan will help avoid costly mistakes. If you do handle these materials, your personnel will need additional training and a medical monitoring program.

2.0.0 ◆ CHEMICAL-SPECIFIC PROGRAMS

Some environmental programs were created to deal with just one chemical. The most common are asbestos, lead, and PCBs. Asbestos and PCBs were considered so harmful that they were banned for most uses. Any current use of them is

Ensure that you have everything you need to teach the course. Check the Materials and Equipment list in the front of this module.

Explain that terms shown in bold are defined in the Glossary at the back of this module.

Review the goals of the course and explain what will be expected of the participant.

Show Transparency 1, Objectives.

Discuss the environmental issues that affect construction trades.

Explain why knowing the rules and regulations for proper handling and disposal of hazardous materials can save a company from legal and financial problems.

Discuss the importance of including environmental issues in pre-bid planning.

Explain why some environmental programs were created to deal with just one chemical.

Explain how asbestos was used in construction. Refer the participants to Appendix A in the Participant Module.

Explain the two federal laws that regulate asbestos.

Describe the types of asbestos.

Explain the health effects of asbestos.

Explain when and how asbestos was determined to be hazardous.

Describe the diseases asbestos causes.

Identify the types of existing materials that may contain asbesots. Refer the participants to Figure 1.

Discuss the requirements and regulations for new products made with asbestos.

very restricted. Unfortunately, these compounds were once widely used in construction. You may encounter them during the renovation or demolition of older buildings.

2.1.0 Asbestos

Asbestos removal is a very common environmental issue in construction. It was once used in many building materials and insulation. *Appendix A* lists products that may contain asbestos. You can create a health hazard if you break **asbestos containing materials (ACM)** during demolition or renovation.

> **WARNING!**
> Asbestos containing materials discovered during work must be immediately reported to a supervisor, safety technician, or other qualified individual. Craftworkers must be warned not to handle any materials they suspect may contain asbestos without proper training and equipment.

Two federal laws regulate asbestos. The general program is the Asbestos Hazard Emergency Response Act (AHERA). Asbestos exposure is more dangerous for children than adults. An additional law was created to deal with asbestos removal in school buildings: the Asbestos School Hazard Abatement Act (ASHAA).

2.1.1 Types of Asbestos

Asbestos is a type of mineral. Asbestos crystals form long, thin fibers. It is a good insulator and has been used in building materials as a fire retardant for many years. There are six types of asbestos:

- Chrysolite (white asbestos)
- Amosite (brown asbestos)
- Crocidolite (blue asbestos)
- Tremolite
- Actinolite
- Anthophyllite

Chrysolite makes up 90% to 95% of all asbestos found in buildings in the United States. The second most common type is Amosite. Crocidolite is only found in very high-temperature applications. The other types are rare.

2.1.2 Health Effects of Asbestos

Asbestos is hazardous when it comes into contact with your lungs by breathing dust. Asbestos that can crumble is known as friable asbestos. When asbestos material is broken, dust is created. The dust contains tiny asbestos fibers. The long, thin fibers are easily airborne. If you breathe this dust without respiratory protection, the fibers get into your lungs and can cause damage.

The known health effects are based on very large exposures to asbestos. In the 1940s, insulators would spray the interiors of ships and tanks with asbestos. The air was thick with asbestos dust. The workers did not have any respiratory protection. They inhaled a lot of asbestos fibers into their lungs. Many later died from **asbestosis**. There is not much information on the effects of low-level exposure.

Asbestos is a known carcinogen. A small exposure to a carcinogen can trigger cancer many years later. Because of this, and the well-documented asbestosis deaths, asbestos handling and removal is tightly regulated.

The primary diseases caused by asbestos exposure are asbestosis and cancer. Asbestosis is a scarring of the lung tissue. It develops slowly over time. The effects may not be seen for 15 to 30 years. At that point it is usually fatal. Asbestos exposure can also cause lung cancer and mesothelioma, another type of cancer. However, both have latency periods of over 20 years. Smoking increases the risk of disease. Tar from cigarettes sticks to the asbestos fibers in a smoker's lungs. This makes it more difficult for your body to get rid of it by natural processes.

2.1.3 Identifying Asbestos Projects

Asbestos or ACM (*Figure 1*) can be found in most buildings built before 1980. The U.S. government banned production of most asbestos products in the 1970s. Installation of building materials containing asbestos continued into the early 1980s. Asbestos can still be found in existing buildings. Existing materials that may contain asbestos include:

- HVAC duct insulation
- Boiler and pipe insulation
- Acoustical ceiling tiles
- Asphalt or vinyl floor tiles
- Exterior siding

New products made with asbestos must be clearly labeled. Building owners must identify existing asbestos and ACM through an asbestos survey. The owner must give this information to employees and contractors working in the building. Contractors must find out if there is asbestos or ACM in the pre-bid job analysis. If the building owner has not done an asbestos survey, you should request a walk-through of the area by someone trained in asbestos identification.

Instructor's Notes:

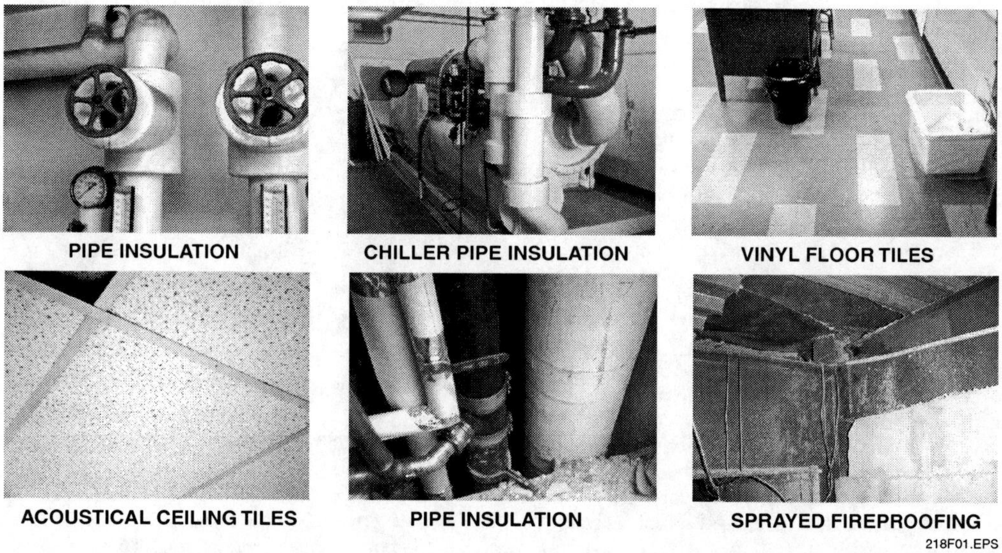

Figure 1 ◆ Typical ACM found in buildings.

Explain OSHA's requirements/standards for the removal of asbestos. Refer the participants to Figure 2.

Show Transparency 2, Asbestos Removal Requirements.

Show Transparency 3 (Figure 3).

2.1.4 Asbestos Removal Projects

An asbestos removal project must follow strict rules for work-zone safety. The OSHA Construction Standard for asbestos is *29 CFR 1926.1101.* It covers all phases of asbestos handling and removal. These include: demolition, removal, alteration, repair, maintenance, installation, cleanup, transportation, disposal, and storage.

Some of the requirements include:

- A control zone must be created as a barrier to contain dust.
- Only trained workers may enter the control zone.
- The air in the control zone must be tested.
- Workers in the control zone must wear respiratory protection *(Figure 2).*
- A decontamination zone must be used when leaving the control zone.
- Decontamination procedures must be used.
- Access to the area must be limited.
- Warning signs must be posted *(Figure 3).*

Figure 2 ◆ Typical control zone for an asbestos removal project.

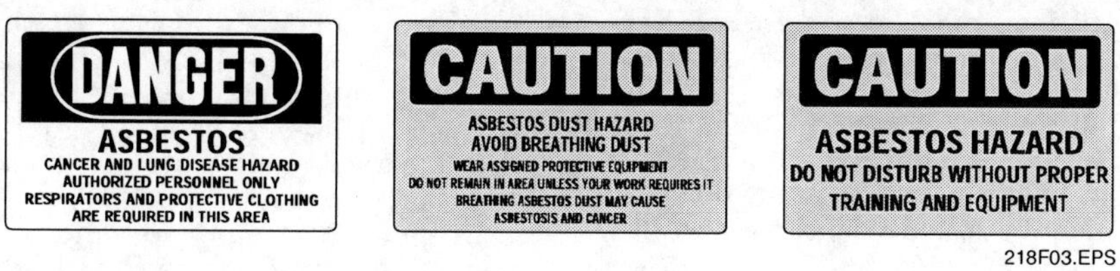

Figure 3 ◆ Asbestos warning signs and labels.

ENVIRONMENTAL AWARENESS 18.3

Describe how lead is used in construction.

Explain its properties and why it is a hazardous substance.

Describe the physical effects of human exposure to lead.

Explain federal legislation regarding the use of lead in workplaces. Refer the participants to Figure 4.

Explain how PCB chemicals are used and why they were banned.

Define the term dieletrics.

Explain in detail the health effects PCBs have on the human body.

Strict procedures must be followed when removing asbestos or ACM. All asbestos waste must be bagged in heavy plastic. It must be tagged with specific hazard labels. The waste must be transported and disposed of by a certified firm. Because of the complex procedures, many general construction companies sub-contract with a certified asbestos abatement firm to remove asbestos.

2.2.0 Lead

Lead is a common metal that is often used for pipes. Lead was also used in the past as an additive to paints, gasoline, and other chemicals. Lead-based paints were banned in 1978. Lead has also been phased out of gasoline. Lead pipes are no longer used for drinking water.

Lead is a toxic metal that can cause serious health problems, especially in children. Exposure to lead in young children can damage the brain and kidneys, impair hearing, and cause learning and behavioral problems. It can also cause less severe problems including vomiting, headaches, and appetite loss. Adults are also affected by lead exposure. It can lead to kidney damage, muscle and joint pain, high blood pressure, nerve disorders, and mood changes.

People are exposed to lead in several ways. Very small children have been exposed to lead by eating paint chips that contain lead. Small lead dust particles can be inhaled or swallowed. Lead can seep into water from lead pipes, lead solder on copper pipes, and brass faucets. Once lead is swallowed, it is carried by the bloodstream and will damage your brain, kidneys, and other organs.

The Federal Residential Lead-Based Paint Reduction Act was enacted to inform homebuyers of possible lead hazards. Sellers and landlords are required to give a lead paint disclosure form to the buyer or renter. In some states, banks will not approve loans for rental units unless lead paints are removed or covered *(Figure 4)*. Due to the risk associated with lead, some people want lead pipes and paint out of their homes.

You may run into lead-based paint during demolition or renovation of buildings built before 1978. Dust created from sanding lead-based paints is hazardous. An approved dust mask must be worn to protect your lungs. Any wastes contaminated with lead may be considered hazardous waste. This includes wood or metal painted with a lead-based paint. Lead piping and solder is usually not considered a hazardous waste. The waste will be regulated as a hazardous waste if it fails a specific test. Any waste that may contain lead must be tested. Handling and disposal of hazardous waste is detailed in the following sections.

218F04.EPS

Figure 4 ◆ Lead paint abatement project.

> **NOTE**
> Fines for improper disposal of hazardous waste start at $25,000. Your company could become a party to a **Superfund** lawsuit if you dispose of hazardous waste improperly.

2.3.0 PCB

Polychlorinated biphenyls (PCBs) are a group of chemicals manufactured by Monsanto from the 1930s to late the 1970s. PCBs are also known by their trade names, for example Aroclor 1260. More than 1.5 billion pounds of PCBs were manufactured in North America.

PCBs were used in many industrial materials, including sealing and caulking compounds, inks, pesticides, and paint additives. Because of their heat resistance and electrical characteristics, PCBs were used as an additive in hydraulic fluids. They were also used to make coolants, lubricants, and **dielectrics** for certain kinds of electrical equipment, including transformers and capacitors. They are now banned for most uses, but more than a million tons remain in use in electrical transformers and **light ballasts**.

2.3.1 Health Effects of PCB

PCBs are not as toxic as many other chemicals. They are persistent, which means they do not break down or degrade easily. Due to widespread use and persistence, PCBs can be found in almost every living organism on Earth, including humans. PCBs were banned because they do not degrade easily and would eventually reach toxic levels in humans.

In the 1970s, many soil and water samples were tested. PCBs were found in many areas, including many rivers and game fish. PCBs are

Instructor's Notes:

easily transported through groundwater, ocean currents, or by evaporation. If swallowed, PCBs are stored in the body's fat tissue. PCBs **bioaccumulate** in the food chain. This means that when one animal eats another, it also eats the PCB. As larger animals eat many smaller animals, the PCBs become concentrated. Because of this, many areas have warnings against eating too many fish from PCB-contaminated waters.

PCBs may cause damage to the immune, nervous, and reproductive systems. Workplace exposure to high levels of PCB over a long time may increase a worker's chance of getting cancer, especially cancer of the liver and kidneys. Some of the short-term health effects are known because people were exposed to high levels of PCB due to accidents on the job site. A concentrated exposure may cause the following conditions:

- A severe form of acne called chloracne
- Swelling of the upper eyelids
- Numbness in the arms and/or legs
- Discoloring of the nails and skin
- Muscle spasms
- Chronic bronchitis
- Problems with the nervous system

2.3.2 PCB Regulations

PCB use was very widespread. When it was banned, it was impossible to remove all PCBs at the same time. PCBs were allowed to remain in some existing electrical equipment. Construction workers may encounter PCBs when decommissioning large electrical equipment or during demolition or renovation of buildings built before 1978.

The U.S. Environmental Protection Agency (EPA) under the Toxic Substances Control Act (TSCA) regulates the handling, transportation, storage, and disposal of highly concentrated PCBs. Many restrictions apply. The waste can only be accepted at approved facilities. Some states regulate smaller concentrations of PCBs, such as those found in light ballasts. You must find out if your project has PCBs and how your state regulates them.

2.3.3 Identifying PCB Transformers

PCBs were used in electrical transformers as a fire retardant from 1929 through 1977. The majority of these PCB transformers were installed in apartments, residential and commercial buildings, industrial facilities, campuses, and shopping centers built before 1978. *Figure 5* shows a transformer that contains PCB oils.

A transformer must have a nameplate attached to one side with the trade name of the dielectric fluid,

218F05.EPS

Figure 5 ◆ A transformer that contains PCB oils.

the approximate weight in pounds, and the amount of fluid (usually in gallons). Since PCBs were marketed under different trade names, the nameplate on a PCB transformer may not carry the specific term "PCB." Trade names for PCB could include:

- Abestol, Aroclor, Askarel, Chlophen
- Chlorextol, DK, EEC-18, Fenclor
- Inerteen, Kennechlor, No-Flamol, Phenoclor
- Pyralene, Pyranol, Saf-T-Kuhl, Solvol
- Non-Flammable Liquid

If the nameplate says "PCB" or any of the names on this list, then the transformer contains PCB. If the nameplate does not carry any of these labels, or if the label is missing or illegible, the utility company may be able to tell you if the transformer contains PCB. Otherwise, the only way to be certain is to have the electrical fluid tested by qualified personnel.

2.3.4 Identifying PCB Light Ballasts

Fluorescent light ballasts installed before to 1978 may also contain PCB. A small capacitor inside the ballast probably contains PCB. If it does not contain PCB, it must be labeled "no PCB." It may be marked with the percentage of PCB in the fluid.

If the label is marked >50 ppm (more than 50 parts per million), or the if ballast is damaged or leaking, then it is regulated under TSCA. If it is not leaking or is <50 ppm (less than 50 parts per million) it is not regulated under TSCA. *Table 1* shows the federal disposal requirements. You must also check state disposal regulations, which may be more restrictive.

Explain the role of the EPA in regulating the use of PCBs.

Describe how to identify PCB transformers. Refer the participants to Figure 5.

Identify the trade names for PCBs.

Describe the characteristics of PCB light ballasts, including labeling requirements.

Show Transparency 4 (Table 1).

Explain the treatment process for unmarked ballasts.

Explain that in addition to laws for specific chemicals, the government has established several general environmental laws. Refer the participants to Appendix B in the Participant Module.

Identify the environmental areas that are impacted by the Clean Water Act (CWA). Explain how these regulations affect the construction industry.

Explain how the storm water run-off provisions in the CWA affect the construction industry.

Table 1 TSCA Disposal Requirements for Fluorescent Light Ballasts

PCB CAPACITOR	PCB FLUIDS	LABELING, TRANSPORTING AND MANIFESTING FOR DISPOSAL	DISPOSAL OPTIONS
"No PCB" label	Does not contain PCB	Not regulated under TSCA	Not regulated under TSCA. State laws may apply.
Intact and non-leaking	<50 ppm	No labeling or manifesting required.	Municipal solid waste under TSCA. State laws may apply.
Intact and non-leaking	>50 ppm	Is a PCB bulk product waste. No labeling is required. Manifesting may be required.	TCSA incinerator, TCSA/RCRA incinerator, secure landfill or alternate destruction method, risk-based approval.
Leaking	Either <50 ppm or >50 ppm	Is a PCB bulk product waste. No labeling is required. Manifesting may be required.	TCSA incinerator, TCSA/RCRA incinerator, secure landfill or alternate destruction method, risk-based approval.

218T01.EPS

Any unmarked ballasts must be treated as if they contained PCB. Testing each ballast is not practical or recommended. Testing for PCB requires destroying the ballast and removing the fluid, creating a waste that is more dangerous. The cost of testing usually outweighs the cost of proper disposal. Improper disposal can lead to very large fines and lawsuits. Federal laws may not regulate disposal of light ballasts, but many state laws do. You need to find out the local laws and dispose of PCB items properly. You can find information on federal PCB regulations online at www.epa.gov. Contact your state Department of Environmental Conservation to see how PCBs are regulated in your state.

3.0.0 ◆ GENERAL ENVIRONMENTAL LAWS

In addition to laws that deal with specific chemicals, there are many general environmental laws to control pollution. These include: the Clean Water Act (CWA); the Federal Insecticide, Fungicide, and Rodenticide Act (FIFRA); and the Comprehensive Environmental Response, Compensation, and Liability Act (CERCLA). A list of environmental laws and other common acronyms are listed in *Appendix B*.

3.1.0 Clean Water Act

The Clean Water Act (CWA) is a federal law that protects all surface waters. The program includes: lakes, rivers, estuaries, oceans, and **wetlands**. The five major sections are:

- A permit program
- National effluent standards
- Water quality standards
- Oil and hazardous substances
- Storm water discharges

The CWA sets certain water quality standards and prohibits pollution. Any discharge into a body of water must be permitted. These permits are known as NPDES permits, for the National Pollutant Discharge Elimination System.

The first two sections of the CWA apply to companies that have a fixed-pipe discharge into a body of water. A fixed-pipe discharge is known as a point source and must have a permit. The CWA also protects wetlands. Excavation or fill in wetlands or swamp areas is controlled by the CWA. Permits must be obtained for these activities.

3.1.1 Storm Water Management

The CWA also covers **storm water run-off**. Storm water that carries any pollutants from a construction site to a body of water requires a permit. This includes suspended particles, even clean soil, that is mixed in the storm water. There are national and state programs for storm water management. Construction firms must control storm water runoff in accordance with local laws.

Construction sites that disturb more than five acres of land must obtain a NPDES permit. As of 2003, construction sites that disturb one to five acres also need a permit. The permit requires that the site prepare a storm water pollution prevention plan. The plan must include best management practices. These practices include erosion and sediment control and site management or housekeeping practices. The first part requires you to prevent storm water, which picks up dirt, from reaching

Instructor's Notes:

storm drains or water bodies. The second part requires you to prevent possible contaminants, like oil and chemicals, from getting in the dirt.

Erosion and sediment controls include:

- Silt fences *(Figure 6A)* and earthen berms
- **Swales** and collection basins
- Storm drain protection *(Figure 6B)*
- Stabilized site entrances

Housekeeping and site management concerns may include:

- Truck washout areas
- Waste collection areas
- Temporary fueling areas
- On-site storage of hazardous substances

You need to be aware of your local storm water rules. If you are excavating more than one acre of land you may need a NPDES permit. The permit will require a pollution prevention program.

(A)

(B)

218F06.EPS

Figure 6 ◆ Tools for control of storm water run-off on a construction site.

3.2.0 Federal Insecticide, Fungicide, and Rodenticide Act

The Federal Insecticide, Fungicide, and Rodenticide Act (FIFRA) covers the use of pesticides in the United States. The EPA is responsible for controlling the use of pesticides. They make sure that pesticides are used properly and do not pose a threat to humans or wildlife.

All pesticides must be registered with the EPA by the manufacturer. The EPA can limit the use of the pesticide to protect health and the environment. Pesticides must be marked and used only as approved. The person applying pesticides must be licensed. You need to make sure that only licensed professionals use pesticides on your site.

3.3.0 Comprehensive Environmental Response, Compensation, and Liability Act

There are hundreds of areas across the United States where chemicals have been dumped, spilled, or leaked out of storage. The Comprehensive Environmental Response, Compensation, and Liability Act (CERCLA), commonly known as the Superfund, was created in 1980 to clean up these areas.

The government can force private parties to clean up these areas. In many cases, the original property owner no longer exists or cannot afford the clean up. They can find other firms who may have added to the pollution, known as **potentially responsible parties (PRP)**. The key is that any company, no matter how much or little they are at fault, can be held responsible for the entire cost of the cleanup. In most cases, the EPA and the PRPs litigate the cleanup and costs.

The very high cost of Superfund actions is a big incentive to make sure that you do not become a PRP. Your firm could become a PRP if you do not properly dispose of hazardous materials or waste. You need to make sure that your waste is not hazardous, or that any hazardous waste is disposed of properly.

In addition to Superfund actions, CERCLA requires that all spills must be reported. If you spill or intentionally pour out a hazardous substance into the water or soil, you must file a report with the National Response Center. The minimum amount of a chemical that must be spilled before a report is required is called a **reportable quantity (RQ)**.

4.0.0 ◆ CHEMICAL MANAGEMENT

Proper chemical handling is the best defense against chemical spills, exposures, and accidents. A good chemical management program includes

Explain how erosion and sediment controls are implemented. Refer the participants to Figures 6(A) and 6(B).

Identify housekeeping concerns related to storm-water rules on a site.

Identify the act that regulates pesticides in the United States. Explain the role of the EPA in regulating the use of these pesticides.

Explain the purpose and requirements of the Superfund.

ENVIRONMENTAL AWARENESS 18.7

Describe the basics of chemical management.

Identify how to properly store and handle chemicals, including the general rules for chemical management.

Explain the concept of secondary containment. Refer the participants to Figure 7.

Explain the importance of pollution prevention in relation to chemical management.

Identify and explain the act (RCRA) that was created to handle all aspects of hazardous waste management.

Explain how these regulations affect the construction industry.

information on safety, MSDSs, personal protective equipment, storage, spills, and disposal. The program should cover all products that contain hazardous chemicals.

4.1.0 Chemical Handling

Chemicals should be stored safely. This includes keeping them in approved containers. All chemical containers should be properly labeled. If the chemicals must be transferred to another container, it should also be an approved container. The new container should be labeled with any warnings. Storing chemicals in temporary containers like coffee cans, creates a hazard and may lead to improper disposal.

Chemicals should be stored in a central location, if possible. This area should be marked with hazard warnings. Spill prevention and containment tools must be available. Some general rules for good chemical management are:

- Keep hazards separate. For example, keep flammables and caustics in separate cabinets.
- Keep container away from weather or heat extremes.
- Protect containers from accidental puncture.
- Have spill-cleanup materials at the storage area.
- Post appropriate warning signs.

Secondary containment *(Figure 7)* is an additional barrier between the chemical and the soil. This means that if a chemical is spilled, it falls into a basin or container. It is much easier to clean the basin than the soil. Any areas where chemicals are transferred should have secondary containment. This includes temporary fueling areas.

When chemicals are spilled, all cleanup materials may be hazardous waste. This includes rags, absorbents, and broken containers. It can also include soil onto which the chemical is spilled. Spill prevention is usually cheaper than hazardous waste disposal.

4.2.0 Pollution Prevention

Pollution prevention and waste minimization are parts of a best practices plan for chemical management. One way to reduce possible pollution is to use less toxic materials. There are many nontoxic solvents available. Good housekeeping will also minimize waste. Separate chemical or hazardous wastes from general construction debris. Remember, even a small amount of a hazardous substance in a container of waste makes the whole container a hazardous waste.

218F07.EPS

Figure 7 ◆ Chemical storage area and secondary containment.

5.0.0 ◆ RESOURCE CONSERVATION AND RECOVERY ACT

In 1976, the Resource Conservation and Recovery Act (RCRA) was created to handle all aspects of hazardous waste management. Hazardous wastes are tracked from when they are created to when they are finally destroyed. RCRA covers hazardous waste **generators**, transporters, and treatment, storage, and disposal facilities.

All firms that create waste must determine if the waste is hazardous. Firms that create hazardous waste are known as generators. Firms that only produce a small amount of hazardous waste are allowed some exemptions from the rules. They are known as conditionally exempt small quantity generators and small quantity generators. Conditionally exempt small quantity generators (CESQGs) create less than 220 pounds of hazardous waste per month. Small quantity generators (SQGs) create 200 to 2,000 pounds per month. Large quantity generators (LQGs) are not exempt and produce more than 2,200 pounds per

Instructor's Notes:

month. How much hazardous waste your firm creates each month determines which generator category governs your firm.

The U.S. EPA offers a free booklet: *Managing Your Hazardous Wastes: A Guide for Small Businesses.* This guide explains all of the rules and exemptions for all generators. It is available on the EPA Web site or by phone.

The general RCRA requirements for generators include:

- Generators must identify hazardous waste.
- All parties must obtain an EPA identification number.
- Certain records must be kept and reported to EPA.
- Personnel must be trained.
- Hazardous waste can be stored for a limited time.
- Hazardous waste containers must be labeled *(Figure 8)*.
- Containers used for hazardous waste must meet international specifications.
- Only permitted carriers and disposal facilities may be used.
- All wastes must be shipped using the **hazardous waste manifest** system.
- Generators must have a waste minimization plan.

5.1.0 Identifying Hazardous Wastes

The EPA regulates hazardous waste disposal. Most states have programs for handling hazardous wastes. The state rules must be at least as tough as the federal standards. The state rules may differ from the federal program explained here. Some state programs include additional chemicals such as nickel, PCB, and oil-contaminated water and soil. You need to be familiar with what wastes are regulated in your state. If you ship waste to other states for disposal, you need to be familiar with the regulations for those states. Contact your state Department of Environmental Conservation for state regulations. Most disposal facilities will assist you in properly identifying the waste and completing paperwork.

Identification of hazardous waste can be complex. The EPA rules for properly identifying waste are listed at *40 CFR 261: Identification and Listing of Hazardous Waste.* These rules will help you determine the proper EPA codes to use on the manifest.

Generally, a waste is hazardous if it has hazardous properties or it is listed as a hazardous waste. The hazardous characteristics are:

- Flammable
- Corrosive
- Reactive
- Toxic

A drum of gasoline and water would be a flammable hazardous waste. In addition, many chemicals are listed wastes. Cadmium and lead are examples. Any container of waste, even if it has a small amount of a listed chemical, would be a hazardous waste. For example, a discarded bucket of paint that contains cadmium or lead is a hazardous waste.

> **NOTE**
> If you put a small amount of a hazardous waste into a container of general waste, the whole container is considered hazardous waste. Separate your hazardous waste for proper disposal.

Explain the RCRA requirements for hazardous waste generators. Refer the participants to Figure 8.

Explain the state and federal EPA requirements for identifying hazardous waste.

Identify hazardous waste characteristics.

Figure 8 ◆ Shipping hazardous waste.

Explain the DOT and EPA requirements for transporting hazardous waste.

Describe how the hazardous waste manifest is used as a shipping document. Identify the contents of the manifest.

Show Transparency 5 (Figure 9).

Explain who completes the manifest and how it is completed.

Explain OSHA's requirements for medical monitoring. Identify the elements of a medical program.

Show Transparency 6 (Table 2).

Explain the types of medical testing an employee may be required to undergo.

In addition to EPA identification, you must also classify the waste according to DOT rules when you ship it. The DOT rules for shipping hazardous materials *(49 CFR 170)* apply to hazardous waste transportation. You must choose the proper DOT shipping name and hazard class. The containers must be labeled with both EPA and DOT labels and markings.

Most disposal companies that accept hazardous waste will assist you in determining the correct EPA codes and DOT shipping name for your waste.

5.2.0 Shipping Hazardous Waste

A hazardous waste manifest *(Figure 9)* is a shipping document. It must be used to ship all hazardous wastes. The waste is tracked from where it is produced until it reaches the disposal facility. Each step in the process is documented. This is known as cradle-to-grave tracking. Both DOT and EPA require the manifest for shipping hazardous waste.

> **NOTE**
> Some states have their own hazardous waste manifest and do not use the federal form. You need to find out which manifest to use.

The manifest is prepared by the generator. It lists the type and quantity of each waste. The form has multiple copies. Initially, the generator must send a copy to the EPA. Each party that handles the waste signs the manifest and keeps a copy. When the waste reaches the disposal facility, that facility returns a signed copy of the manifest to the generator. This confirms that the waste has been received. The disposal facility also sends a copy to the EPA. The EPA matches the initial copy from the generator with the copy from the disposal facility.

5.2.1 Completing a Hazardous Waste Manifest

The firm that creates the waste must obtain a Generator number from the EPA. This number is limited to a particular site. If the firm creates waste in several places, each site must have its own Generator EPA ID number. Transportation firms that carry hazardous waste must have an EPA ID number. The facility that accepts the waste for disposal must also have an EPA ID number. These are all listed on the top of the manifest.

The second section of the form lists each type of waste. You must use a proper shipping name. All proper shipping names are listed in the *DOT Hazardous Materials Table* at *49 CFR 171.101*.

In Section I of the manifest, you must list the EPA codes associated with each waste. These are the same EPA codes you found when you initially identified your waste. The RQ and emergency response information must also be listed for each waste.

Finally, the manifest must be signed. This signature is a certification that you have fully and accurately described and packaged the waste. Your signature also certifies that the company has a waste minimization plan. This certification is a legal document. You can be held responsible for improper disposal.

6.0.0 ◆ MEDICAL MONITORING PROGRAMS

Many OSHA and environmental programs require medical monitoring. Workers must be tested periodically to make sure that their health has not been affected by exposure to hazardous materials. A medical program consists of these elements:

- Surveillance, including pre-employment screening, annual screening, exposure-related testing, and termination examination
- Treatment, both emergency and non-emergency
- Recordkeeping
- Program review

Table 2 is the recommended medical program from NIOSH, OSHA, EPA, and the U.S. Coast Guard for hazardous waste site operations.

6.1.0 Medical Testing

The employee should have a medical evaluation at critical stages in his or her employment. The first screening takes place during pre-employment. The second is done at annual checkups. Screening is also done if an accident occurs, and finally, when the employee stops working for the company.

The pre-employment screening is a series of medical tests that workers must have before they can start the job. These tests serve two purposes. The first is to make sure that the worker is physically qualified to perform the work. This is especially important if the worker will be wearing a respirator. Some firms require extensive testing for work in extreme environments or where medical services are not readily available.

Second, pre-employment screening will document the initial health of the worker. This is known as establishing a **baseline**. Any changes in the worker's health can be compared to the baseline. Without the baseline, it will be unclear if any health issues arose during the person's employment. All people are different. Although there are general norms, there are no right answers on medical tests.

Instructor's Notes:

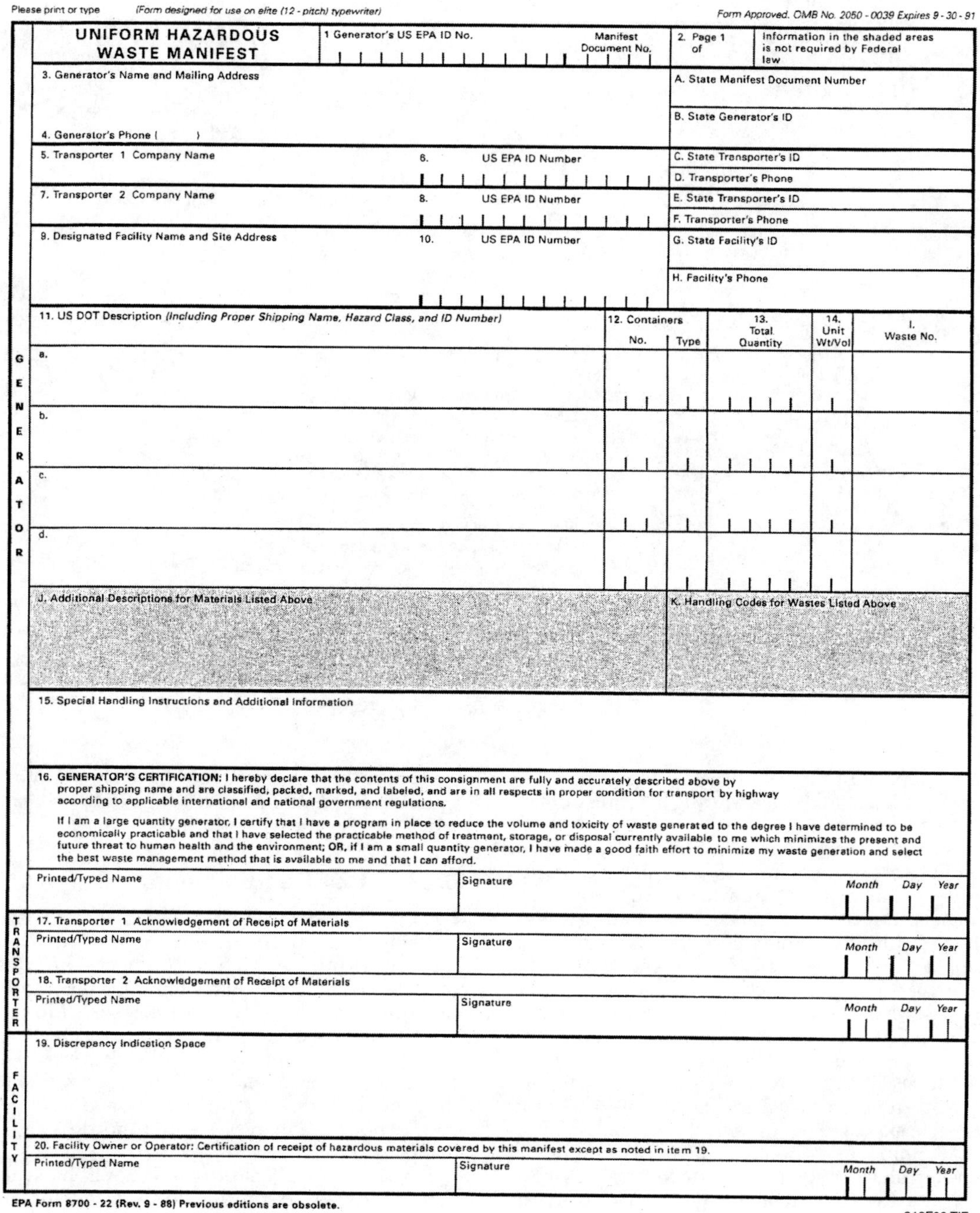
Please print or type (Form designed for use on elite (12 - pitch) typewriter)

Form Approved. OMB No. 2050 - 0039 Expires 9 - 30 - 91

UNIFORM HAZARDOUS WASTE MANIFEST

1 Generator's US EPA ID No.

Manifest Document No.

2. Page 1 of

Information in the shaded areas is not required by Federal law

3. Generator's Name and Mailing Address

A. State Manifest Document Number

B. State Generator's ID

4. Generator's Phone ( )

5. Transporter 1 Company Name

6. US EPA ID Number

C. State Transporter's ID

D. Transporter's Phone

7. Transporter 2 Company Name

8. US EPA ID Number

E. State Transporter's ID

F. Transporter's Phone

9. Designated Facility Name and Site Address

10. US EPA ID Number

G. State Facility's ID

H. Facility's Phone

GENERATOR

11. US DOT Description (Including Proper Shipping Name, Hazard Class, and ID Number)	12. Containers No.	Type	13. Total Quantity	14. Unit Wt/Vol	I. Waste No.
a.					
b.					
c.					
d.					

J. Additional Descriptions for Materials Listed Above

K. Handling Codes for Wastes Listed Above

15. Special Handling Instructions and Additional Information

16. GENERATOR'S CERTIFICATION: I hereby declare that the contents of this consignment are fully and accurately described above by proper shipping name and are classified, packed, marked, and labeled, and are in all respects in proper condition for transport by highway according to applicable international and national government regulations.

If I am a large quantity generator, I certify that I have a program in place to reduce the volume and toxicity of waste generated to the degree I have determined to be economically practicable and that I have selected the practicable method of treatment, storage, or disposal currently available to me which minimizes the present and future threat to human health and the environment; OR, if I am a small quantity generator, I have made a good faith effort to minimize my waste generation and select the best waste management method that is available to me and that I can afford.

Printed/Typed Name | Signature | Month Day Year

TRANSPORTER

17. Transporter 1 Acknowledgement of Receipt of Materials

Printed/Typed Name | Signature | Month Day Year

18. Transporter 2 Acknowledgement of Receipt of Materials

Printed/Typed Name | Signature | Month Day Year

FACILITY

19. Discrepancy Indication Space

20. Facility Owner or Operator: Certification of receipt of hazardous materials covered by this manifest except as noted in item 19.

Printed/Typed Name | Signature | Month Day Year

EPA Form 8700 - 22 (Rev. 9 - 88) Previous editions are obsolete.

218F09.TIF

Figure 9 ◆ Uniform hazardous waste manifest.

Explain in detail the contents of a uniform hazardous waste manifest. Provide the participants with a similar manifest and have them identify specific areas of the manifest in order to become more familiar with reading manifests.

ENVIRONMENTAL AWARENESS 18.11

Explain the special circumstances for which an employee may be required to have a medical exam.

Describe the recordkeeping requirements for medical programs.

Explain the importance of regular evaluation of medical programs.

Table 2 Recommended Medical Program

Component	Recommended	Optional
Pre-Employment Screening	• Medical history • Occupational history • Physical examination • Determination of fitness to work wearing protective equipment • Baseline monitoring for specific exposures	• Freezing pre-employment specimen for later testing (limited to specific situations
Periodic Medical Examinations	• Yearly update and occupational history • Yearly physical examination • Testing based on 1) examination results, 2) exposures, and 3) job class and task	• Yearly testing with routine medical tests
Emergency Treatment	• Provide first aid on site • Develop liaison with local hospital and medical specialties • Arrange for decontamination of victims • Arrange in advance for transport of victims • Transfer medical records; give details of incident and medical history to next care provider	
Non-Emergency Treatment	• Develop mechanism for non-emergency health care	
Recordkeeping and Review	• Maintain and provide access to medical records in accordance with OSHA and state regulations • Report and record occupational injuries and illnesses • Review Site Safety Plan regularly to determine if additional testing is needed • Review program periodically. Focus on current site hazards, exposures, and industrial hygiene standards	

Source: NIOSH/OSHA/USCG/EPA

It is important to have a baseline to see how an employee's health changes during the course of his or her employment.

The next element of medical monitoring is annual screening. Workers who handle many types of hazardous chemicals must have an annual physical. Chemical exposure will be tested through blood or other samples. Medical tests cannot directly check for a specific chemical exposure. Instead, the tests check the status of **target organs**. Most chemicals affect a particular organ, such as the lungs or liver, known as a target organ. The annual results will be compared to the baseline to see if there are any significant changes. General trends in a worker's health will also show possible exposure or stress.

All employees who are injured, become ill, or develop signs or symptoms due to possible overexposure to a hazardous substance must receive a medical exam as soon as possible. This includes illness or symptoms due to health hazards from an emergency response or hazardous waste operation. Follow-up treatment and medical evaluations may be required.

Finally, the worker should have an exit physical. This serves to document the worker's health when he or she leaves the job.

6.2.0 Other Requirements

Records must be kept of any employment-related physical exams. OSHA and state regulations require certain restrictions on access to these records. OSHA also requires reporting and recordkeeping for accidents and illnesses. Medical records on any exposed worker must be kept for the duration of employment plus 30 years. This requirement is waived if the employee works for the contractor for less than one year and receives a copy of his or her medical records upon leaving the company.

Finally, the medical program should be evaluated regularly. It should be updated for any changes in legal requirements. It should also be effective in protecting worker health.

Instructor's Notes:

7.0.0 ◆ TRAINING REQUIREMENTS

Many of the laws covered in this module require training. The training requirements are listed in the regulations. Some programs have a set number of hours of training, for example the HAZWOPER (Hazardous Waste Operations) 40-hour initial course. Most training programs include annual or periodic refresher courses. You must keep accurate records of worker training. OSHA, insurance companies, and other government agents may inspect these training records.

There are many professional safety firms that offer training. These programs come in many formats, from standard lecture format to on-line courses. You need to be sure that the training meets the specific program requirements. Some training may overlap.

There are different training needs, depending on job tasks. Most programs have several levels of training. The first level is for employees whose daily tasks are not directly related to the hazard. The second level is for those who may encounter the hazard. The third level is for those who work around the hazard every day. The final level is for those responsible for safety or emergency response.

Several programs offer certification or licenses. The following are some of the training programs you may need:

- Asbestos Handlers: *29 CFR 1926.1101*
- Confined Space Entry: *29 CFR 1910.146*
- DOT HM-126: Hazardous materials handling and transportation, *49 CFR 172*
- Hazard Communication: *29 CFR 1910.1200*
- HAZWOPER: Hazardous waste operations and emergency response, *29 CFR 1910.120*
- Commercial Pesticide Applicators: FIFRA laws managed by state agencies.
- Respiratory Protection: *29 CFR 1910.134*

Summary

There are many environmental issues that confront the construction industry. You need to be aware of these issues. If they are handled poorly, they could become very expensive headaches. Your company may choose to pass the handling and abatement of hazardous materials and wastes to a qualified sub-contractor. If your firm must handle these issues, you may need additional training. Most programs require worker training and medical monitoring. The safety technician may be responsible for managing training and medical monitoring programs.

The laws are always changing. New programs are developed or expanded to protect the health of the community and preserve the environment. You need to be aware of new developments that will affect your job site.

Discuss the training requirements for environmental rules and laws.

Identify the training programs that are required for certification or licensure.

Review the material covered in the module.

Have the participants complete the Review Questions and go over the answers prior to administering the Module Examination.

Administer the Module Examination. Record the results on Craft Training Report Form 200 and submit the results to the Training Program Sponsor.

ENVIRONMENTAL AWARENESS 18.13

Review Questions

1. The most common type of asbestos found in buildings is _____.
   a. amosite
   b. brown asbestos
   c. blue asbestos
   d. white asbestos

2. Lead paint is often found in buildings built before _____.
   a. 1990
   b. 1983
   c. 1978
   d. 1970

3. Light ballasts should be treated as a PCB waste unless labeled "No PCB."
   a. True
   b. False

4. Best Management Practices for a Storm Water Pollution Prevention Plan include _____.
   a. medical monitoring and training
   b. sediment control and site management
   c. a generator ID number and a manifest
   d. a dust barrier and a decontamination zone

5. A company that may be responsible for clean up in a Superfund action is known as a(n) _____.
   a. PRP
   b. SWPP
   c. RQ
   d. CERCLA

6. Spill prevention is achieved through using _____.
   a. sediment control
   b. secondary containment
   c. reportable quantities
   d. dust barriers

7. The firm that creates a hazardous waste is known as a(n) _____.
   a. polluter
   b. transporter
   c. generator
   d. applicator

8. When you sign a hazardous waste manifest as the generator you certify that you have a(n) _____.
   a. medical monitoring program
   b. waste minimization program
   c. best management plan
   d. NPDES permit

9. A medical monitoring program should include _____.
   a. best management practices and site maintenance
   b. annual screening and recordkeeping
   c. respiratory protection and secondary containment
   d. baseline and target organs

10. Medical records for an employee who has had a chemical exposure must be kept for the duration of employment plus _____ years.
    a. 5
    b. 10
    c. 20
    d. 30

Instructor's Notes:

GLOSSARY

Trade Terms Introduced in This Module

Asbestos: A natural mineral that forms long crystal fibers, used in the past as a fire retardant. It is a known carcinogen.

Asbestos containing material (ACM): An object that is comprised of asbestos and other compounds.

Asbestosis: Scarring of the lung tissue caused by inhaled asbestos fibers. This terminal condition is caused by asbestos exposure.

Baseline: In medical monitoring programs, this refers to the initial health status of the person. Subsequent medical reports are compared to the baseline.

Bioaccumulate: The natural process by which chemicals become concentrated in higher levels of the food chain as larger animals consume many smaller contaminated animals or plants.

Dielectric: A nonconductor of electricity, especially a substance with electrical conductivity of less than one millionth (10^{-6}) of a siemens.

Generator: A firm that creates hazardous waste.

Hazardous waste: A discarded material that has dangerous properties; it may be ignitable, corrosive, toxic, and/or reactive.

Hazardous waste manifest: A manifest is similar to a bill of lading. It is a shipping document that must be used for shipping waste that is considered hazardous by DOT and EPA standards.

Light ballasts: A part of a fluorescent light which contains an electric capacitor that may contain PCB.

Polychlorinated biphenyls (PCB): A group of man-made chemicals, which were widely used as dielectric fluids or additives.

Potentially responsible party (PRP): An individual or firm who may be liable for paying the costs of a Superfund cleanup; a defendant in a Superfund lawsuit.

Reportable quantity (RQ): The amount of a chemical that, when spilled, must be reported to the National Response Center.

Secondary containment: A barrier that collects chemical overflow or spills from their original containers.

Storm water run-off: Rain that is not absorbed by the soil. Uncontrolled rainwater flows over land and picks up dirt and other contaminants and carries these to the nearest water body.

Superfund: The common name for the Comprehensive Environmental Response, Compensation, and Liability Act.

Swale: A shallow trough-like depression that carries water mainly during rainstorms or snow melts.

Target organ: A specific organ in the human body most affected by a particular chemical.

Wetlands: A lowland area, such as a marsh or swamp, which is saturated with moisture, especially when regarded as the natural habitat of wildlife.

ENVIRONMENTAL AWARENESS 18.15

Instructor's Notes:

APPENDIX A

Asbestos Containing Materials

Note: The following list does not include every product or material that may contain asbestos. It is intended as a general guide to show which types of materials may contain asbestos.

Acoustical Plaster
Adhesives
Asphalt Floor Tile
Base Flashing
Blown-in Insulation
Boiler Insulation
Breaching Insulation
Caulking and Putties
Ceiling Tiles and Lay-In Panels
Cement Pipes
Cement Siding
Cement Wallboard
Chalkboards
Construction Mastics (floor tile, carpet, ceiling tile, etc.)
Cooling Towers
Decorative Plaster
Ductwork Flexible Fabric Connections
Electrical Cloth
Electrical Panel Partitions
Electrical Wiring Insulation
Elevator Brake Shoes
Elevator Equipment Panels
Fire Blankets
Fire Curtains
Fire Doors
Fireproofing Materials
Flooring Backing
Heating and Electrical Ducts
High Temperature Gaskets
HVAC Duct Insulation
Joint Compound
Laboratory Gloves
Laboratory Hoods or Table Tops
Pipe Insulation (corrugated air-cell, block, etc.)
Roofing Felt
Roofing Shingles
Spackling Compounds
Spray-Applied Insulation
Taping Compounds (thermal)
Textured Paints or Coatings
Thermal Paper Products
Vinyl Floor Sheeting
Vinyl Floor Tile
Vinyl Wall Coverings
Wallboard

ENVIRONMENTAL AWARENESS 18.17

APPENDIX B

Common Environmental Laws and Acronyms

Most environmental programs are known by their acronyms. These are a some of the commonly used terms and programs.

ACM – Asbestos Containing Material
AHERA – Asbestos Hazard Emergency Response Act
ASHAA – Asbestos School Hazard Abatement Act
BMP – Best Management Practices
CAA – Clean Air Act
CAS – Chemical Abstract Service
CERCLA – Comprehensive Environmental Response, Compensation, and Liability Act
CWA – Clean Water Act
DOT – US Department of Transportation
EPA – US Environmental Protection Agency
EPCRA – Emergency Planning and Community Right-to-Know Act
FIFRA – Federal Insecticide, Fungicide and Rodenticide Act
HMIS – Hazardous Materials Information System
HMTUSA – Hazardous Materials Transportation Uniform Safety Act
HazCom – Hazard Communication Standard
HSWA – Hazardous and Solid Waste Amendments
LEPC – Local Emergency Planning Committee
MSDS – Material Safety Data Sheet
NFPA – National Fire Protection Association
NPDES – National Pollution Discharge Elimination System
OSHA – Occupational Safety and Health Administration
PCB – Polychlorinated Biphenyls
POTW – Publicly Owned Treatment Works
PRP – Potentially Responsible Party
RCRA – Resource Conservation and Recovery Act
RQ – Reportable Quantity
SARA – Superfund Amendments and Reauthorization Act
SIC – Standard Industrial Classification
SWPPP – Storm Water Pollution Prevention Plan
TSCA – Toxic Substances Control Act
EPA ID – United States Environmental Protection Agency Identification Number

Instructor's Notes:

Additional Resources

This module is intended to present thorough resources for task training. The following reference works are suggested for further study. These are optional materials for continued education rather than for task training.

www.osha.gov

www.epa.gov

www.asse.org

Hazardous Materials Behavior and Emergency Response Operations, 2000. Denis E. Zeimet, Ph.D., CIH and David N. Ballard. Des Plaines, IL: The American Society of Safety Engineers (ASSE).

Hazardous Materials Management Desk Reference, 2000. Edited by Doye B. Cox, P.E., CHMM. New York, NY: McGraw-Hill.

Figure Credits

APL Environmental, Inc.	218F01
Midwest Environmental Control, Inc.	218F03, 218F04, 218F05, 218F08
Ultra Tech International	218F06, 218F07

ENVIRONMENTAL AWARENESS 18.19

MODULE 75218-03 — ANSWERS TO REVIEW QUESTIONS

Answer	Section
1. d	2.1.1
2. c	2.2.0
3. a	2.3.4
4. b	3.1.1
5. a	3.3.0
6. b	4.1.0
7. c	5.0.0
8. b	5.2.1
9. b	6.1.0
10. d	6.2.0

CONTREN™ LEARNING SERIES — USER UPDATES

The NCCER makes every effort to keep these textbooks up-to-date and free of technical errors. We appreciate your help in this process. If you have an idea for improving this textbook, or if you find an error, a typographical mistake, or an inaccuracy in NCCER's Contren™ textbooks, please write us, using this form or a photocopy. Be sure to include the exact module number, page number, a detailed description, and the correction, if applicable. Your input will be brought to the attention of the Technical Review Committee. Thank you for your assistance.

Instructors – If you found that additional materials were necessary in order to teach this module effectively, please let us know so that we may include them in the Equipment/Materials list in the Instructor's Guide.

Write: Curriculum Revision and Development Department
National Center for Construction Education and Research
P.O. Box 141104, Gainesville, FL 32614-1104

Fax: 352-334-0932

E-mail: curriculum@nccer.org

Craft ______________________ Module Name ______________________

Copyright Date ______________ Module Number ______________ Page Number(s) ______________

Description ____

____

____

____

(Optional) Correction ____

____

____

(Optional) Your Name and Address ____

____

____

Index

Index

NCCER's Contren Instructor Resource Center

This Web site is passcode-protected and is only meant for instructor use. The access code below provides you with access to all levels of the craft area. In order for you to access other craft areas, you will need to have an access code that is specific to that craft. Be sure to check back with the IRC on a regular basis for any updates to the exam banks or the TestGen software.

This site contains the following resources for instructor use!

TestGen – This product contains software to create the secure paper-based module exams that trainees must pass in order to acquire NCCER credentials. Instructors may use this software to scramble the order of exam questions and create additional questions as desired. Note that according to the NCCER Accreditation Policies and Guidelines, existing questions may not be removed or edited from any module examination. Doing so may seriously jeopardize either the accreditation status of the training program sponsor or any recognition of trainees, instructors, and trainers through the NCCER National Registry.

Performance Profile Sheets – Included on the Web site is a downloadable document containing all the hands-on tasks that are necessary for trainees to complete to receive credentials from NCCER.

Getting started is easy!

1. Visit **www.NCCERContrenIRC.com**. You will need a valid email address that you check on a regular basis as well as the access code that is on this sheet. You are allowed to redeem this access code only once.
2. Select the appropriate craft area by either using the drop down menu or by selecting "Login" next to the book cover.
3. Login using your username and password every time you return to **www.NCCERContrenIRC.com**.
4. If you are a new TestGen user, you will need to download the software. Go to: **www.pearsoned.com/testgen** and use the login information below:

   Login Name: tginstructor
   Password: testgen

Note: Additional resources may be added to this site as they become available.
For 24/7 technical support, please call **1-888-433-8435** or go to www.247pearsoned.com.

Access Code: Safety Tech

PEEL HERE →

STORE IN A SECURE AREA!

PEARSON

nccer